SAUNDERS COMPLETE PACKAGE FOR TEACHING GENERAL PHYSICS

CONCEPTS, PROBLEMS AND SOLUTIONS IN GENERAL PHYSICS

VOLUME TWO

A STUDY GUIDE FOR STUDENTS OF ENGINEERING AND SCIENCE

RAYMOND A. SERWAY

CLARKSON COLLEGE OF TECHNOLOGY

1975 **W. B. SAUNDERS COMPANY · Philadelphia · London · Toronto**

W. B. Saunders Company: West Washington Square
Philadelphia, PA 19105

12 Dyott Street
London, WC1A 1DB

833 Oxford Street
Toronto, Ontario, M8Z 5T9, Canada

Library of Congress Cataloging in Publication Data

Serway, Raymond A.

Concepts, problems, and solutions in general physics.

1. Physics. I. Title.

QC21.2.S38 530 74–4590

ISBN 0–7216–8066–6 (v. 2)

Concepts, Problems and Solutions in General Physics – VOLUME II ISBN 0–7216–8066–6

Last digit is the print number: 9 8 7 6 5 4 3 2 1

PREFACE

The purpose of this book is to provide students of science and engineering with a review of basic concepts and fundamental principles in general physics, together with a collection of illustrative examples and programmed exercises. The book contains two volumes. The first volume covers Newtonian Mechanics, Heat, Thermodynamics and Fluids. This second volume covers Waves, Electricity and Magnetism, Geometrical and Wave Optics. It is a widely acknowledged fact that physics is best understood through the experience of application of fundamental concepts to a variety of physical situations. Therefore, methods and techniques of problem solving are emphasized which, when mastered, should provide the student with the background and confidence required for more complex situations.

Each chapter is divided into several parts. The first part is a presentation of basic concepts and definitions, together with a collection of illustrative examples. The second part is a set of programmed exercises, which serves as a review of the fundamental principles in that chapter, as well as being a self-test of techniques in problem solving. The last part contains a summary of important definitions and equations, and a collection of unsolved problems. Most of these problems are closely related to the "theory" portion of that chapter. Problems and examples which involve calculus are labeled with a dagger (†), so that they can be avoided in the event that the book is used in a non-calculus physics course. In general, problems range in difficulty from those of the confidence building variety to complex problems involving several thought processes. The more complex problems are labeled with an asterisk (*). Hints to solving these complex problems are usually given. Answers to all problems are given at the end of the book, together with a collection of useful tables and an index.

This book should be most useful in an introductory physics course where problem solving and concepts are emphasized. The added feature of the programmed exercises will hopefully provide some guidelines of "self-learning" for the student. A cross-index is provided in the event that the book is used as a supplement to other texts such as *Fundamentals of Physics* by Halliday and Resnick, *University Physics* by Sears and Zemansky, *Physics for Students of Science and Engineering* by Bueche and *Elementary Classical Physics* by Weidner and Sells. Since the order of topics is somewhat standard, the book can be used in conjunction with other introductory physics texts.

The author would be grateful to the users of this book if they would point out errors by writing to him directly.

RAYMOND A. SERWAY

TO THE STUDENT

It is difficult to teach good study habits to a student. However, my experience with incoming students has shown that the major difficulty in learning physics is the "wrong" approach, namely, memorization of textbook and lecture material. Memorizing sections of a text, including definitions, derivations and the like *does not* necessarily mean the student *understands* the material. Understanding the basic concepts and formalism is possible only through efficient study habits, many hours of problem solving, and discussions with other students and instructors. Therefore, my first word of advice is *reduce memorization* of material (including basic equations and definitions) to a *minimum.* (It has been my own policy to list basic equations and constants on the front of exams to encourage students to "learn" physics through "thinking" rather than memorizing.) It is very important, in fact essential, that you understand basic concepts and principles *first,* before attempting to solve assigned problems.

Second, try to solve as many problems at the end of the chapter as possible. This can be done only after carefully reading through the text material, examples and exercises. Keep in mind that very few people are able to absorb the full meaning of scientific writing after one reading. Lectures should provide some clarification of troublesome points, but several readings of notes and text material are usually necessary. When solving problems, try to find alternate solutions to the same problem. For example, many problems in mechanics can be treated by solving the motion or by the more direct energy method. This book provides a variety of examples and step-by-step problems (programmed exercises) which should be of value in this regard.

The method of solving certain problems, especially those requiring the use of several concepts, should be *carefully* planned. Always read the problem through a few times until you are confident that you understand what is being asked. Next, read the problem through with more thought, with special attention to the available information. Finally, write down the basic structure of the method (or methods) you feel would be applicable to the problem, and proceed with the solution. Be careful not to misinterpret the problem. The ability to properly interpret what is being asked is an integral part of solving the problem.

Finally, it is always useful to supplement the study of material with models and experiments. Whenever it is possible, try to set up simple experiments at home or in the laboratory to substantiate ideas and models discussed in class or in the text. For example, the common "slinky" toy is invaluable for demonstrating traveling waves; an old pair of Polaroid sunglasses and some discarded lenses and magnifying glass are central components of various experiments in optics; collisions between billiard balls can be conveniently studied in the pool room, with the addition of a paper covered table to provide a permanent record of the collisions. The list is endless. When physical models are not available, try to develop "mental" models and devise thought experiments to improve your understanding of the concepts or the situation at hand.

A few words of advice pertaining to the use of this book are in order. Note that each chapter is divided into three major parts. The first part is a discussion of basic concepts, definitions and principles, together with a collection of solved example problems. The second is a collection of programmed exercises which should be read with care. Some of these exercises serve as a review of the concepts and definitions in that chapter. Others are step-by-step solutions to new problems, some of which involve more than one concept discussed in that chapter. Finally, a summary of important concepts and equations is presented, followed by a collection of unsolved problems, with answers at the end of the text. The student should attempt to solve these problems only after reading through and, hopefully, "understanding" the previous portions of that chapter. Some of the more difficult problems are labeled with an asterisk (*), while those requiring the use of calculus are labeled with a dagger (†). It would be to the student's advantage to use the programmed exercises in an "honest" manner. By this I mean that the answers to each part of the exercise in the *right column* of the page should be covered up, while reading the questions in the *left column.* Obviously, the exercises will prove most effective when used in this manner. I suggest blocking the right column with a blank sheet of paper (which you can use for calculations). After writing what you think is the correct answer to that part of the exercise, check the answer in the right column by sliding your blank paper down one frame. If your answer is correct, go on to the next part of the exercise. If your answer is incorrect, reread the section of the text applicable to that exercise and go over it a second time. Repeat this until you are confident that the answer in the text is correct. If there is still disagreement, check your work with the instructor.

Cross Index to Other Texts

Given by chapter numbers*

LM	S	HR	S	SZ	S	B	S	WS	S
1, 2	1	16, 17	1	21, 22, 23	1	18, 19	2	16, 17	1
3, 4	8, 9	23, 24	2	24, 25	2	20	3	22, 23, 24	2
5, 7	2	25, 26	3	26, 27	3	21, 22	3, 4	25, 26	3
6, 8	3	27, 28	4	28, 29	4	23	3	27, 28	4
9	4	29, 30	5	30, 31, 32	5	24, 25	5	29, 30	5
10, 11	5, 7	31, 32, 33	6	33, 34	6	26	6	31, 32	6
12	6	34, 35	7	35, 36	7	27	5	33	5
13	7	36	8	37-40	8	28, 29	7	34, 35	7
		37, 38	9	41, 42	9	30	1	36, 37	8
						31	8	38, 39, 40	9
						32, 33	9		
						35	1		

**Note:* This index is only a *rough* guide, and the symbols used are as follows:

LM – Lobkowicz & Melissinos – *Physics for Scientists and Engineers (Vol. II)*
HR – Halliday & Resnick – *Fundamentals of Physics*
SZ – Sears & Zemansky – *University Physics*
B – Bueche – *Introduction to Physics for Scientists and Engineers*
WS – Weidner & Sells – *Elementary Classical Physics*
S – Serway – *(This Text)*

ACKNOWLEDGMENTS

I am grateful to a number of people who, in various ways, assisted me in the preparation of the final manuscript. Mrs. Agatha Hollister did an outstanding job of typing the manuscript. Helpful criticisms and suggestions were provided by Professors A. Czanderna, H. Helbig, D. Kaup, D. Larsen, F. Lobkowicz, J. Love, F. Otter, J. Marion, and Messrs. G. Anton, B. Davis, A. Miller and A. Serway. I thank the hundreds of students who used this manuscript in its original "uncut" version for pointing out various errors and misleading statements.

Finally, I dedicate this book to my wife, Elizabeth, and my children Mark, Michele and David. They were a constant source of inspiration and eventually learned to cope with my unusual hours.

CONTENTS

1

WAVE PHENOMENA

1.1 WAVE TYPES

Various types of disturbances in nature exhibit wavelike characteristics. *Mechanical waves,* such as waves on strings, water waves and sound waves, require the presence of a medium through which the disturbance propagates. Light waves and radio waves are electromagnetic in origin and do not require the presence of a medium for their propagation. In this chapter, we will confine our attention to *mechanical waves.*

In general, energy is transferred from one part of the medium to another as the result of the motion of a large number of particles. If the particles of the medium move *perpendicular* to the direction of wave propagation, the wave is called a *transverse* wave. On the other hand, if the particles of the medium move *along* the direction of wave propagation, the wave is called a *longitudinal* wave. A pulse moving along a string is an example of a transverse wave [Figure 1–1(*a*)], whereas a wave on a spring (for example a slinky) is an example of a longitudinal wave [Figure 1–1(*b*)].

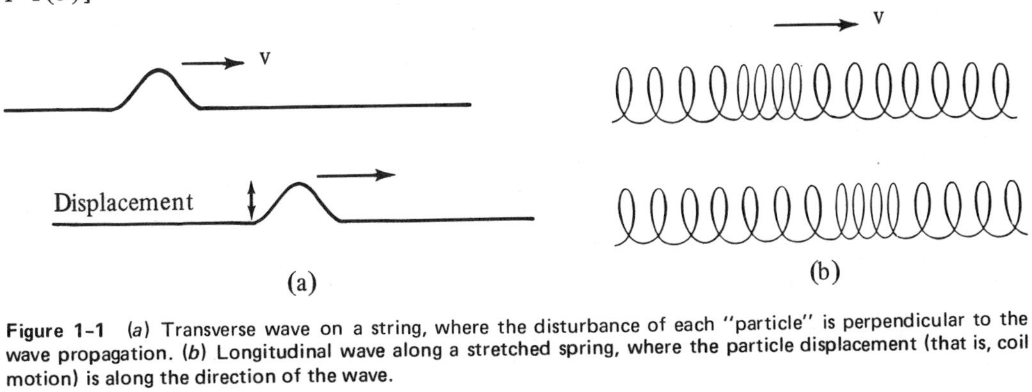

Figure 1–1 (*a*) Transverse wave on a string, where the disturbance of each "particle" is perpendicular to the wave propagation. (*b*) Longitudinal wave along a stretched spring, where the particle displacement (that is, coil motion) is along the direction of the wave.

1.2 TRAVELING WAVES

A wave which travels to the *right* along the x axis with a velocity v can be represented by the expression

$$y(x, t) = f(x - vt) \tag{1.1}$$

where $y(x)$ represents the shape of the curve at $t = 0$, as in Figure 1-2. At some time t, the curve has traveled to the right a distance vt, but the shape of the curve $y(x,t)$ hasn't changed. If the wave travels to the *left* with a velocity, v, Equation (1.1) would then read $y(x, t) = f(x + vt)$.

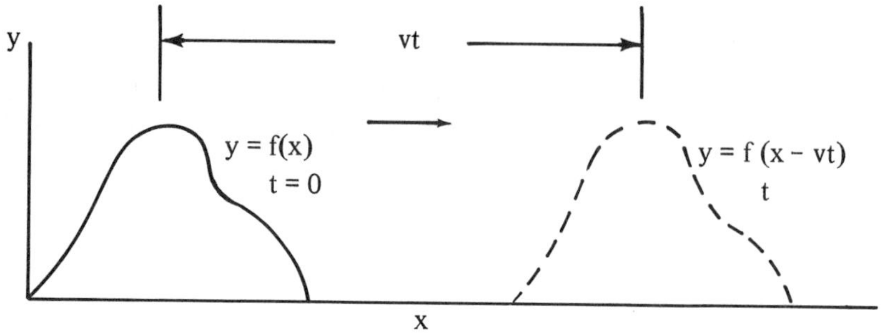

Figure 1-2 Graphical description of a traveling wave of arbitrary shape. The wave moves to the right with a velocity v.

The *harmonic wave,* that is, one which behaves sinusoidally, is especially interesting. We can represent such a wave as

$$y(x,t) = A\sin(kx - vt) \tag{1.2}$$

where k is called the *wave number* and A is the amplitude of the wave. If x is replaced by $x + 2\pi/k$, the function $y(x, t)$ must *repeat* itself, since the wave is periodic. If we define the *wavelength* λ as the distance through which the curve repeats itself, as in Figure 1–3, then we have

$$\lambda = \frac{2\pi}{k} \tag{1.3}$$

or,

$$k = \frac{2\pi}{\lambda} \tag{1.4}$$

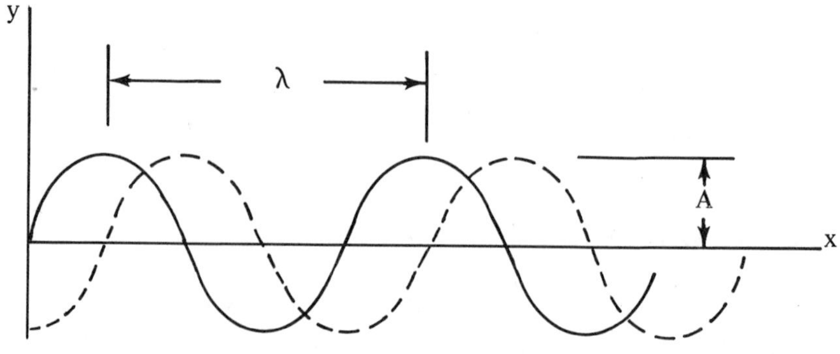

Figure 1-3 Schematic representation of a harmonic wave. The solid curve is the wave at t = 0; the dotted curve is the displaced wave at some time t.

The *angular frequency* of the harmonic wave, from Equation (1.2), is given by

$$\omega = kv = \frac{2\pi}{\lambda} v \tag{1.5}$$

The angular frequency can also be written in terms of the frequency f (the number of oscillations per second) and period T (the time for one oscillation) as

$$\omega = 2\pi f = \frac{2\pi}{T} \tag{1.6}$$

Comparing Equation (1.5) with (1.6) gives the important expression relating the wave velocity to the wavelength and frequency:

$$v = \lambda f \tag{1.7}$$

Therefore, the sinusoidal wave traveling to the right can be written in the equivalent forms

$$y(x, t) = A\sin(kx - \omega t) = A\sin 2\pi\left(\frac{x}{\lambda} - \frac{t}{T}\right) \tag{1.8}$$

The student should note that the harmonic wave has two periodicities, one which is *spatial* (given by λ) and one in *time* (given by the period T).

The form of y(x, t) as given by Equation (1.8) is such that y(0, 0) = 0, as in Figure 1-3. This need not be the case. If the transverse displacement of the wave $y(0, 0) \neq 0$ as in Figure 1-4(*a*), we must introduce a *phase constant* ϕ into the expression for y(x, t), and write

$$y(x, t) = A\sin(kx - \omega t - \phi) \tag{1.9}$$

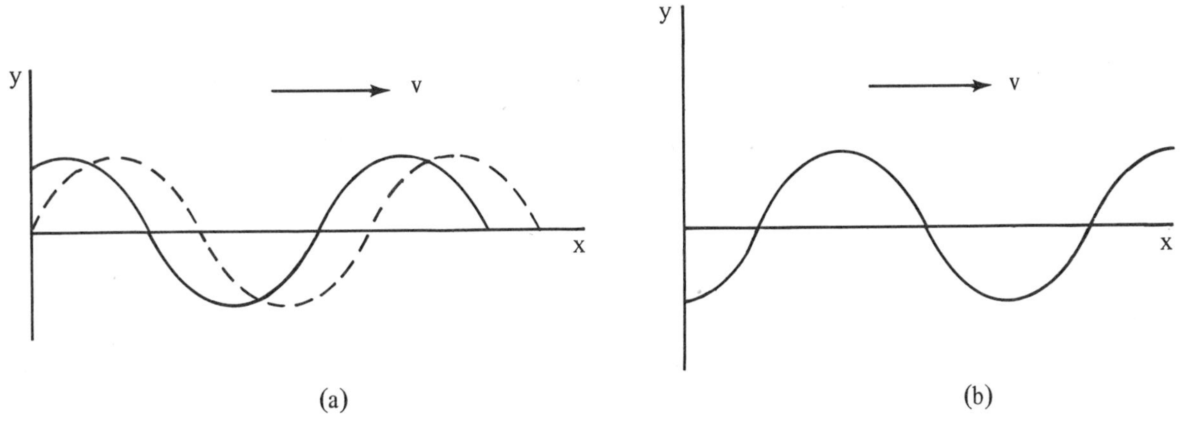

Figure 1-4 (*a*) A harmonic wave traveling to the right, according to Equation (1.9), with $\phi \neq 0$. The solid curve represents the wave at t = 0; the dotted curve represents the wave at some later time t. (*b*) A traveling wave, where $\phi = \pi/2$.

Example 1.1

Suppose a traveling wave moving to the *right* has transverse displacement $y(0, 0) = -A$, as in Figure 1-4(*b*). What is the phase angle ϕ, and what is the form of $y(x, t)$?

In this case, from Equation (1.9) we *require* that

$$y(0, 0) = -A = A \sin(-\phi)$$

or,

$$\sin(-\phi) = -1$$

But $\sin(-\phi) = -\sin\phi = -1$, so we get

$$\phi = \pi/2$$

Furthermore, since $\sin(a - b) = \sin a \cos b - \cos a \sin b$, the traveling wave can be expressed in the equivalent form

$$y = A \sin\left(kx - \omega t - \frac{\pi}{2}\right) = A\left[\sin(kx - \omega t)\cos\frac{\pi}{2} - \cos(kx - \omega t)\sin\frac{\pi}{2}\right]$$

or,

$$y(x, t) = -A\cos(kx - \omega t)$$

1.3 TRAVELING WAVES ON A STRETCHED STRING

Consider a flexible string of uniform mass per unit μ length under constant tension F, as in Figure 1–5(*a*). A small segment of the string is disturbed, forming a pulse which travels to the right. This small segment can be considered to form a circular arc of radius R if its curvature is small.

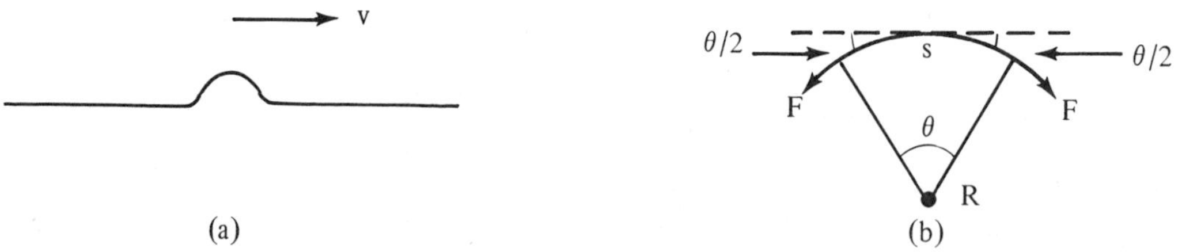

Figure 1-5 (*a*) A pulse on a flexible string traveling to the right with a speed v. (*b*) A small segment of the pulse, where the string is under constant tension F.

From Figure 1–5(*b*), we see that the resultant force on the segment is directed toward the center of curvature, and is given by $2F\sin\frac{\theta}{2}$. Therefore, from Newton's second law, we have

$$\Sigma F_r = 2F\sin\frac{\theta}{2} = ma_r = m\frac{v^2}{R} \tag{1.10}$$

But $m = \mu s = \mu R\theta$, and for small θ, $\sin\frac{\theta}{2} \cong \frac{\theta}{2}$. Therefore, Equation (1.10) reduces to

$$2F\frac{\theta}{2} = \mu R\theta\,\frac{v^2}{R}$$

or,

$$v = \sqrt{\frac{F}{\mu}} \qquad (1.11)$$

That is, the speed of propagation of traveling waves on a stretched string for *small* transverse displacements depends only on the tension and the mass per unit length. The student should show that Equation (1.11) is dimensionally correct.

Example 1.2

A uniform string of length 15 m and mass 0.3 kg is stretched horizontally with a 4 kg mass suspended at one end over a frictionless pulley, as in Figure 1-6. (a) Determine the speed of a pulse on this string and (b) find the time it takes to travel from A to B.

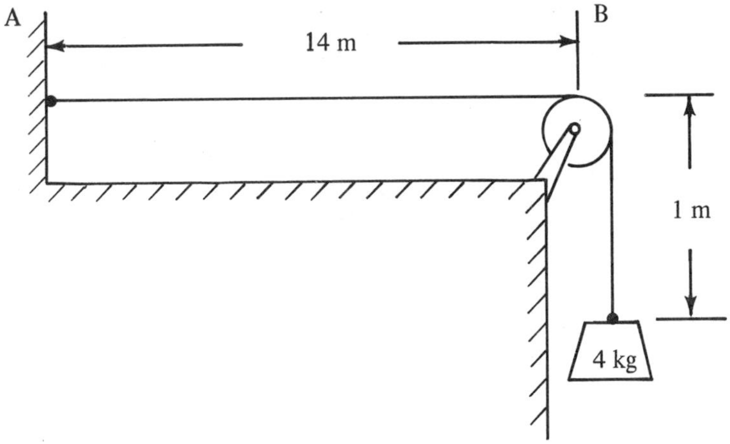

Figure 1-6

Solution

(a) The tension $= F = mg = 4 \times 9.8\text{ N} = 39\text{ N}$.

The mass per unit length $= \mu = \frac{0.3\text{ kg}}{15\text{ m}} = \frac{1\text{ kg}}{50\text{ m}}$

$$\therefore\quad v = \sqrt{\frac{F}{\mu}} = \sqrt{39 \times 50\,\frac{m^2}{sec^2}} = 44\text{ m/sec}$$

(b) $t = \frac{d}{v} = \frac{14\text{ m}}{44\text{ m/sec}} = 0.32\text{ sec}$

If one end of a continuous string is driven up and down in *simple harmonic motion,* a periodic traveling wave will propagate down the string, as in Figure 1–7. Each particle of the string oscillates in the y direction with simple harmonic motion, assuming the amplitude of the vibration is small. This is illustrated by the point B in Figure 1–7. Therefore, each part of the string can be treated as a simple harmonic oscillator.

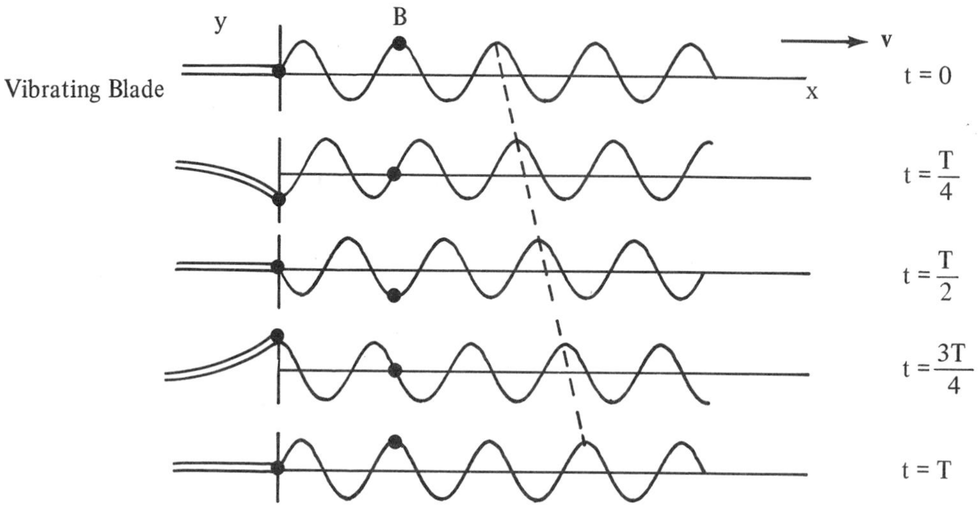

Figure 1–7 A harmonic wave traveling down a continuous string being driven at one end. Note that each particle of the string oscillates in the y direction in SHM, with a period T.

Note that although the particles oscillate in the y direction, the wave (or disturbance) propagates in the x direction with a velocity v; this is what is meant by a transverse wave. We say that the traveling wave transfers energy from one side of the string to the other, the energy being supplied by some external oscillator. (In the laboratory, it is common to use a tuning fork as the driving source.) Each particle of the string has the same period T. If the wave is as shown in Figure 1–7 at t = 0, then we can write

$$y = A\sin(kx - \omega t)$$

Now the point B shown in Figure 1–7 (or any point on the string) moves in such a way that its x coordinate *remains constant.* Therefore, we can find the *transverse velocity* and *transverse acceleration* of a point on the string (that is, the components of velocity and acceleration in the y direction).

$$v_y = \left.\frac{dy}{dt}\right|_{x = \text{constant}} = -A\omega\cos(kx - \omega t) \tag{1.12}$$

$$a_y = \left.\frac{dv_y}{dt}\right|_{x = \text{constant}} = -A\omega^2 \sin(kx - \omega t) \tag{1.13}$$

Therefore, the *maximum* transverse velocity and *maximum* transverse acceleration are given by

$$(v_y)_{max} = \omega A \tag{1.14}$$

$$(a_y)_{max} = \omega^2 A \tag{1.15}$$

Example 1.3

Suppose the string described in Example 1.2 is driven at the left end with a tuning fork whose natural frequency of oscillation is 400 Hz, producing a traveling wave of maximum amplitude 2 mm. (a) Determine the period and wavelength of the wave. (b) Calculate the maximum values of the transverse velocity and transverse acceleration of a point on the string.

Solution

(a) $T = \frac{2\pi}{\omega} = \frac{1}{f} = \frac{1}{400}\text{ sec} = 2.5 \times 10^{-3}\text{ sec.}$

From Example 1.2, v = 44 m/sec; therefore,

$$\lambda = \frac{v}{f} = \frac{44\text{ m/sec}}{400\text{ sec}^{-1}} = 0.11\text{ m}$$

(b) $(v_y)_{max} = \omega A = 2\pi f A = 2\pi \times 400 \times 2 \times 10^{-3}\,\frac{\text{m}}{\text{sec}}$

$$(v_y)_{max} = 1.6\pi\text{ m/sec}$$

$(a_y)_{max} = \omega^2 A = 4\pi^2 f^2 A = 4\pi^2 \times (400)^2 \times 2 \times 10^{-3}\,\frac{\text{m}}{\text{sec}^2}$

$$(a_y)_{max} = 1.28\,\pi^2 \times 10^3\text{ m/sec}^2$$

In a traveling wave, such as a pulse traveling down a string, energy is transmitted through the medium. This energy is supplied by some external agent which initiates the disturbance, such as a vibrating reed at the end of a stretched string. The rate at which energy is transferred past a given point on the string, or the *power*, is given by

$$P = \frac{1}{2}\omega^2 A^2 \mu v \tag{1.16}$$

For example, the student should show that the power transmitted down the string described in Examples 1.2 and 1.3 is 11 watts.

(See Programmed Exercises 1, 2 and 3 for a review and other applications.)

1.4 SUPERPOSITION AND INTERFERENCE OF WAVES

When two or more traveling waves propagate along the same medium, such as a string, the resultant displacement of the particles of that medium is the vector sum of the displacements that each wave produces. This is called the *superposition principle* and is valid for mechanical waves for *small* displacements from equilibrium. This is equivalent to assuming a linear restoring force for the displaced particles.

The term *wave interference* is used to describe the result of superimposing two or more waves. For example, consider two sinusoidal waves traveling to the right, having the same frequency and amplitude, but with a phase difference ϕ.

$$y_1 = A \sin (kx - \omega t) \tag{1.17}$$

$$y_2 = A \sin (kx - \omega t - \phi) \tag{1.18}$$

The superposition principle says that the resultant displacement, y, is given by the sum $y_1 + y_2$. Recalling the trigonometric identity,

$$\sin a + \sin b = 2\cos\left(\frac{a - b}{2}\right)\sin\left(\frac{a + b}{2}\right)$$

we get

$$y = y_1 + y_2 = 2A \cos\frac{\phi}{2} \sin\left(kx - \omega t - \frac{\phi}{2}\right) \tag{1.19}$$

From this result, we conclude that the resultant wave has the same frequency as the original waves, but its amplitude is given by $2A\cos \phi/2$. If the phase difference ϕ is given by $0, 2\pi, 4\pi, \ldots$, the resultant wave has an amplitude of 2A, and the waves are said to be *in phase* everywhere. Such waves *interfere constructively*. In this case, the crests and troughs of both waves occur at the same positions, as in Figure 1–8(*a*). On the other hand, if ϕ is given by $\pi, 3\pi, 5\pi, \ldots$, the resultant wave has *zero* amplitude, and the waves *interfere destructively*. In this case, the crest of one wave "cancels" the trough of the second, as in Figure 1–8(*b*). For an arbitrary value of ϕ, the resultant wave has an amplitude lying somewhere between 0 and 2A, as in Figure 1–8(*c*).

By inspection of Figure 1–8(*a*), we see that the *path difference* between y_1 and y_2 for *constructive interference* (the dotted curves) is $0, \lambda, 2\lambda$, and so on. Likewise, from Figure 1–8(*b*) we see that the path difference between y_1 and y_2 for *destructive interference* is $\lambda/2, 3\lambda/2, 5\lambda/2$, and so on. The general relation between the path difference δ and the phase difference ϕ is

$$\delta = \frac{\phi}{k} = \frac{\lambda\phi}{2\pi} \tag{1.20}$$

(See Programmed Exercise 4 for a review.)

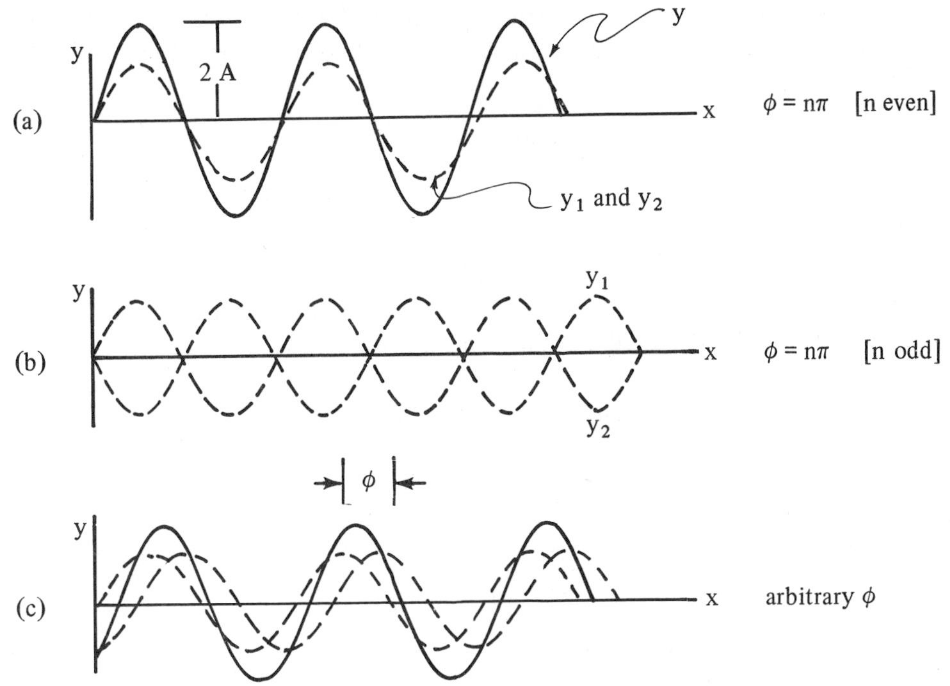

Figure 1-8 Superposition of two waves with the same amplitude and frequency, with a phase difference ϕ of (a) $n\pi$ (n even), (b) $n\pi$ (n odd) and (c) arbitrary ϕ. The dotted curves represent y_1 and y_2; the solid curves represent $y = y_1 + y_2$.

1.5 STANDING WAVES IN A STRING

If a stretched string is fixed at both ends and traveling waves are produced, say, by plucking the string, waves will reflect from the fixed ends, creating waves traveling in both directions. These waves superimpose and interfere with each other. If two sinusoidal waves with the same amplitude and frequency travel in opposite directions, the resultant displacement is given by

$$y = y_1 + y_2 = A\sin(kx - \omega t) + A\sin(kx + \omega t)$$

But $\sin(a \pm b) = \sin a \cos b \pm \cos a \sin b$; therefore, this reduces to

$$y = 2A \sin kx \cos \omega t \qquad (1.21)$$

This expression represents the equation of a *standing wave*. Note that it satisfies the boundary condition that $y = 0$ at $x = 0$, as in Figure 1–7. The points of *zero amplitude* of the resultant wave are called nodes, labeled N in Figure 1–9. These are spaced one-half wavelength apart, and occur at values of kx given by

$$kx = n\pi \qquad (n = 1, 2, 3, \ldots)$$

Condition for Nodes

$$x = \frac{n\lambda}{2} \qquad (n = 1, 2, 3, \ldots)$$

Likewise, there are points on the string which have a *maximum* amplitude given by 2A. These points are called *antinodes*, labeled A, and occur at values of kx given by

$$kx = \frac{n\pi}{2} \quad (n \text{ odd})$$

Condition for Antinodes

$$x = \frac{n\lambda}{4} \quad (n \text{ odd})$$

We also must require that y = 0 and x = *l*, since the string is fixed at both ends. Using this condition, together with the condition on x for nodes, gives

$$\frac{n\lambda}{2} = l \quad (n = 1, 2, 3, \dots)$$

or,

$$\lambda = \frac{2l}{n} \quad (n = 1, 2, 3, \dots) \tag{1.22}$$

Fundamental

$$l = \frac{\lambda}{2}$$

$$f_1 = \frac{1}{2l}\sqrt{\frac{F}{\mu}}$$

First Overtone

$$l = \lambda$$

$$f_2 = 2f_1 = \frac{1}{l}\sqrt{\frac{F}{\mu}}$$

Second Overtone

$$l = \frac{3\lambda}{2}$$

$$f_3 = 3f_1 = \frac{3}{2l}\sqrt{\frac{F}{\mu}}$$

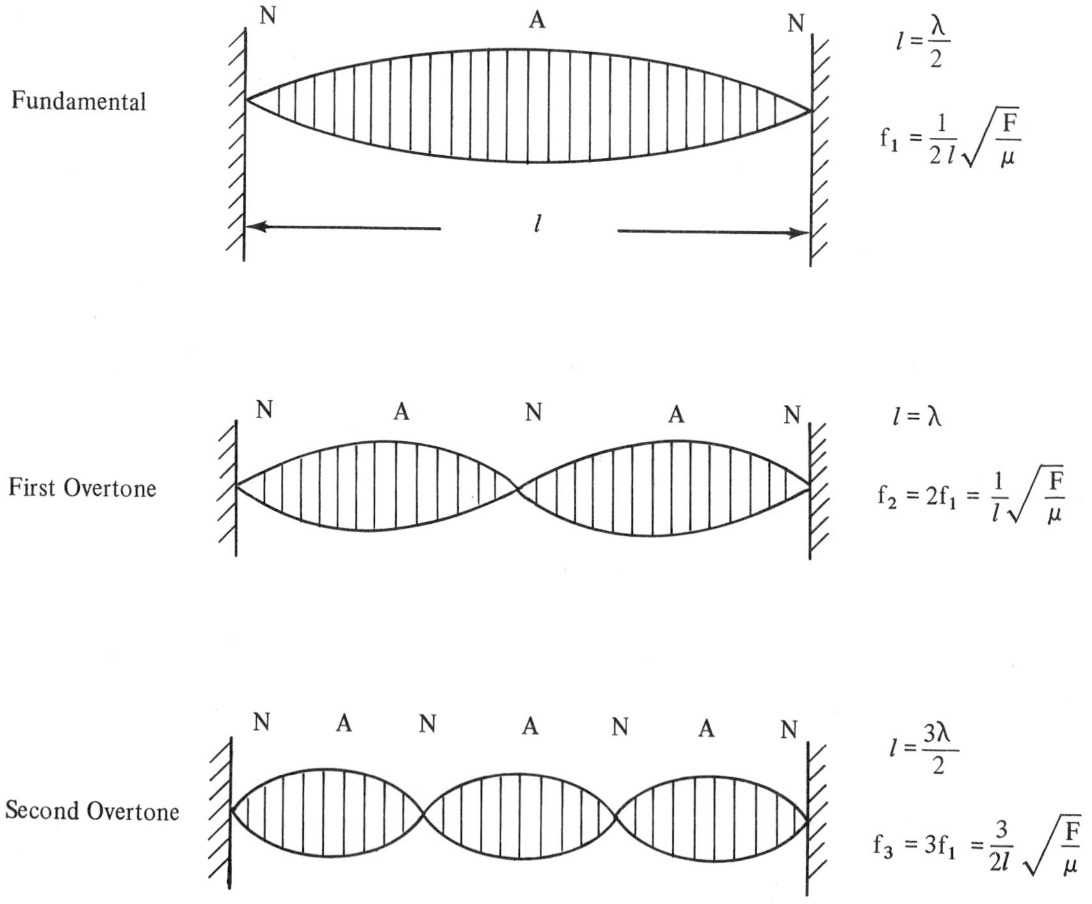

Figure 1-9 Schematic representation of standing waves on a stretched string of length *l*, where the envelope represents many successive vibrations. The points of zero displacement are called *nodes;* the points of maximum displacement are called antinodes.

If the string is driven at one of the fixed ends by some mechanical vibrator, and the frequency of the vibrator is near one of the natural frequencies of the stretched string, given by $f = v/\lambda$, the string will vibrate with a large amplitude and a *resonance* condition can be established. The *resonance frequencies* can be obtained using Equation (1.2):

$$f_n = \frac{v}{\lambda} = \frac{n}{2l}\sqrt{\frac{F}{\mu}} \qquad (n = 1, 2, 3, \ldots) \qquad (1.23)$$

In practice, one can fix the tension F and vary the vibrator frequency until one (or more) loops appear between the fixed ends. One loop corresponds to $n = 1$ in Equation (1.2), and a frequency f_1. Two loops correspond to $n = 2$, and a frequency $f_2 = 2f_1$, and so forth. On the other hand, one can also fix f and vary the tension. If F_1 corresponds to the tension for one loop ($n = 1$), then the tension for two loops is $F_2 = F_1/4$, and, in general, $F_n = F_1/n^2$.

Example 1.4

The C note on the C major scale of a piano has a fundamental frequency of 264 Hz, while the A note has a fundamental frequency of 440 Hz. (a) Calculate the frequencies of the first two overtones of the A note. (b) If the two piano strings for the A and C notes have the same mass per unit length, and the same tension, determine the ratio of lengths of the two strings.

Solution

(a)

$$f_1 = \frac{v}{2l} = 440 \text{ Hz}$$

$$f_2 = 2f_1 = 880 \text{ Hz}$$

$$f_3 = 3f_1 = 1320 \text{ Hz}$$

(b)

$$f_{1A} = \frac{1}{2l_A}\sqrt{\frac{F}{\mu}}$$

$$f_{1C} = \frac{1}{2l_C}\sqrt{\frac{F}{\mu}}$$

$$\frac{f_{1A}}{f_{1C}} = \frac{440}{264} = \frac{l_C}{l_A}$$

or

$$\frac{l_C}{l_A} = 1.67$$

(See Programmed Exercise 5 for a review.)

1.6 LONGITUDINAL WAVES

As was mentioned earlier, longitudinal waves are waves which propagate in such a way that the particle displacements are parallel to the direction of propagation of the wave. One example of a longitudinal wave is the deformed spring illustrated in Figure 1-1(*b*), where the compressed (or elongated) portion propagates along the spring. Other longitudinal waves include any compressional elastic wave propagating through a solid, liquid or gas. These compressional waves can be described by a density or pressure variation in the medium. The medium is said to be *elastically deformed* when a longitudinal wave propagates.

Sound waves are a common example of longitudinal waves which fall into three categories: (1) *audible* waves are sound waves which lie within the range of stimulation of the human ear, typically 20 Hz to 20,000 Hz; (2) *infrasonic* waves have frequencies below the audible range; and (3) *ultrasonic* waves fall above the audible range. The source of the sound waves, such as the diaphragm of a loudspeaker, causes air molecules to oscillate about their equilibrium positions parallel to the direction of propagation of the wave. Unlike transverse waves, the displacement, velocity and acceleration of the particles of the medium are parallel to the direction of the wave. However, if the source vibrates sinusoidally, the displacements of the air molecules will also vary sinusoidally. Therefore, the mathematical description of the longitudinal wave is identical to that of the transverse wave. The speed of a longitudinal, compressional wave in a medium of density ρ is given by

$$v = \sqrt{\frac{B}{\rho}} \qquad (1.24)$$

where B is the *bulk modulus* of the medium, which is a measure of the resistance of the medium to compression. If the pressure change arising from the longitudinal stress is Δp, and the change in volume arising from the deformation is ΔV, the bulk modulus is given as

$$B = -\frac{\Delta p}{\Delta V/V} \qquad (1.25)$$

Note that Equation (1.24) is of the same form as Equation (1.11) in that both are equal to the *square root of an elastic property divided by an inertial property.* The negative sign in Equation (1.25) assures that B is always positive, since the ratio $\Delta p/\Delta V$ is always negative (for example, for an increase in pressure, Δp is positive, but ΔV is negative since the volume decreases). The speed of sound is generally large in solids (for example 5100 m/sec in aluminum) and small in gases (for example, (331 m/sec in air at STP). For the case of extended solids, the velocity of longitudinal waves will also depend on the so-called *shear modulus,* which is a measure of the resistance of the medium to shearing forces. Therefore, Equation (1.24) applies to all fluids which do not sustain shear forces. The correct expression for the *velocity of longitudinal waves in an extended noncrystalline solid* is given by

$$v = \sqrt{\frac{B + \frac{4}{3}S}{\rho}} \qquad (1.26)$$

where S is the *shear modulus* of the material. If the solid is in the form of a long, thin rod, the velocity of longitudinal waves along the rod is given by $\sqrt{Y/\rho}$, where Y is *Young's modulus* for that material.

Example 1.5

Calculate the velocity of longitudinal waves in copper. Use the fact that $B = 14 \times 10^{10}\ N/m^2$, $S = 4.1 \times 10^{10}\ N/m^2$ and $\rho = 8.89 \times 10^3\ kg/m^3$ for copper at room temperature.

Applying Equation (1.26) gives

$$v = \sqrt{\frac{[14 \times 10^{10} + \frac{4}{3}(4.1 \times 10^{10})]\ N/m^2}{8.89 \times 10^3\ kg/m^3}} = \sqrt{22 \times 10^6 \frac{m^2}{sec^2}} \cong 4700 \frac{m}{sec}$$

When sound waves propagate through a *gas*, the variations in pressure occur so rapidly that the changes are essentially *adiabatic*. It is left as an exercise to show that under this assumption, together with Equation (1.24), the speed of sound in a gas is given by

$$v = \sqrt{\frac{\gamma p_0}{\rho}} \tag{1.27}$$

where p_0 is the equilibrium pressure and γ is the ratio C_p/C_v which depends on the gas in question. It is left as an exercise to show that v for an ideal gas is actually *independent* of the pressure, but is only a function of the *temperature* of the gas.

Example 1.6

(a) Calculate the velocity of sound in hydrogen gas at atmospheric pressure and at 273°K. Use the fact that $\gamma = 1.41$ for hydrogen, and $\rho = 9 \times 10^{-2}\ kg/m^3$. (b) Determine the wavelength of sound waves in hydrogen gas at a frequency of 300 Hz.

Solution

(a) $$v = \sqrt{\frac{\gamma p_0}{\rho}} = \sqrt{\frac{1.41 \times 1.01 \times 10^5\ N/m^2}{9 \times 10^{-2}\ kg/m^3}} = \sqrt{1.58 \times 10^6 \frac{m^2}{sec^2}} = 1260 \frac{m}{sec}$$

(b) $$\lambda = \frac{v}{f} = \frac{1260\ m/sec}{300\ sec^{-1}} = 4.2\ m$$

(See Programmed Exercises 6 and 7 for further detail and review.)

1.7 STANDING LONGITUDINAL WAVES IN HOLLOW PIPES

In section 1.6 we found that a stretched string fixed at both ends has a discrete set of natural frequencies of vibration, determined by the tension, mass and length of the string. These modes of vibration are common to such stringed musical instruments as the violin, piano and guitar. Other musical instruments make use of natural frequencies of sound waves in hollow pipes, such as the flute and organ. In these cases, standing sound waves are established in the hollow pipe, whose natural frequencies depend on the length of the pipe and on whether one end is open or closed. Reflections of waves at the open and closed ends set up standing waves. Boundary conditions are such that *open ends are antinodes* (where air molecules move with the largest amplitude), whereas the *closed ends are nodes* (air molecules have no net longitudinal motion at these points).

The first three natural modes of vibration of a hollow pipe open at both ends are shown in Figure 1–10. Note that the fundamental frequency $f_1 = v/2l$, the first overtone is $2f_1$, and so on. That is, *all* harmonics are present, and we can write

$$f_n = n \frac{v}{2l} \qquad (n = 1, 2, 3, \ldots) \qquad (1.28)$$

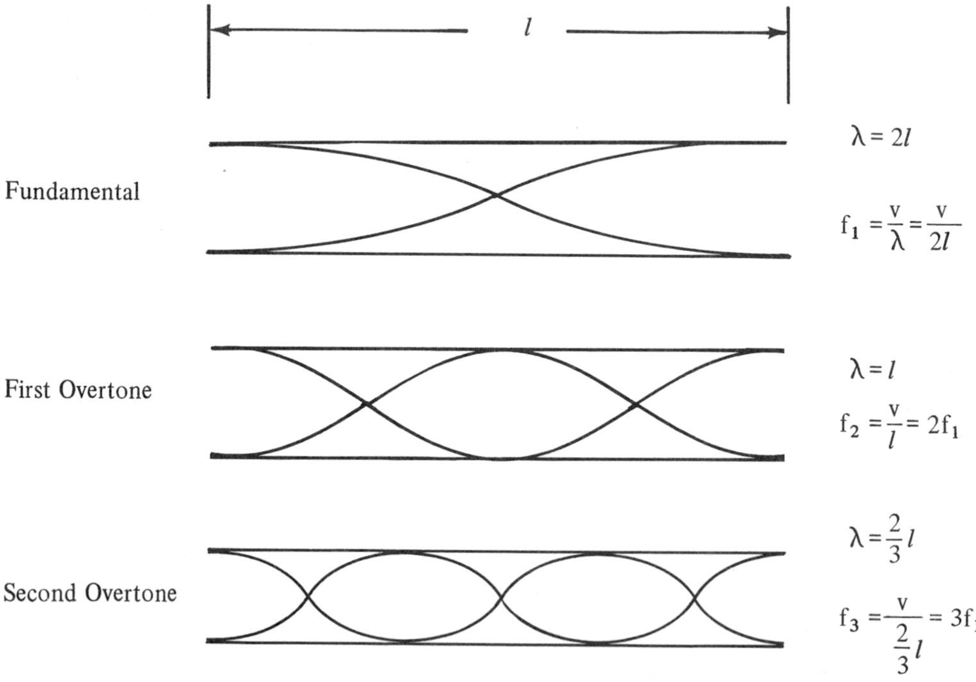

Figure 1–10 Natural modes of vibration in a hollow pipe open at each end. All harmonics are present.

The first three natural modes of vibration for a hollow pipe closed at one end are illustrated in Figure 1–11. In this case, the fundamental frequency is $f_1 = v/4l$, the first overtone is at $3f_1$, and so on. That is, in a pipe closed at one end, only *odd* harmonics are present and are given by

$$f_n = (2n + 1) \frac{v}{4l} \qquad (n = 0, 1, 2, \ldots) \qquad (1.29)$$

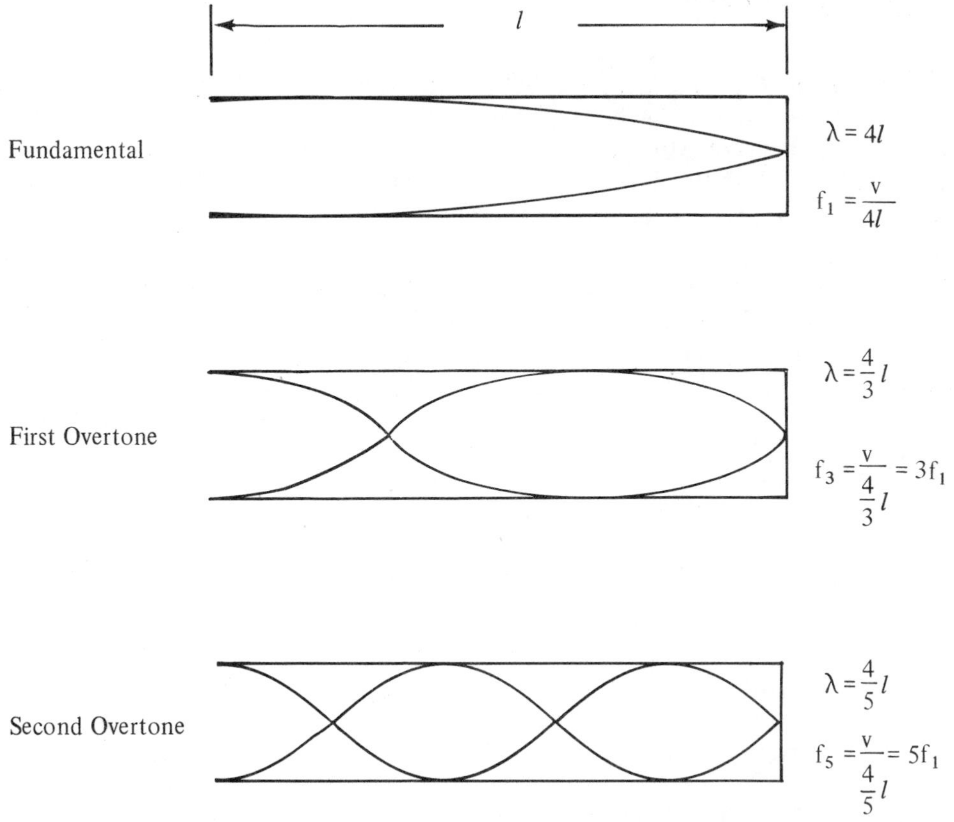

Figure 1–11 Natural modes of vibration for a hollow pipe closed at one end. Only odd harmonics are present.

Example 1.7

A hollow pipe has a length of 2 m. Determine the frequencies of the fundamental and first two overtones if (a) the pipe is open at each end, and (b) the pipe is closed at one end.

Assume $v = 330$ m/sec for the velocity of sound in air.

Solution

(a)

$$f_1 = \frac{v}{2l} = \frac{330}{2(2)} = 82.5 \text{ Hz}$$

$$f_2 = 2f_1 = 165 \text{ Hz}; \quad f_3 = 3f_1 = 248 \text{ Hz}$$

(b)

$$f_1 = \frac{v}{4l} = \frac{330}{4\,(2)} = 41.3 \text{ Hz}$$

$$f_3 = 3f_1 = 124 \text{ Hz}; \quad f_5 = 5f_1 = 207 \text{ Hz}$$

1.8 DOPPLER EFFECT

A common observation involving sound waves is the apparent pitch (or frequency) heard by an observer when the source of the sound is moving towards or away from the observer (or when the observer is moving towards or away from the source). This is called the *Doppler effect,* and anyone who has listened to a passing car sounding its horn has experienced the phenomenon.

First, consider the source of sound moving in the +x direction towards a stationary observer A, as in Figure 1–12(*a*). For simplicity, we will assume that the air is stationary. To observer A, the wavelength is *decreased* by the distance which the source travels in one cycle of the sound vibration. This apparent decrease in wavelength is given by $\Delta\lambda = v_s/f$, where v_s is the velocity of the source and f is the frequency of the source. If v is the velocity of sound, then the *apparent* frequency heard by observer A is

$$f' = \frac{v}{\lambda'} = \frac{v}{\lambda - \Delta\lambda} = \frac{v}{\frac{v}{f} - \frac{v_s}{f}}$$

$$f' = f\left(\frac{v}{v - v_s}\right) \qquad (1.30)$$

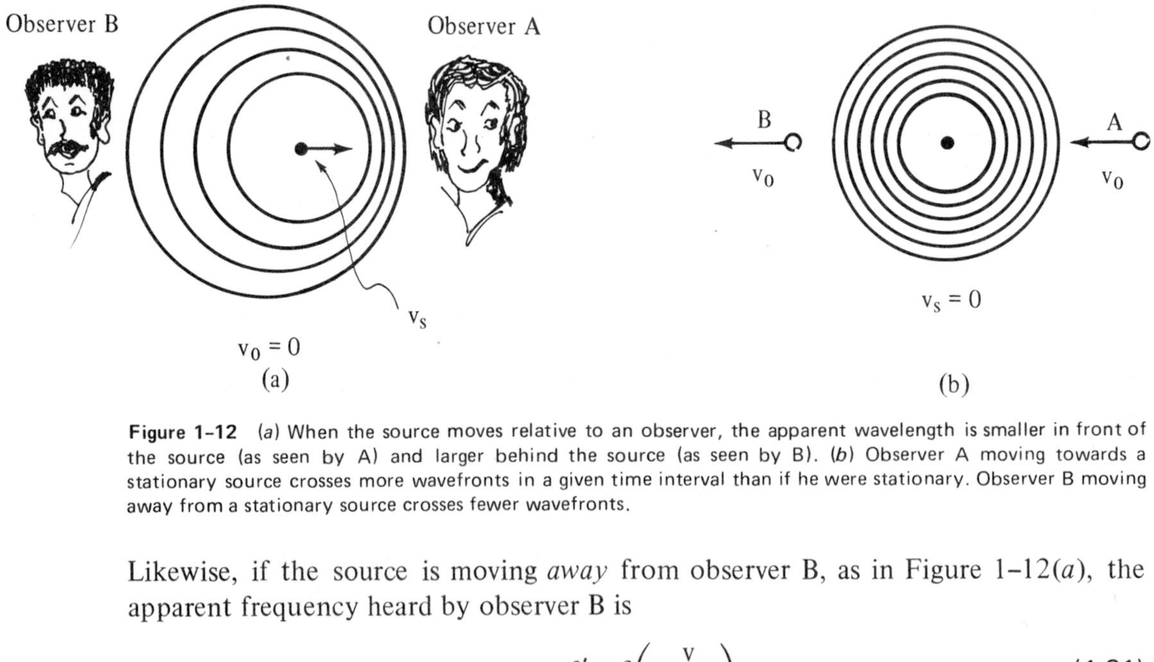

Figure 1–12 (*a*) When the source moves relative to an observer, the apparent wavelength is smaller in front of the source (as seen by A) and larger behind the source (as seen by B). (*b*) Observer A moving towards a stationary source crosses more wavefronts in a given time interval than if he were stationary. Observer B moving away from a stationary source crosses fewer wavefronts.

Likewise, if the source is moving *away* from observer B, as in Figure 1–12(*a*), the apparent frequency heard by observer B is

$$f' = f\left(\frac{v}{v + v_s}\right) \qquad (1.31)$$

Finally, if the source is stationary, and the observer moves towards or away from the source with a velocity v_0, as in Figure 1–12(*b*), the apparent frequency heard by the observer is

$$f' = f\left(\frac{v \pm v_0}{v}\right) \qquad (1.32)$$

where the plus sign refers to observer A moving towards the source and the minus sign refers to observer B moving away from the source. If *both* the source and the observer are in motion, the apparent frequency is the combination given by

$$f' = f\left(\frac{v \pm v_0}{v \mp v_s}\right) \tag{1.33}$$

Example 1.8

A train traveling parallel to a highway at a speed of 30 mph sounds its whistle, whose true frequency is 300 Hz. What is the frequency heard by a passenger in a car traveling in the opposite direction to the train at a speed of 60 mph as (a) the car approaches the train, and (b) the car has passed the train and they recede from each other? Take the velocity of sound in air to be 1080 ft/sec.

Solution

First, we note that 30 mph ≅ 44 ft/sec, and 60 mph ≅ 88 ft/sec.

(a) $f' = f\left(\frac{v + v_0}{v - v_s}\right) = 300\left(\frac{1080 + 88}{1080 - 44}\right) = 300\left(\frac{1168}{1036}\right) \cong 340 \text{ Hz}$

(b) $f' = f\left(\frac{v - v_0}{v + v_s}\right) = 300\left(\frac{1080 - 88}{1080 + 44}\right) = 300\left(\frac{992}{1124}\right) \cong 260 \text{ Hz}$

Therefore, the change in frequency as heard by the passenger as the train passes by is 340 - 260 = 80 Hz.

If the observer is at rest, but the source moves with a speed $v_s > v$, that is, its speed is greater than the speed of sound in that medium, the resulting wave is a *shock wave.* The wave front has a *conical* shape with a half angle given by $\sin \alpha = v/v_s$, as in Figure 1–13. The ratio v_s/v is referred to as the *Mach number.* The shock wave caused by the supersonic speed carries a great deal of energy in the form of a highly compressed "wall" of air, which in turn can cause a great deal of damage, for example, to structures, or to the ear. Similar wave fronts are observed in the wakes of a speed boat, when the boat travels faster than the speed of surface water waves.

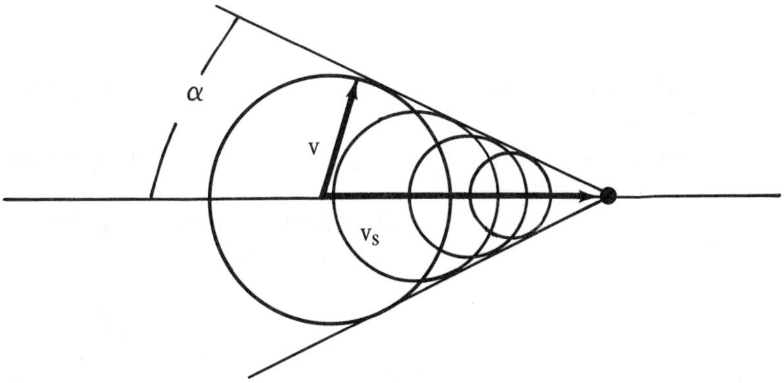

Figure 1–13 Schematic of a shock wave.

1.9 PROGRAMMED EXERCISES

1.A

A pulse travels down a stretched string as shown. What kind of traveling wave is this? Describe the motion of the string "particles."

This is a *transverse wave.* The string particles move *perpendicular* to the direction of motion of the pulse.

1.B

If the pulse moves to the right with a speed v (called the phase velocity), what general expression gives the amplitude of the pulse, y, as a function of x and t?

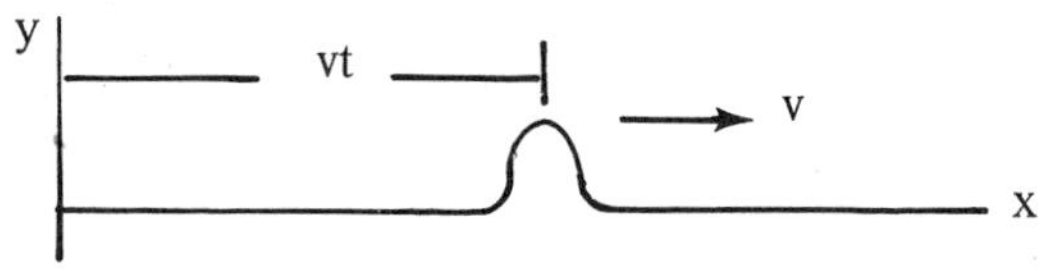

$$y = f(x - vt) \qquad (1)$$

Note that this function gives the shape of the pulse. The shape doesn't change and an observer moving along with the pulse at the speed v would also "see" the same shape.

1.C

If the pulse moves to the *left* with a speed v, what is the form of y?

$$y = f(x + vt)$$

1.D

The argument x - vt is called the *phase* of the traveling wave. Since the shape of the wave, that is, y(x, t), remains constant, what can we say about the phase velocity?

$$y = f(x - vt) = \text{constant}$$

$$\therefore \qquad x - vt = \text{constant}$$

Differentiating this expression with respect to time gives

$$v = \frac{dx}{dt}$$

We could have guessed this from the definition of velocity.

1.E

A sinusoidal wave traveling to the right on a string gives rise to a transverse displacement

$$y = A\sin\frac{2\pi}{\lambda}(x - vt)$$

Sketch this function at t = 0 and some later time t. Define the parameters A and λ in the expression.

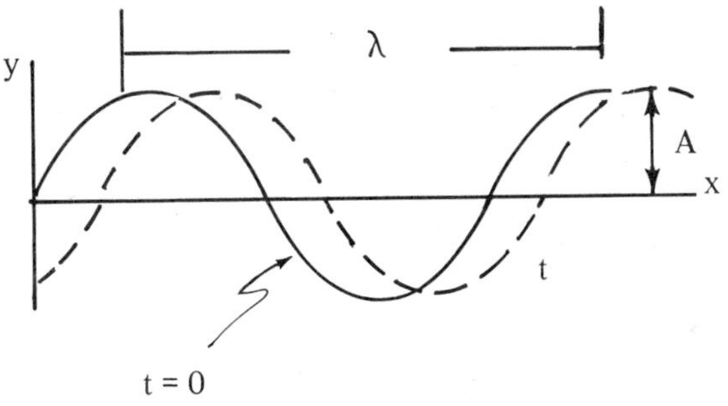

A is the *maximum* displacement of any particle of the string, and λ is the *wavelength* or the distance between two adjacent points having the same phase.

1.F

We can also write the displacement of the harmonic wave in the form

$$y = A\sin(kx - \omega t)$$

Compare this with the expression

$$y = A\sin\frac{2\pi}{\lambda}(x - vt)$$

and obtain expressions for k and ω in terms of λ and v.

$$k = \frac{2\pi}{\lambda} \qquad (3)$$

called the *wave number.*

$$\omega = 2\pi\frac{v}{\lambda} = kv \qquad (4)$$

is the *angular frequency* of the wave.

1.G

Each point on the string exhibits *simple harmonic motion* in the y direction. The period of this motion (or the time for one complete vibration) is T. (a) How is T related to λ? (b) What is the frequency f of the wave, that is, the number of vibrations per second? (c) How is T related to ω?

(a) $\frac{v}{\lambda} = \frac{1}{T}$ (5)

(b) $f = \frac{1}{T}\ \text{sec}^{-1}$ (7)

(c) $\omega = \frac{2\pi}{T}\ \frac{\text{rad}}{\text{sec}}$ (6)

It also follows that

$$v = \lambda f$$

and

$$\omega = 2\pi f$$

1.H

A sinusoidal wave has the form

$$y = 5.0\sin(2x - 3t)\ \text{cm},$$

where x is measured in cm. Determine A, λ, T, f and v for this wave.

Comparing this with the expression

$$y = A\sin(kx - \omega t)$$

gives A = 5.0 cm;

$$k = \frac{2\pi}{\lambda} = 2, \quad \therefore \lambda = \pi\ \text{cm};$$

$$\omega = \frac{2\pi}{T} = 3, \quad \therefore T = \frac{2\pi}{3}\ \text{sec};$$

$$f = \frac{1}{T} = \frac{3}{2\pi}\ \text{Hz};$$

$$v = \lambda f = \pi\left(\frac{3}{2\pi}\right) = 1.5\ \text{cm/sec}.$$

1.I

Determine the *transverse velocity* v_y of a point on the sinusoidal wave given by

$$y = 5.0\sin(2x - 3t)\ \text{cm}.$$

Obtain a numerical value of v_y at the point x = 2 cm for t = 1 sec. Note that the argument (2x - 3t) is in *radians.* Recall that 1 rad = 57.3°.

$$v_y = \frac{\partial y}{\partial t} = 5.0\frac{\partial}{\partial t}\sin(2x - 3t)$$

$$v_y = 5.0(-3)\cos(2x - 3t)\ \frac{\text{cm}}{\text{sec}}$$

$$v_y\ (x = 2, t = 1) = -15\cos(4 - 3)$$

$$= -15\cos(1)$$

$$= -8.1\ \frac{\text{cm}}{\text{sec}}$$

1.J

Calculate the *transverse acceleration* a_y of a point on the sinusoidal wave given by

$$y = 5.0\sin(2x - 3t)\ \text{cm}.$$

Obtain a numerical value for a_y at x = 2 cm and t = 1 sec. (Note that v_y and a_y are in the y direction, that is, *perpendicular* to the direction of propagation of the wave.)

$$a_y = \frac{\partial v_y}{\partial t} = -15\frac{\partial}{\partial t}\cos(2x - 3t)$$

$$a_y = -15(-3)\sin(2x - 3t)\ \frac{\text{cm}}{\text{sec}^2}$$

$$a_y\ (x = 2, t = 1) = 45\sin(1)$$

$$= 38\ \frac{\text{cm}}{\text{sec}^2}$$

1.K

What are the maximum values of v_y and a_y for *any* point on the sinusoidal wave described in 1.I and 1.J?

By inspection of the expressions for v_y and a_y, we see that

$$(v_y)_{max} = 15\ \text{cm/sec}$$

$$(a_y)_{max} = 45\ \text{cm/sec}^2$$

2 The wave equation corresponding to traveling waves on a stretched string is given by $\frac{\partial^2 y}{\partial t^2} = \frac{F}{\mu}\frac{\partial^2 y}{\partial x^2}$, where F is the tension in the string and μ is its mass per unit length. We will show that a *harmonic* wave is one solution to this equation.

2.A

The wave equation is used to describe many wave phenomena. What is the main assumption one has to make in applying the wave equation?

It must be assumed that the disturbance of the medium from equilibrium is *small,* that is, the wave amplitude is small.

2.B

A harmonic traveling wave on a string has the form $y = A\sin(kx - \omega t)$. Evaluate $\partial^2 y/\partial t^2$. Note that the partial derivative operation $\partial y/\partial t$ implies that the derivative is taken with respect to t. All other quantities are taken to be constant. Therefore,

$$\frac{\partial}{\partial t}(kx - \omega t) = -\omega.$$

$$y = A \sin(kx - \omega t)$$

$$\frac{\partial y}{\partial t} = A\frac{\partial}{\partial t}\sin(kx - \omega t)$$

$$= A\cos(kx - \omega t)\frac{\partial}{\partial t}(kx - \omega t)$$

$$\frac{\partial y}{\partial t} = -A\omega\cos(kx - \omega t)$$

$$\frac{\partial^2 y}{\partial t^2} = -A\omega\frac{\partial}{\partial t}\cos(kx - \omega t)$$

$$= +A\omega\sin(kx - \omega t)\frac{\partial}{\partial t}(kx - \omega t)$$

$$\frac{\partial^2 y}{\partial t^2} = -A\omega^2\sin(kx - \omega t) \qquad (1)$$

2.C

Now evaluate the quantity $\partial^2 y/\partial x^2$ which appears in the wave equation, assuming $y = A\sin(kx - \omega t)$. Again, note that

$$\frac{\partial}{\partial x}(kx - \omega t) = k.$$

$$\frac{\partial y}{\partial x} = A\frac{\partial}{\partial x}\sin(kx - \omega t)$$

$$\frac{\partial y}{\partial x} = A\cos(kx - \omega t)\frac{\partial}{\partial x}(kx - \omega t)$$

$$\frac{\partial y}{\partial x} = Ak\cos(kx - \omega t)$$

$$\frac{\partial^2 y}{\partial x^2} = Ak\frac{\partial}{\partial x}\cos(kx - \omega t)$$

$$\frac{\partial^2 y}{\partial x^2} = -Ak^2\sin(kx - \omega t) \qquad (2)$$

2.D

Substitute (1) and (2) in the wave equation

$$\frac{\partial^2 y}{\partial t^2} = \frac{F}{\mu}\frac{\partial^2 y}{\partial x^2}$$

and obtain a relation between ω and k.

$$-A\omega^2 \sin(kx - \omega t) = \frac{F}{\mu}[(-Ak^2 \sin(kx - \omega t)]$$

$$\omega^2 = \frac{F}{\mu}k^2$$

or,

$$\omega = \sqrt{\frac{F}{\mu}}\,k \qquad (3)$$

2.E

We have verified that the harmonic wave is a solution of the wave equation. Recall that for any periodic wave ω and k are related by the expression $\omega = kv$, where v is the speed of the wave. Use this fact and (3) to obtain the speed of a traveling wave on a stretched string. Show that the result for v has dimensions of velocity.

$$\omega = \sqrt{\frac{F}{\mu}}\,k = kv$$

$$\therefore \quad v = \sqrt{\frac{F}{\mu}} \qquad (4)$$

$$[v] = \left[\frac{F}{\mu}\right]^{1/2} = \left[\frac{\text{Force}}{\text{Mass/Length}}\right]^{1/2}$$

$$= \left[\frac{ML/T^2}{M/L}\right]^{1/2} = \left[\frac{L^2}{T^2}\right]^{1/2}$$

$$[v] = \frac{L}{T}$$

3 A uniform rope of mass M, length l hangs vertically under its own weight as shown below. We wish to find the speed of transverse waves on this rope and the time it takes a pulse to travel the length of the rope.

3.A

Is the tension in the rope uniform? Explain.

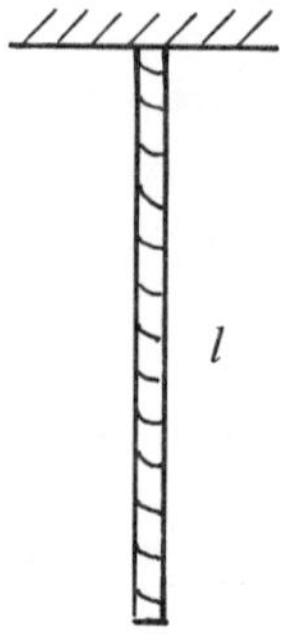

No. The upper part of the rope supports more weight than the lower part, so the tension will be a *maximum* at the top where it supports all its weight, and *zero* at the bottom.

3.B

What is the tension at the point where the rope is supported at the top? [*Hint:* Draw a free body diagram for the knot at the top.]

Since the entire weight of the rope is supported at this point, $F_m = Mg$, the *maximum* tension.

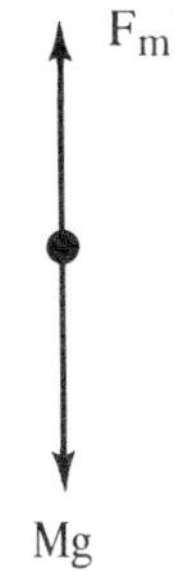

3.C

Let y = 0 be the coordinate of the lowest point on the rope. Determine the mass M′ of a segment of the rope whose length is y.

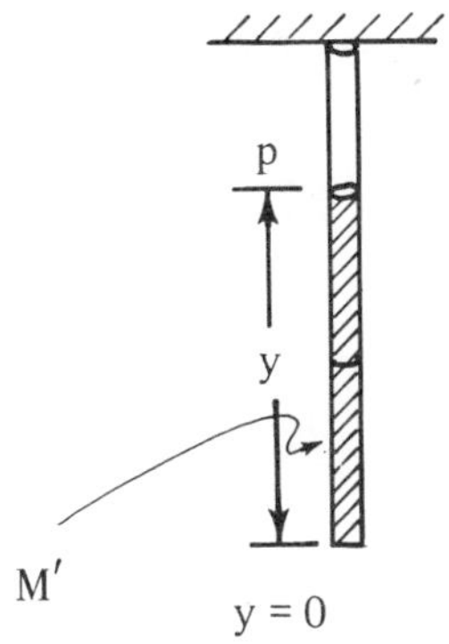

Since the rope is uniform,

$$M' = \frac{y}{l} M \qquad (1)$$

That is, for y = 0, M′ = 0. For y = l, M′ = M.

3.D

Now determine the tension F in the rope at the point p, a distance y above the bottom. (We need F in order to obtain the speed of transverse waves.)

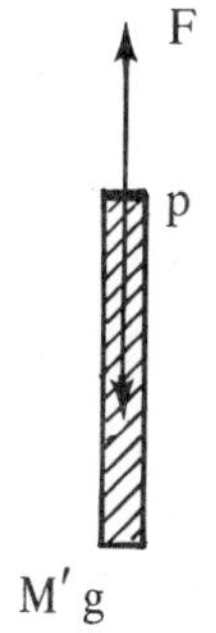

If we isolate the point p, we see that this point must only support *part* of the rope, namely M′g.

$$\therefore \quad F = M'g = M \frac{y}{l} g \qquad (2)$$

3.E

Note that (2) tells us that F = 0 at y = 0 (the bottom) and F = Mg at y = l (the top). Now determine the speed of transverse waves on the rope, using the relation $v = \sqrt{F/\mu}$.

$$v = \sqrt{\frac{F}{\mu}} = \sqrt{\frac{F}{M/l}}$$

$$v = \sqrt{\frac{M\frac{y}{l}g}{M/l}}$$

$$v = \sqrt{gy} \qquad (3)$$

Note that v depends on the coordinate y. v is a maximum at the top and zero at the bottom!

3.F

Suppose a transverse wave starts from the top of a hanging rope which is 10 m in length. What is the speed of the wave at the top and at the midpoint of the rope?

At the top, y = 10 m, and using (3) gives

$$v = \sqrt{9.8 \times 10} = 9.9 \text{ m/sec}$$

At the midpoint, y = 5 m, so

$$v = \sqrt{9.8 \times 5} = 7 \text{ m/sec}$$

3.G

Use the fact that v = dy/dt, together with (3), and determine the time it takes a transverse wave to travel from the top to the bottom of the vertical rope of length l. You must integrate the expression, noting that v is a function of y and $v = -\sqrt{gy}$, since the wave travels in the *negative* y direction.

$$v = \frac{dy}{dt} = -\sqrt{gy}$$

$$\therefore \quad dt = -\frac{dy}{\sqrt{gy}}$$

At t = 0, y = l. When the wave reaches the bottom at time t, y = 0.

$$\therefore \quad \int_0^t dt = -\frac{1}{\sqrt{g}}\int_l^0 \frac{dy}{\sqrt{y}}$$

$$t = \frac{1}{\sqrt{g}}\int_0^l y^{-1/2}\, dy = \frac{1}{\sqrt{g}} 2y^{1/2}\Big]_0^l$$

$$t = 2\sqrt{\frac{l}{g}} \qquad (4)$$

Note that t is independent of M!

3.H

Obtain a numerical value for t if l = 25 m.

$$t = 2\sqrt{\frac{25}{9.8}} = 2\sqrt{2.55}$$

$$t = 3.2 \text{ sec}$$

The student should verify the dimensions of (4).

3.I

Suppose a mass m is suspended at the end of the rope as shown. What is the tension at the point p? What is the speed of transverse waves in this case?

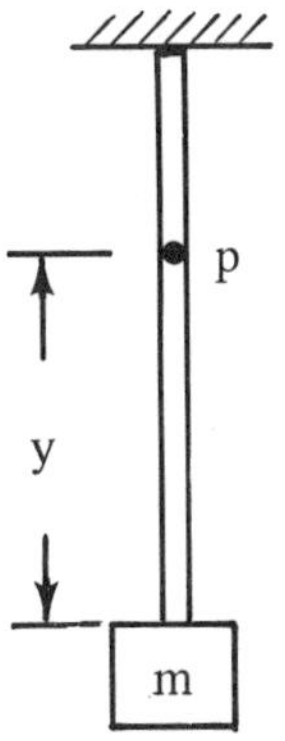

The point p now supports the rope segment of length y and the mass m. Using (2) we get

$$F = M'g + mg = Mg\frac{y}{l} + mg$$

$$\therefore\ v = \sqrt{\frac{F}{M/l}} = \sqrt{\frac{Mg\,y/l + mg}{M/l}}$$

$$v = \sqrt{gy + \frac{m}{M}gl} \qquad (5)$$

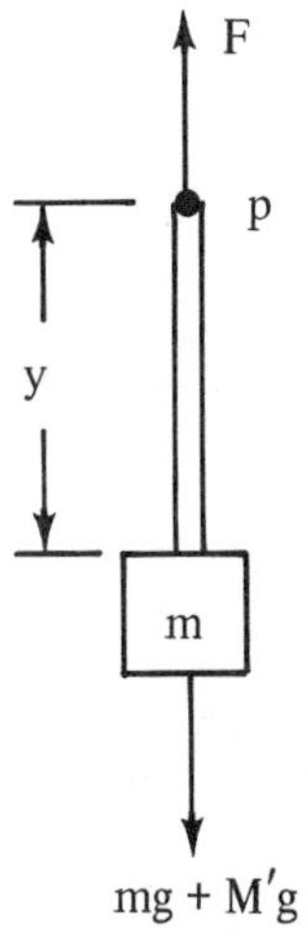

3.J

If a pulse travels from the top to the bottom when m is suspended as in 3.I, would the time of travel be greater or less than that given by (4)? Explain. (In this case, the tension is nonzero everywhere on the rope so v is never zero, and the pulse will reflect from the lower end and propagate upwards.)

Since v given by (5) is *greater* than the speed of transverse waves without m, ($v = \sqrt{gy}$), the time of travel would be *less* when m is present. The time can be found in the same manner as in frame 5.G, that is, integrate $v = \frac{dy}{dt}$ using (5) for v. Try it!

4.A

Discuss the *superposition* principle as applied to traveling waves. What is the condition under which the principle is valid?

The principle states that when two or more traveling waves propagate in a medium, the resultant displacement is the *vector sum* of the displacements that each wave produces. It is valid for *small* displacements of the medium from equilibrium.

4.B

If two pulses of equal amplitude move as shown below, sketch the resultant wave when they meet at x = 0 at t = 0. Sketch the pulses at some later time t > 0.

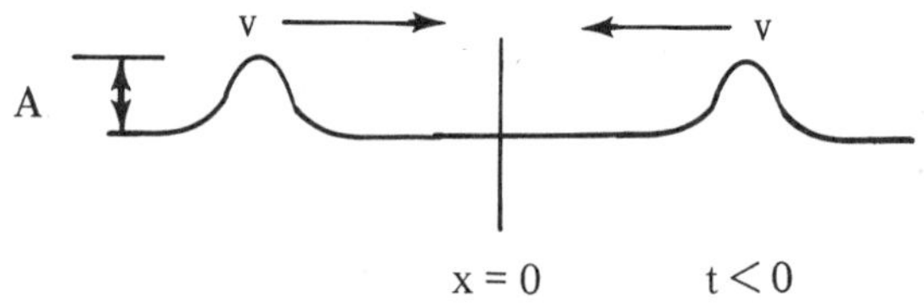

The two pulses superimpose to give a pulse of amplitude 2A.

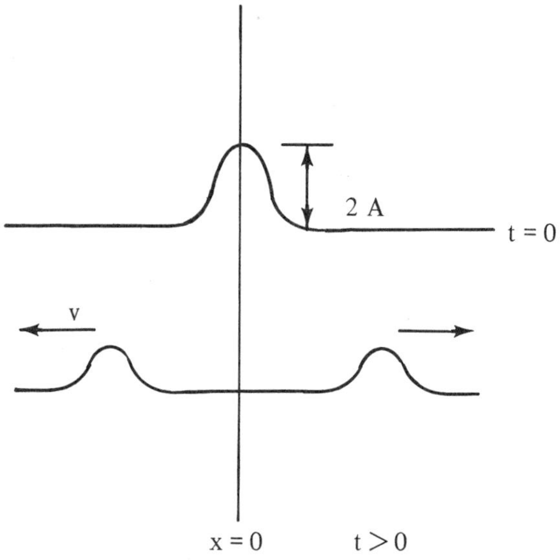

4.C

Suppose one pulse is inverted, as shown below. Sketch the resultant wave when they meet at t = 0, x = 0, and at some later time t > 0.

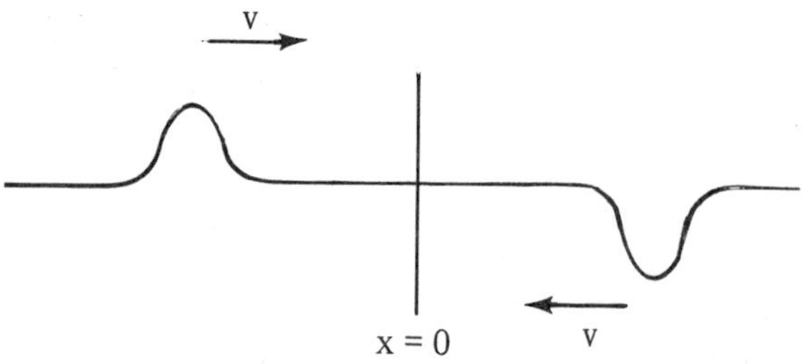

The pulses cancel each other when they meet at t = 0.

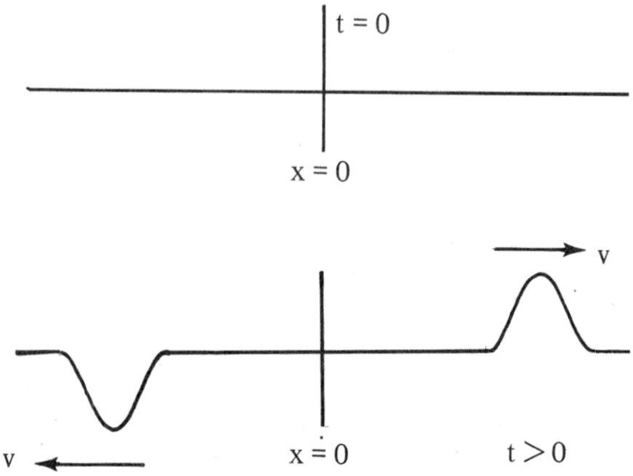

4.D

Two waves are represented by the equations

$$y_1 = A\sin\omega t \tag{1}$$

$$y_2 = A\sin(\omega t - \phi) \tag{2}$$

What is common to both waves? Define ϕ.

The two waves have the *same* amplitude A and angular frequency ω. ϕ is a phase angle, that is, the phase difference between the waves measured in *radians.*

4.E

If the two waves given by (1) and (2) propagate in the same medium, determine the *resultant amplitude* y using the *superposition* principle. Note that

$$\sin a + \sin b = 2\cos\left(\frac{a-b}{2}\right)\sin\left(\frac{a+b}{2}\right)$$

$$y = y_1 + y_2 = A\sin\omega t + A\sin(\omega t - \phi)$$

$$y = 2A\cos\left(\frac{\omega t - \omega t + \phi}{2}\right)\sin\left(\frac{\omega t + \omega t - \phi}{2}\right)$$

$$y = 2A\cos\left(\frac{\phi}{2}\right)\sin\left(\omega t - \frac{\phi}{2}\right) \tag{3}$$

4.F

The amplitude of the resultant wave given in (3) is $2A\cos(\phi/2)$. What values of ϕ give a *maximum* amplitude? What kind of interference is this called?

The constant $2A\cos(\phi/2)$ has a *maximum* value of 2A for $\phi = 0, 2\pi, 4\pi, \ldots$. These values of ϕ correspond to *constructive* interference.

4.G

For what values of ϕ is the resultant amplitude *zero*? What kind of interference is this called?

$2A\cos(\phi/2)$ has zero values for $\phi = \pi, 3\pi, 5\pi, \ldots$. These values of ϕ correspond to *destructive* interference.

5 *Standing waves on a stretched string* are formed as the result of the superposition of traveling waves traveling in opposite directions on the string.

5.A

Assume that the string has a length l and is fixed at each end as shown below. If the string is driven at one end with a sinusoidal source, why are waves produced traveling in opposite directions?

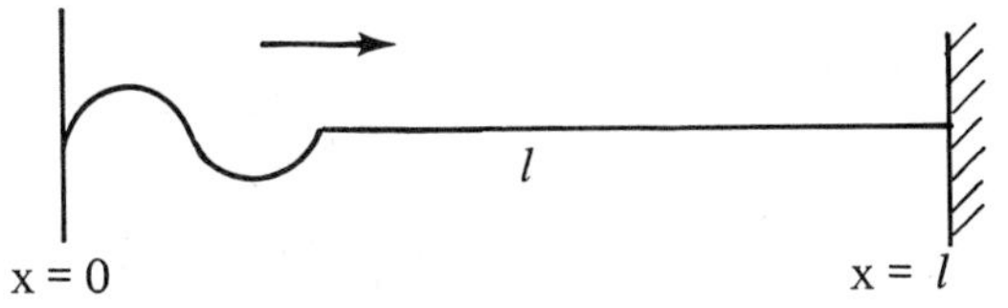

The initial disturbance will propagate to the right until it meets the fixed support. It will then be reflected to the left with practically the same amplitude. Therefore, sinusoidal waves are produced traveling in opposite directions.

5.B

If the waves traveling to the right and left are given by $y_1 = A \sin(kx - \omega t)$ and $y_2 = A \sin(kx + \omega t)$, respectively, find the resultant amplitude by using the superposition principle.

$$y = y_1 + y_2$$

$$y = A\sin(kx - \omega t) + A\sin(kx + \omega t)$$

But $\sin(a \pm b) = \sin a \cos b \pm \cos a \sin b$

$$\therefore y = A\sin kx \cos\omega t - A\cos kx \sin\omega t$$
$$+ A\sin kx \cos\omega t + A\cos kx \sin\omega t$$

$$y = 2A\sin kx \cos\omega t \qquad (1)$$

5.C

Note that (1) is *not* of the form $y = f(x \pm vt)$; therefore, it is *not* a traveling wave. The resultant wave is a *standing wave.* What is its amplitude?

By inspection of (1), we see that the wave's amplitude is given by $2A\sin kx$. That is, each point on the string has an amplitude dependent on its x coordinate measured along the length of the string.

5.D

Describe the motion of each particle making up the string.

Each particle undergoes *simple harmonic motion* in the transverse (y) direction according to Equation (1).

5.E

What are the *boundary conditions* on the amplitude of the wave? That is, what is y at $x = 0$ and $x = l$?

Since the string is fixed at each end, we require that $y = 0$ at $x = 0$ *and* $x = l$.

5.F

Show that (1) satisfies the condition that $y = 0$ at $x = 0$.

Since the coefficient of (1) contains $\sin kx$, we see that for $x = 0$, $\sin 0 = 0$, so $y = 0$.

5.G

The function $\sin kx$ goes through an infinite number of zeroes. The corresponding amplitude y is zero at these points, called *nodes.* What are the values of x such that the amplitude is zero? Recall that $k = 2\pi/\lambda$, where λ is the wavelength.

$$\sin kx = 0$$

$$\therefore \quad kx = n\pi \quad (n = 1, 2, 3, \ldots)$$

$$x = \frac{n\pi}{k} = \frac{n\pi}{2\pi/\lambda}$$

$$x = \frac{n\lambda}{2} \quad (n = 1, 2, 3, \ldots) \qquad (2)$$

The *nodes* are therefore separated by $\lambda/2$, that is, one-half of a wavelength.

5.H

The amplitude of the resultant wave is 2Asinkx, and has a maximum value of 2A. For what values of x is the amplitude a maximum? (These points are called *antinodes.*) Determine these values of x in terms of λ.

$|2A\sin kx|$ is a maximum for $|\sin kx| = 1$ or,

$$kx = n\frac{\pi}{2} \quad (n = 1, 3, 5, \ldots)$$

Since $k = 2\pi/\lambda$, we get

$$x = \frac{n\lambda}{4} \quad (n = 1, 3, 5, \ldots) \qquad (3)$$

The *antinodes* are also separated by $\lambda/2$ and lie halfway between nodes.

5.I

Use the condition that y = 0 at x = l to obtain the *allowed* frequencies, or natural frequencies, of vibration for standing waves on the stretched string. Recall that $v = \sqrt{F/\mu}$ for traveling waves on a string.

Using (2), we can write, at x = l,

$$n\frac{\lambda}{2} = l$$

$$\lambda = \frac{2l}{n} \qquad (4)$$

where n is 1, 2, 3, ... But $v = f\lambda$, $\therefore$

$$f_n = \frac{v}{\lambda} = \frac{n}{2l}v$$

$$f_n = \frac{n}{2l}\sqrt{\frac{F}{\mu}} \quad (n = 1, 2, 3, \ldots) \qquad (5)$$

5.J

The fundamental frequency is f_1, which occurs for n = 1. Sketch this *fundamental* vibration. Label the nodes and antinodes. Note that $f_1 = v/2l = v/\lambda$, so $l = \lambda/2$. Sketch the second harmonic corresponding to n = 2, where $f_2 = v/l = v/\lambda$, and $l = \lambda$.

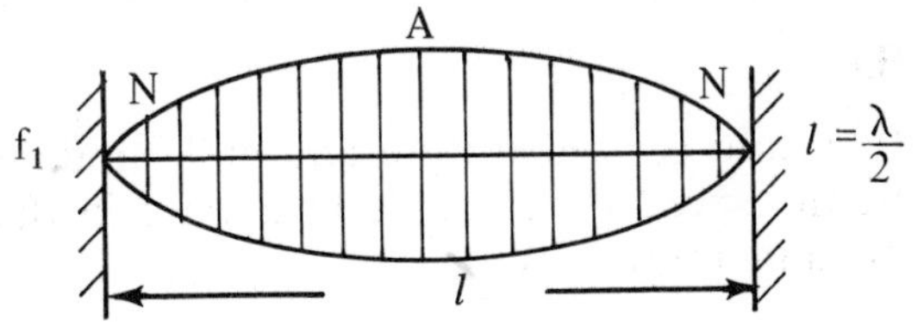

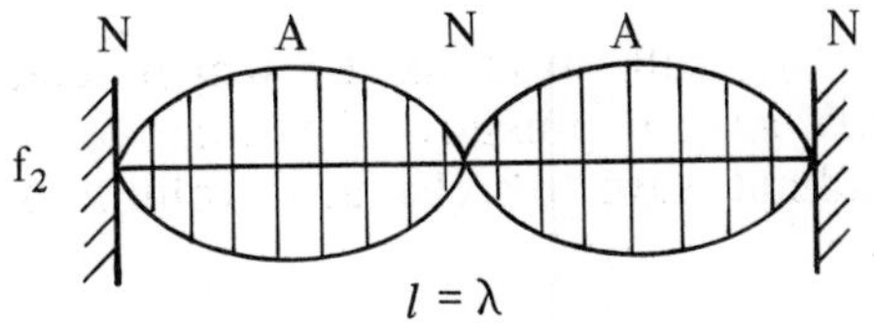

5.K

Calculate the fundamental frequency (first harmonic) and second harmonic of a string of length 2 m, mass 50 g, which is under a tension of 20 newtons.

$$\mu = \frac{50 \times 10^{-3}\ \text{kg}}{2\ \text{m}} = 2.5 \times 10^{-2}\ \text{kg/m}$$

$$f_1 = \frac{1}{2l}\sqrt{\frac{F}{\mu}} = \frac{1}{2(2)}\sqrt{\frac{20}{2.5 \times 10^{-2}}}$$

$$f_1 = \frac{1}{4}\sqrt{8 \times 10^2} = 7.1\ \text{Hz}$$

$$f_2 = 2f_1 = 14.2\ \text{Hz}$$

5.L

How long does it take a pulse to make one complete round trip for the string described in 5.K?

The speed of the pulse is

$$v = \sqrt{\frac{F}{\mu}} = \sqrt{\frac{20}{2.5 \times 10^{-2}}} = 28\ \text{m/sec}$$

Since it must travel a distance of $2l = 4$ m, the time required is

$$t = \frac{2l}{v} = \frac{4\ \text{m}}{28\ \text{m/sec}} = 0.14\ \text{sec}$$

Note that this is equivalent to $1/f_1$. Why?

6 *Longitudinal waves* are mechanical waves which propagate in solids, liquids or gases.

6.A

In general, longitudinal waves differ from transverse waves in that the motion of the medium for longitudinal waves is __________ to the direction of the wave. One common example of a longitudinal wave is a __________.

parallel

sound wave

6.B

If a fluid has a density ρ and bulk modulus B, the speed of a longitudinal wave in that medium is given by__________.

$$v = \sqrt{\frac{B}{\rho}} \qquad (1)$$

6.C

Define the bulk modulus B, which relates the *change* in pressure Δp due to the longitudinal wave to a change in volume ΔV.

$$B = -\frac{\Delta p}{\Delta V/V} \qquad (2)$$

Note that B is always positive, since the ratio $\Delta p/\Delta V$ is always negative.

6.D

According to (2), B has units of pressure, that is, Force/Area. Use this fact and the units of ρ to verify that v has units of L/T. Note that

$$v \sim \sqrt{\text{Elastic Property/Inertial property}}.$$

From (1), we get

$$[v] = \left[\frac{B}{\rho}\right]^{1/2} = \left[\frac{\text{Force/Area}}{\text{Density}}\right]^{1/2}$$

$$= \left[\frac{ML/T^2}{L^2 \cdot M/L^3}\right]^{1/2}$$

$$[v] = \left[\frac{L^2}{T^2}\right]^{1/2} = \frac{L}{T}$$

6.E

Consider a compressional wave traveling along the x axis, where the *displacements* of the air molecules from equilibrium are given by

$$\xi = \xi_m \sin(kx - \omega t) \qquad (3)$$

What is the direction of the displacement ξ?

ξ is the displacement *along the x axis*. That is, the air molecules oscillate about their equilibrium position with an amplitude ξ_m.

6.F

Show that (3) represents a solution to the wave equation

$$\frac{\partial^2 \xi}{\partial t^2} = \frac{B}{\rho}\frac{\partial^2 \xi}{\partial x^2} \qquad (4)$$

provided that $v = \omega/k = \sqrt{B/\rho}$. (Note that (4) is of the same form as the wave equation for a stretched string given in Programmed Exercise 2 of this chapter.)

$$\frac{\partial \xi}{\partial t} = -\,\xi_m\,\omega \cos(kx - \omega t)$$

$$\frac{\partial^2 \xi}{\partial t^2} = -\xi_m\,\omega^2 \sin(kx - \omega t) \qquad (5)$$

Likewise,

$$\frac{\partial^2 \xi}{\partial x^2} = -\,\xi_m\,k^2 \sin(kx - \omega t) \qquad (6)$$

Substituting (5) and (6) into (4) gives

$$\omega^2 = k^2\,\frac{B}{\rho}$$

or,

$$v = \frac{\omega}{k} = \sqrt{\frac{B}{\rho}}$$

6.G

If a portion of the fluid of cross-sectional area A has an initial volume $V = A\Delta x$, and the width of that portion changes by an amount $\Delta\xi$ during the compression, the change in volume in $\Delta V = A\Delta\xi$. Use this, together with (2), to obtain an expression for the *pressure variation* in the fluid. What does the expression reduce to as $\Delta x \to 0$?

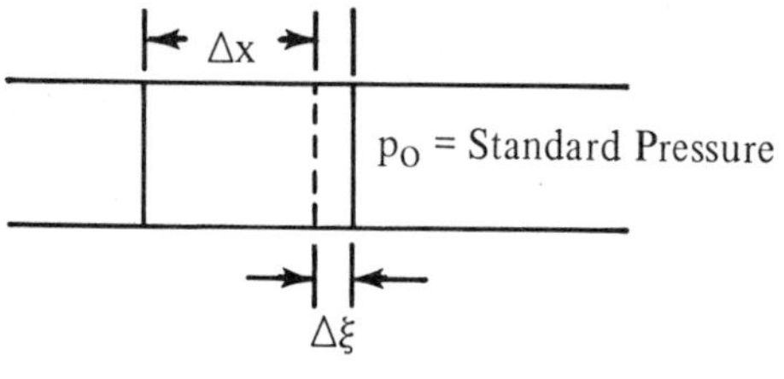

From (2) we get

$$\Delta p = -B\frac{\Delta V}{V}$$

$$\Delta p = -BA\frac{\Delta\xi}{\Delta x}$$

as $\Delta x \to 0$,

$$\Delta p = -BA\frac{\partial\xi}{\partial x} \quad (7)$$

Note that we must use *partial* derivatives since ξ is a function of x and t.

6.H

Use the fact that the particle displacement ξ is harmonic, given by (3), and show that the pressure variation Δp is also harmonic.

$$\Delta p = -BA\frac{\partial}{\partial x}[\xi_m \sin(kx - \omega t)]$$

$$\Delta p = -BA\,\xi_m k\cos(kx - \omega t) \quad (8)$$

Note that the pressure variation is a maximum when the particle displacement ξ is zero. Likewise, when ξ is a maximum, Δp is zero.

6.I

If a compressional wave propagates in a *gas,* the pressure variations are so rapid that little heat transfer takes place. We then can assume an adiabatic process for which pV^γ = constant. Use this assumption, together with (1) and (2), and verify that $v = \sqrt{\gamma p/\rho}$ for a gas.

In the limit $\Delta V \to 0$, we can write (2) as

$$B = -V\frac{dp}{dV}$$

But since pV^γ = constant, we have

$$p = \text{constant } V^{-\gamma}$$

$$\frac{dp}{dV} = \text{constant }(-\gamma)\,V^{-\gamma-1}$$

$$= -\gamma p V^{-1}$$

∴ B can be written as

$$B = -V(-\gamma p V^{-1}) = \gamma p$$

and the speed v is

$$v = \sqrt{\frac{B}{\rho}} = \sqrt{\frac{\gamma p}{\rho}} \quad (9)$$

6.J

For an *ideal gas*, recall that $pV = \mu kT$, where μ is the number of moles of gas. Use this, together with (9), and show that

$$v = \sqrt{\frac{\gamma RT}{M}}$$

where M = the mass of one mole of gas given by $M = \frac{m}{\mu}$, where m is the total mass.

$$pV = \mu RT$$

$$p = \frac{\mu RT}{V}$$

$$\frac{p}{\rho} = \frac{\mu RT}{V\rho} = \frac{\mu RT}{Vm/V} = \frac{\mu RT}{m}$$

$$\therefore \quad \frac{p}{\rho} = \frac{RT}{M}$$

and

$$v = \sqrt{\frac{\gamma p}{\rho}} = \sqrt{\frac{\gamma RT}{M}} \qquad (10)$$

6.K

The speed of sound waves in air at 0°C is 330 m/sec. Use this fact and (10) to determine the speed of sound in air at 50°C.

Since 0°C = 273°K, and 50°C = 323°K

$$v_0 = \sqrt{\frac{\gamma R(273)}{M}} \quad \text{at } 273°\text{K}$$

$$v = \sqrt{\frac{\gamma R(323)}{M}} \quad \text{at } 323°\text{K}$$

$$\frac{v}{v_0} = \left(\frac{323}{273}\right)^{1/2} = 1.09$$

$$v = 1.09\,(330) \cong 360 \text{ m/sec}$$

6.L

The natural modes of sound vibration in a pipe of length l open at both ends is given by

$$f_n = n\frac{v}{2l}\,(n = 1, 2, 3, \ldots).$$

If the fundamental has a frequency of 150 Hz at 0°C, determine the *shift* in frequency when the air is heated to 50°C.

At 0°C,

$$f_1 = \frac{v_0}{2l} = 150 \text{ Hz}$$

At 50°C,

$$f_1' = \frac{v}{2l}$$

$$\therefore \quad \frac{f_1'}{f_1} = \frac{v}{v_0} = 1.09$$

$$f_1' = 1.09\,(150) = 163 \text{ Hz}$$

$$\Delta f = f_1' - f_1 = 13 \text{ Hz}$$

7 A steel rail on a railroad track is struck with a hammer by a worker. A listener some distance d from the worker with one ear on the rail hears *two* sounds separated in time by 2.0 sec. One sound wave corresponds to waves propagated through the air, the other through the rail. From this information, we wish to find the separation of the worker and listener, assuming the rail is straight.

7.A

First let us find the speed of longitudinal waves in steel. Use the fact that Young's modulus for steel is 2.0×10^{12} dynes/cm^2 and the density of steel if 7.8 g/cm^3.

For a long rod, we have

$$v = \sqrt{\frac{Y}{\rho}}$$

$$v_s = \sqrt{\frac{2.0 \times 10^{12}\ \text{dynes/cm}^2}{7.8\ \text{g/cm}^3}}$$

$$v_s \cong 5.1 \times 10^5\ \text{cm/sec}$$

$$v_s = 5100\ \text{m/sec} \qquad (1)$$

7.B

Use the fact that the speed of sound in air is 330 m/sec at STP and write relations for the *time* it takes the sound to travel the distance d in air (t_a) and in steel (t_s).

In air, we have

$$t_a = \frac{d}{v_a} = \frac{d}{330}\ \text{sec} \qquad (2)$$

where d is in meters. In steel,

$$t_s = \frac{d}{v_s} = \frac{d}{5100}\ \text{sec} \qquad (3)$$

Note that $t_a > t_s$, so the wave traveling in the rail reaches the listener first.

7.C

From the information stated in the exercise, $t_a - t_s = 2.0$ sec. Use this fact together with (1) and (3) to find the separation d between the listener and the worker.

$$t_a - t_s = \frac{d}{330} - \frac{d}{5100} = 2.0\ \text{sec}$$

Solving for d gives

$$d \cong 710\ \text{m}$$

1.10 SUMMARY

A harmonic wave traveling in the positive x direction can be represented by the expression

$$y(x, t) = A\sin(kx - \omega t - \phi) \tag{1.9}$$

where A is the maximum amplitude of the wave, ϕ is a phase constant, k is the wave number, and ω is the angular frequency given by the expressions

$$k = \frac{2\pi}{\lambda}$$

$$\omega = 2\pi f = 2\pi \frac{v}{\lambda} = kv$$

If the displacement of the particles y is perpendicular to the direction of propagation, the wave is *transverse.* If the displacement is parallel to the direction of propagation, the wave is *longitudinal.* A traveling wave on a *stretched string,* whose tension is F and whose mass per unit length is μ, has a speed given by

$$v = \sqrt{\frac{F}{\mu}} \tag{1.11}$$

Waves on strings are transverse, assuming y is small compared to the length of the string.

The *superposition principle* states that the resultant displacement of a medium when two or more waves propagate is equal to the vector sum of the displacements that each wave produces. Waves are then said to *interfere* with each other. The resultant waveform of the superposition of two traveling waves depends on their amplitudes, frequencies and phase difference. If a string of length l is fixed at both ends, *standing waves* can be set up, resulting from superposition of traveling waves reflected from the fixed ends. The natural frequencies of vibration of the stretched string are given by

$$f_n = \frac{n}{2l}\sqrt{\frac{F}{\mu}} \quad (n = 1, 2, 3, \ldots) \tag{1.23}$$

Sound waves are a common example of longitudinal waves. The disturbance is characterized by variations in the density (or pressure) of the medium. The velocity of longitudinal waves in a fluid of density ρ, bulk modulus B is given by

$$v = \sqrt{\frac{B}{\rho}} \tag{1.24}$$

In general, the velocity of longitudinal waves in an elastic medium is given by the square root of an elastic property divided by an inertial property. Standing sound waves can also be established in such hollow structures as pipes. The resonant frequencies of a hollow pipe are determined by the length of the pipe, the velocity of sound in the medium and the nature of the boundary conditions.

When an observer and the source of sound waves are in relative motion in a medium where the speed of sound is v, the apparent frequency heard by the observer is given by

Doppler Effect
$$f' = f\left(\frac{v \pm v_0}{v \mp v_s}\right) \tag{1.33}$$

where v_0 is the speed of the observer, v_s is the speed of the source and f is the true frequency of the source. The upper signs refer to the case where the separation is decreasing, while the lower signs refer to the case where the separation is increasing.

1.11 PROBLEMS

1. A traveling wave propagates according to the expression

$$y = 4.0\sin(2.0x - 3.0t) \text{ cm}.$$

Determine (a) the amplitude, (b) the wavelength, (c) the frequency and (d) the period of the wave.

2. A traveling wave on a string is harmonic, and its amplitude is given by

$$y = 3.0 \cos(\pi x - 4\pi t) \text{ cm}$$

where x is in centimeters. (a) Determine the wavelength and period of the wave. (b) Find the transverse velocity and transverse acceleration at any time t. (c) Calculate the transverse velocity and transverse acceleration at $t = 0$ for a point located at $x = 0.25$ cm. (d) What are the maximum values of the transverse velocity and transverse acceleration?

3. All electromagnetic waves (light waves, radio waves, radar waves, infrared rays, and others) travel with a speed of 3.0×10^8 m/sec in a vacuum. (a) Determine the frequency of an electromagnetic wave whose wavelength is 5 cm (microwave range). (b) Determine the period of oscillation of an electromagnetic wave whose wavelength is 10 m (radio wavelength).

4. Determine the speed of transverse waves on a stretched string, which is under a tension of 80 newtons, if the string has a length of 2 m and a mass of 5 g.

5. A harmonic traveling wave moving in the positive x direction has an amplitude of 2.0 cm, a wavelength of 4.0 cm and a frequency of 5 Hz. (a) Determine the speed of the wave and (b) write an equation for the amplitude as a function of x and t.

6. Show that the speed of longitudinal waves along a spring of force constant k is

$$v = \sqrt{\frac{kl}{\mu}}$$

where l is the unstretched length of the spring and μ is the mass per unit length.

7. A spring of mass 0.4 kg has an unstretched length of 2 m and a force constant equal to 100 N/m. Determine the speed of longitudinal waves along this spring, using the results to Problem 6.

8. A wave traveling to the right has a transverse displacement given by

$$y = 2.0\sin(5.0x - 20t) \text{ cm}.$$

(a) Determine the phase angle ϕ. (b) What is the phase velocity of the wave, that is, the velocity of the wave in the x direction. (c) Find the transverse velocity and transverse acceleration of a point on the wave at $x = 0$ and $t = 0$. (d) What wave y' when added to y will produce a standing wave of amplitude 4 cm?

9. One string of a guitar has a mass of 2.0 g and a length of 50 cm. To what tension must the string be adjusted to produce a fundamental frequency of 440 Hz?

10. A stretched string fixed at each end has a mass of 40 g and a length of 8 m. The tension in the string is 49 N. (a) Find the speed of transverse waves on this string (b) Determine the frequency of the fundamental (n = 1) and first two overtones (n = 2, 3). (c) Determine the positions of the nodes and antinodes for the second overtone.

11. In Programmed Exercise 5, it is shown that the speed of sound waves in a gas at pressure p and temperature T is given by

$$v = \sqrt{\frac{\gamma p}{\rho}} = \sqrt{\frac{\gamma R}{M} T}$$

(a) Use the fact that the speed of sound in air is 331 m/sec at 0°C, and show that the speed of sound in air at any temperature T is $v = 20.1\sqrt{T}$ m/sec, where T is measured in °K. (b) Make a plot of v *vs* T from 0°C to 200°C.

12. Determine the speed of sound waves in helium gas at 0°C. Compare your result with the speed of sound in air.

13. (a) Calculate the speed of longitudinal waves in a long aluminum rod. Young's modulus for aluminum is 7×10^{11} dynes/cm^2 and the density of aluminum is 2.7 g/cm^3. (b) Determine the time it would take a sound wave to travel the length of an aluminum rod 100 miles long.

14. Water has a compressibility of $49 \times 10^{-11} (\text{Nm}^{-2})^{-1}$ and a density of 1.0 g/cm^3. (a) Recall that the bulk modulus is the inverse of the compressibility, and determine the speed of sound waves in water. (b) How long does it take a sound wave in a sonar system to make a *round* trip between a submarine and a ship 3000 m away?

15. Steel has a bulk modulus of 16×10^{11} dynes/cm^2, a shear modulus of 8×10^{11} dynes/cm^2, and a density of 7.8 g/cm^3. Use this data to calculate the speed of longitudinal waves in steel.

16. Refer to Programmed Exercise 5, and verify that the pressure variations in a medium through which sound waves propagate can be written as

$$\Delta p = -\left[\frac{2\pi\rho v^2}{\lambda}\xi_m\right]\cos(kx - \omega t)$$

where the term in the brackets is the *pressure amplitude* p_m, defined by

$$p_m = \frac{2\pi\rho v^2}{\lambda}\xi_m$$

17. The sensitivity of the ear can be appreciated by calculating the displacement amplitudes ξ_m for the faintest sounds that can be heard and the loudest sounds the ear can tolerate. The *maximum* pressure variations or pressure amplitude for the faintest sounds at a frequency of 1000 Hz is about 2×10^{-5} N/m^2, whereas the corresponding value for the loudest sounds at 1000 Hz is 28 N/m^2. Calculate the range of ξ_m corresponding to these values of p_m. (Use the results of Problem 16.)

18. A long vertical tube partially filled with water is open at the top, and the water level is adjustable, as in Figure 1–14. A tuning fork of unknown frequency excites resonance oscillations in the column when placed near the top. As the water level is lowered, the first peak sound intensity occurs at a water level of 12 cm below the top. (a) Assuming the speed of sound in air is 330 m/sec, determine the frequency of the tuning fork. (b) What are the positions of the water level for the next three maxima in sound intensity?

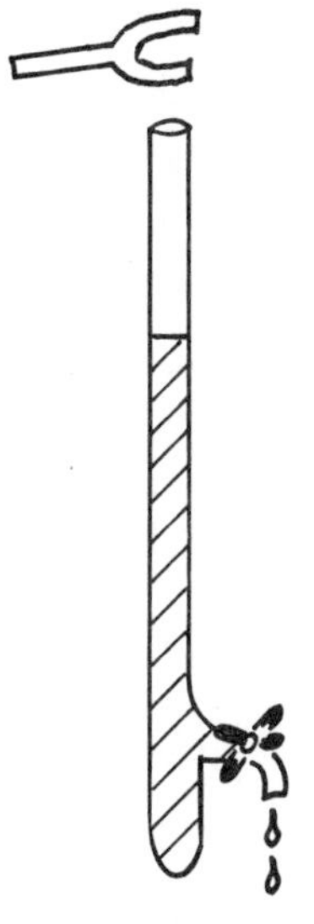

Figure 1–14

19. A hollow pipe open at each end has a fundamental frequency of 300 Hz when filled with air. (a) What is the length of the pipe? (b) Determine the frequencies of the first two overtones.

20. Determine the frequencies of the fundamental and first two overtones of a hollow pipe of length 1 m if the pipe is closed at one end and is filled with hydrogen gas. (See Example 6.)

21. The fundamental frequency of a hollow pipe containing a movable piston at the closed end is 400 Hz when its length is 1.5 m. If the frequency is lowered to 240 Hz by moving the piston, what is the "new length" of the pipe?

22. Suppose two waves having slightly different frequencies f_1 and f_2, but having the same amplitude A, propagate through the same medium. If the individual waves are represented by $y_1 = A\sin\omega_1 t$ and $y_2 = A\sin\omega_2 t$, show that the resultant wave has an amplitude given by

$$y = [2A\cos 2\pi\left(\frac{f_1 - f_2}{2}\right)t]\sin 2\pi\left(\frac{f_1 + f_2}{2}\right)t$$

This result shows that the resultant wave at a given point has an amplitude which *varies* in *time,* where the amplitude is the term in square brackets. This phenomenon in the case of time-varying amplitude of sound waves is known as *beats.* The apparent beat frequency is $f_1 - f_2$, and the average frequency of the resultant wave is $(f_1 + f_2)/2$.

23. Determine the beat frequency and average frequency of two tuning forks of frequencies 800 Hz and 828 Hz. (See Problem 22.)

24. A siren of an ambulance emits sound waves at a frequency of 800 Hz. A car moves at a speed of 50 mi/hr in the same direction as the ambulance, which has a speed of 80 mi/hr. What is the frequency of sound heard by a passenger in the car when (a) the ambulance approaches the car from behind, (b) the ambulance has passed and recedes from the car? (Note that 60 mi/hr = 88 ft/sec, and take the speed of sound in air to be 1100 ft/sec.)

25. A racing car moves in a circular track with a constant speed of 100 mi/hr. A stationary warning whistle located far from the track has a frequency of 620 Hz. What is the *range* of frequencies heard by the driver of the racing car during one revolution of the track?

26. A man waiting at a train station hears a train whistle. The train whistle has a "true" frequency of 500 Hz, and the man hears a frequency of 520 Hz. Determine the speed of the train and its direction relative to the man.

27. (a) Sound waves, emitted at a frequency f from a stationary source, are reflected from a target moving toward the source with a speed v_t, where v_t is much less than the speed of sound v. Show that the frequency of the reflected waves heard by a stationary observer at the source is

$$f' \cong f\left(1 + \frac{2v_t}{v}\right)$$

Hint: Treat the moving target as a *source* of sound moving toward the observer with a speed v_t and note that $v_t \ll v$. (b) If the velocity of the target is 5 m/sec, and f = 500 Hz, what is the frequency of the reflected wave heard by the observer?

28. A photograph shows that a projectile moving at supersonic speeds produces a shock wave whose wave front makes an angle of 33° with the horizontal (the angle α in Figure 1-13). Determine the speed of the projectile. What is the Mach number of the projectile?

2

ELECTROSTATICS

The development of the science of electricity lagged behind the development of mechanics by more than a century. In fact, Coulomb established the inverse square law of force between charged particles in 1785. This is to be compared with Newton's discoveries concerning gravitational forces, starting in 1666. Perhaps part of the reason for the lag is due to the fact that electrical phenomena were not easy to deal with experimentally. In 1747, Benjamin Franklin proposed that electrical charge is conserved. The science of magnetism was thought to be quite separate and distinct from electricity until 1820, when Oersted demonstrated a relation between the two by reporting the deflection of a compass needle when placed near a current-carrying conductor. Quantitative developments of the science of electromagnetism were made a short time later by such investigators as Faraday and Maxwell. Maxwell's equations are basic to *all* electromagnetic phenomena, and are comparable in importance to Newton's laws of motion and gravitation in mechanical phenomena.

We will begin our treatment of the science of electromagnetism by first describing the nature of electric forces between charged particles. Then, the concept of an electric field associated with a charged particle will be introduced, with emphasis on its vector nature and its effect on other charged particles. A simple and powerful tool will be introduced for determining the electric field of highly symmetric charged bodies, namely, Gauss's law. Finally, the motion of a charged particle in a uniform electric field will be discussed.

2.1 PROPERTIES OF ELECTRIC CHARGES AND FORCES

Some basic properties of electric charge can be understood with some relatively simple experiments, whereby electric charge is imparted to some solid substance by rubbing it with some other material. For example, if one takes an ordinary plastic hair comb and rubs it with a piece of wool or fur, the comb will attract a small piece of paper held close to it. In fact, if they are close enough, the paper will adhere to the comb and its weight can be supported by the comb. The same effect can be observed with other materials, such as glass and rubber. When materials behave in this fashion, we say they are *electrified* or have become *electrically charged.* Associated with this electrified state is an electric charge produced by friction. The force of attraction exerted on the piece of paper is due to this electrification which was not present before rubbing. Another simple experiment that is commonly performed is to rub an inflated balloon against one's clothing. The electrified balloon can then adhere to a wall, and sometimes to the ceiling of a room. These experiments generally work best on a dry day, since an excessive amount of moisture in the air can lead to leakage of charge either through the air to earth or through the electrified body to earth by other conducting paths.

Those materials which hold their electrification for long periods of time are called good *insulators.* Materials such as glass, rubber and lucite fall into this category. On the other hand, a material which maintains its electrified state for a very short period of time is known as a good *conductor.* Metals such as copper, aluminum and silver fall into this category. Paper becomes a reasonably good conductor on a humid day, when moisture is readily absorbed, whereas paper is a fairly good insulator on a dry day. If a copper rod is rubbed with wool or fur, and is held in hand, it will not attract a small piece of paper. On the other hand, if the copper rod is held by a lucite handle, and then rubbed, it will become electrified and thereby attract a piece of paper. This is easily explained by noting that when the copper is held directly in hand, the electric charges produced will readily conduct through the copper to the hand and body (a good conductor), and finally to earth. In the second case, when an insulating handle such as lucite is provided, the lucite prevents the flow of charges to the ground. Therefore, on the basis of these simple experiments, we can conclude that all solids may be electrified by friction.

In a more systematic set of experiments, it can be shown that there are *two* types of electric charges, which were given the names *positive* and *negative* by Benjamin Franklin. For example, a rubber rod rubbed with fur obtains a negative charge, while the fur obtains a positive charge. However, the total charge remains constant. This was first suggested by Franklin and is referred to as the law of *charge conservation.* On the other hand, a glass rod rubbed with silk obtains a positive charge, while the silk obtains a negative charge. Clearly, there is a *charge transfer* between the rods and the rubbing material. The distinction in the electrification of a rubbed rubber rod and a rubbed glass rod can be shown by suspending an electrified rubber rod in a stirrup attached to a piece of thread. When a second electrified rubber rod held in hand is brought near the suspended rod, the charged ends will *repel* one another, and we conclude that the force is one of repulsion. Likewise, if the experiment is repeated with two electrified glass rods, the rods will also repel one another. On the other hand, if an electrified rubber rod is brought near an electrified glass rod, the rods will *attract* one another; that is, the force is one of attraction. On the basis of these simple experiments, we can safely conclude that *like charges repel* and *unlike charges attract.* The assignment as to which is positive or negative is completely arbitrary. One should also note that the electrified material (for example, rubber or glass) will always be attracted to the piece of material that was used to electrify it, since they are oppositely charged. Figure 2–1 summarizes the attraction between unlike charges and the repulsion between like charges. In practice, the suspended balls may be made of cork.

An uncharged body can also be electrified by a method called *induction.* In this process, a charged object may impart a charge of opposite sign to another body without losing any of its own charge. For example, suppose a negatively charged rubber rod is brought near a neutral sphere, insulated from ground. The portion of the sphere nearest the charged rod will have an excess of positive charge, whereas the

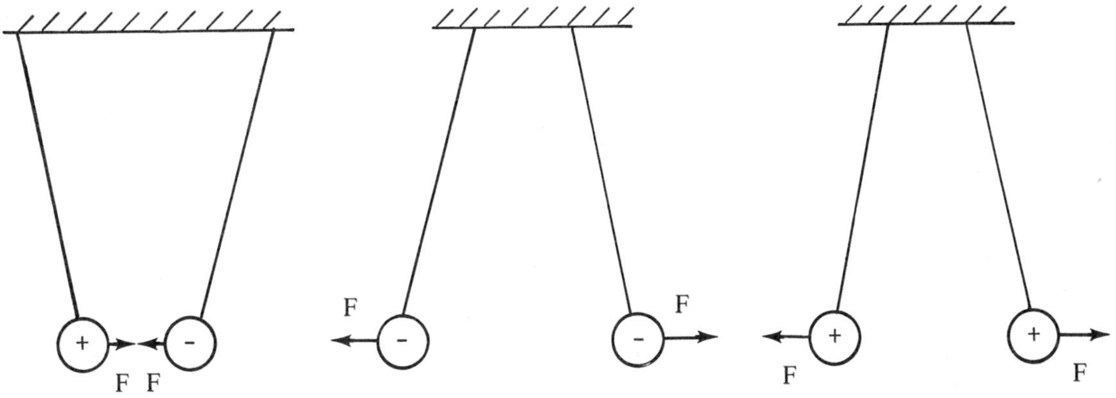

Figure 2-1 Forces between unlike and like charges.

portion opposite the rod will have an excess of negative charge. If a wire is connected from the sphere to earth, the excess negative charges will be repelled to earth. Finally, if the wire is removed, the sphere will contain an excess of *induced* positive charge. In the process, the electrified rubber rod has lost none of its negative charge.

2.2 COULOMB'S LAW

In 1785, Coulomb established the law of electric force between two charged objects. The apparatus he used was a torsion balance, similar to the balance used earlier by Cavendish in establishing the value of G, the universal gravitational constant. First, Coulomb found that the electric force is inversely proportional to the square of the separation, r, between the two charged objects. Second, he found that the force is proportional to the product of the charges q_1 and q_2 on the two objects. Coulomb's law can, therefore, be written as

$$F = k\frac{q_1 q_2}{r^2} \tag{2.1}$$

where k is a constant of proportionality. Coulomb's law serves as a useful definition of a unit charge. In cgs units, $k = 1$, and the unit charge is called an electrostatic unit (esu), or the statcoulomb. In this system of units, two unit charges separated by 1 cm will exert a force of 1 dyne on each other.

The unit of charge in SI units is the *coulomb,* C. The coulomb is defined in terms of a unit current called the *ampere,* A. Since current is charge per unit time, then 1C = 1A sec. In other words, if a wire carries a current of 1A, then 1C of charge conducts through a point in 1 sec. The conversion factor between esu and SI units for charge is

$$1C = 3 \times 10^9 \text{ esu} \tag{2.2}$$

It is standard to write $k = 1/4\pi\epsilon_0$ in SI units, where

$$k = \frac{1}{4\pi\epsilon_0} \cong 9.00 \times 10^9 \text{ Nm}^2/\text{C}^2 \tag{2.3}$$

and

$$\epsilon_0 = 8.8542 \times 10^{-12} \text{C}^2/\text{Nm}^2 \tag{2.4}$$

The constant ϵ_0 is known as the *permittivity* of free space and is related to the speed of light. Usually we will use the SI system of units in our treatment of electricity and magnetism.

One should note that a coulomb is a very large unit of charge. The *smallest* unit of charge known in nature is the charge carried by an electron or proton (part of the "building" blocks of atoms and molecules). *Electrons* have *negative* charge, and *protons* have an equal amount of *positive* charge.* The unit of charge, or quantum of charge e on an electron or proton, has a magnitude

$$e = 1.60207 \times 10^{-19} \text{ C}$$

Therefore, one C of charge is roughly equivalent to the charge of 10^{19} electrons. This can be compared to the number of free electrons in 1 cm^3 of copper, which is of the order of 10^{23}. In typical electrostatic experiments, where a rubber or glass rod is electrified by friction, a net charge of only about 10^{-6} C may be obtained. Thus, only a very small fraction of charges are transferred between the rod and rubbing material.

A list of the basic properties of the electron, proton and neutron is given in Table 2–1.

TABLE 2–1 BASIC PROPERTIES OF THE ELECTRON, PROTON AND NEUTRON

Particle	Charge C	Mass (kg)
Electron	-1.60×10^{-19}	9.11×10^{-31}
Proton	$+1.60 \times 10^{-19}$	1.67×10^{-27}
Neutron	0	1.67×10^{-27}

When dealing with Coulomb's law, as with any force, we must remember that force is a *vector* quantity and must be dealt with carefully. If $\hat{r}$ is a unit vector directed from q_2 to q_1, as in Figure 2–2, then the force acting on particle 1 due to particle 2, written as $\mathbf{F}_{12}$, is given by

$$\mathbf{F}_{12} = k\frac{q_1 q_2}{r^2}\hat{r} \tag{2.5}$$

*No unit of charge smaller than e has yet been detected, although there are current *theories* which have proposed the existence of smaller units of charge in particles called *quarks*.

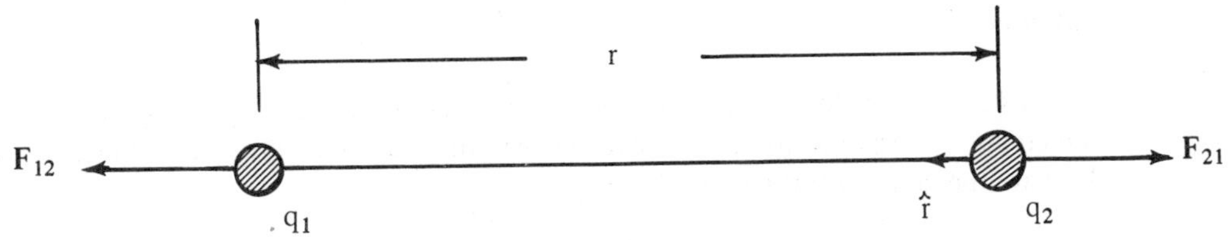

Figure 2-2 Schematic diagram representing the Coulomb force between two charged particles. Note that the force is repulsive if q_1 and q_2 are of the same sign, whereas the force is attractive if q_1 and q_2 are of opposite sign.

In keeping with Newton's third law, it follows that $\mathbf{F}_{21} = -\mathbf{F}_{12}$, that is, the force acting on particle 2 due to particle 1 is equal and opposite to $\mathbf{F}_{12}$. Therefore, it follows that the force $\mathbf{F}_{12}$ is *repulsive* if q_1 and q_2 are of the same sign (as in Figure 2–2), whereas $\mathbf{F}_{12}$ is attractive if q_1 and q_2 are of opposite sign. (In this case, $\mathbf{F}_{12}$ and $\mathbf{F}_{21}$ would be reversed in Figure 2–2.)

Of course, when more than two charges are present, and the force on one of them is required, then the procedure is to apply Equation (2.5) for every *pair* of charges, and add these forces *vectorially*. For example, if there are *four* charges, and the force on particle 1 is required, then

$$\mathbf{F}_1 = \mathbf{F}_{12} + \mathbf{F}_{13} + \mathbf{F}_{14}$$

Likewise, this can be extended to any number of charges. This procedure is known as the *superposition* principle.

Example 2.1

Consider four charges at the corner of a square, as in Figure 2–3, where $q_1 = q_3 = 5\ \mu C$, $q_2 = q_4 = -2\ \mu C$ (one $\mu C = 10^{-6}$ C), and a = 0.1 m. What is the resultant force acting on charge q_3?

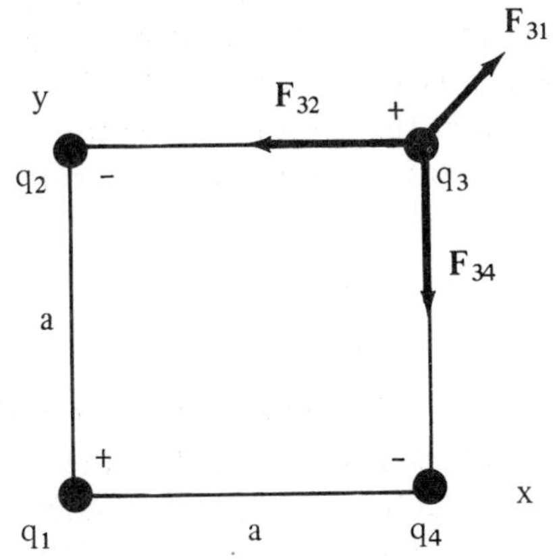

Figure 2-3 The force on q_3 is the vector sum of the forces due to q_1, q_2 and q_4.

Solution

First, note the *directions* of the individual forces on q_3 due to q_1, q_2 and q_4. These directions correspond to the relative charges of the interacting pairs. Since $\mathbf{F}_{32}$ and $\mathbf{F}_{34}$ have the same magnitudes, we have

$$F_{32} = F_{34} = k\frac{q_3 q_2}{a^2} = k\frac{q_3 q_4}{a^2}$$

$$F_{32} = F_{34} = 9 \times 10^9 \frac{\text{Nm}^2}{\text{C}^2} \times \frac{(5 \times 10^{-6}\text{C})(2 \times 10^{-6}\text{C})}{(0.1\text{ m})^2} = 9\text{ N}$$

The magnitude of the force between q_1 and q_3 is given by

$$F_{31} = k\frac{q_3 q_1}{(\sqrt{2}a)^2} = 9 \times 10^9\ \text{N}\frac{\text{m}^2}{\text{C}^2}\ \frac{(5 \times 10^{-6}\text{C})(2 \times 10^{-6}\text{C})}{2\,(0.1\text{ m})^2} = 4.5\text{ N}$$

We can write these in vector form as

$$\mathbf{F}_{32} = -9\mathbf{i}\text{ N}$$

$$\mathbf{F}_{34} = -9\mathbf{j}\text{ N}$$

$$\mathbf{F}_{31} = F_{31x}\mathbf{i} + F_{31y}\mathbf{j} = (3.2\,\mathbf{i} + 3.2\,\mathbf{j})\ \text{N}$$

Therefore, the total force on q_3 is

$$\mathbf{F}_3 = \mathbf{F}_{32} + \mathbf{F}_{34} + \mathbf{F}_{31} = -\,(5.8\,\mathbf{i} + 5.8\,\mathbf{j})\ \text{N}$$

$\mathbf{F}_3$ has a magnitude of 8.2 N and is directed 225° from the +x axis.

2.3 THE ELECTRIC FIELD

The concept of a gravitational field at some point in space has been introduced to describe its effect on a test mass placed at that point. Likewise, we can describe an electric field at some point in space as a vector field which exists if a test charge experiences a force when placed at that point. To be more precise, the electric force $\mathbf{F}$ divided by the test charge q' gives the electric field vector $\mathbf{E}$.

$$\mathbf{E} = \frac{\mathbf{F}}{q'} \tag{2.6}$$

There is one restriction on q', namely, that it must be infinitesimally small. This restriction assures that the charges responsible for the electric field will not be disturbed. Thus, we say that an electric field exists at some point *if* a test charge q' experiences a force when placed at that point. On the basis of the definition of $\mathbf{E}$, if q' is positive, the electric force is in the direction of $\mathbf{E}$, but if q' is negative, $\mathbf{F}$ is opposite to $\mathbf{E}$.

An electric field is a *vector* quantity and in SI units has dimensions N/C. It is an abstract but useful concept because we can use it to describe its effect on *any* charged particle placed at some point in space. Moreover, we can think of an electric field as something which would act if a test charge were present, without actually introducing the test charge.

One useful approach to "visualize" electric fields is a pictorial method, where arrows are used to represent the electric field vectors. We refer to these arrows as *lines of the electric field.* These lines, for a single *positive* charge q, point radially outwards from the charge as in Figure 2–4(*a*). The number of lines per unit area gives the magnitude of the electric field, and the arrow heads point in the direction of **E**. Therefore, as we go further *away* from the charge, the number of lines per unit area decreases; hence, the electric field decreases. The lines of electric field near a negative charge −q are shown in Figure 2–4(*b*). Note that in this case the lines point radially *inwards.* Therefore, if we place a *positive* test charge q′ in the vicinity of a positive charge q [Figure 2–4(*a*)], the force on q′ would be radially outwards, away from q. On the other hand, if q′ is placed near a negative charge [Figure 2–4(*b*)], the force on q′ would be radially inwards, towards −q.

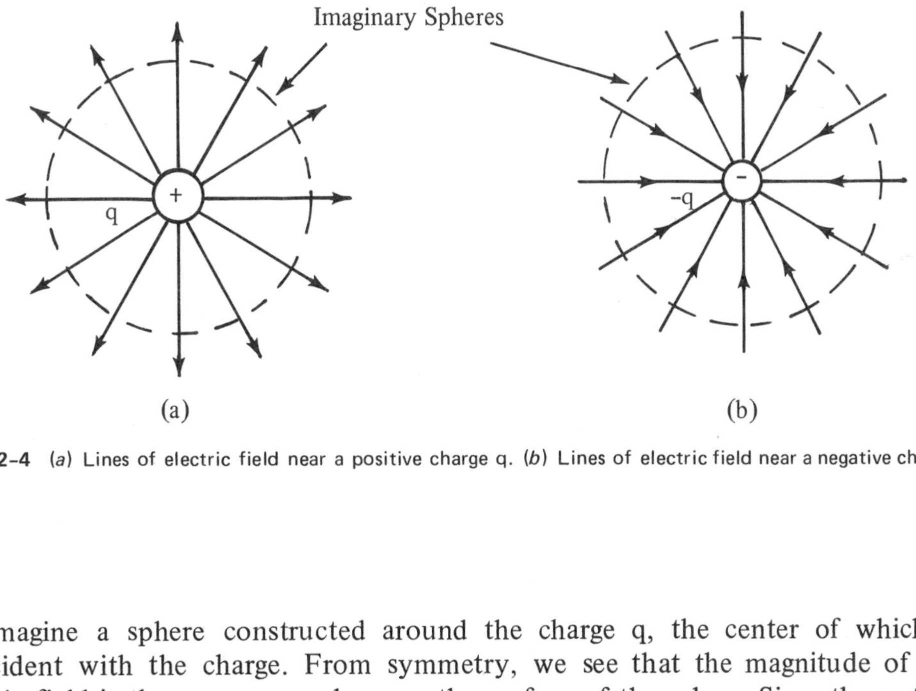

Figure 2–4 (*a*) Lines of electric field near a positive charge q. (*b*) Lines of electric field near a negative charge −q.

Imagine a sphere constructed around the charge q, the center of which is coincident with the charge. From symmetry, we see that the magnitude of the electric field is the same everywhere on the surface of the sphere. Since the surface area of a sphere is $4\pi r^2$, and the electric field is proportional to the number of lines per unit area, we conclude that $E \sim 1/r^2$. This result also follows from Coulomb's law. Therefore, $E \to 0$ as $r \to \infty$, but a singularity in E exists at $r = 0$. As we shall see later, this problem is eliminated when the finite size of the charge is taken into consideration.

Other patterns of electric field lines for two equal positive charges and two equal opposite charges are illustrated in Figure 2–5. Note that the field lines always originate at a positive charge and terminate on a negative charge.

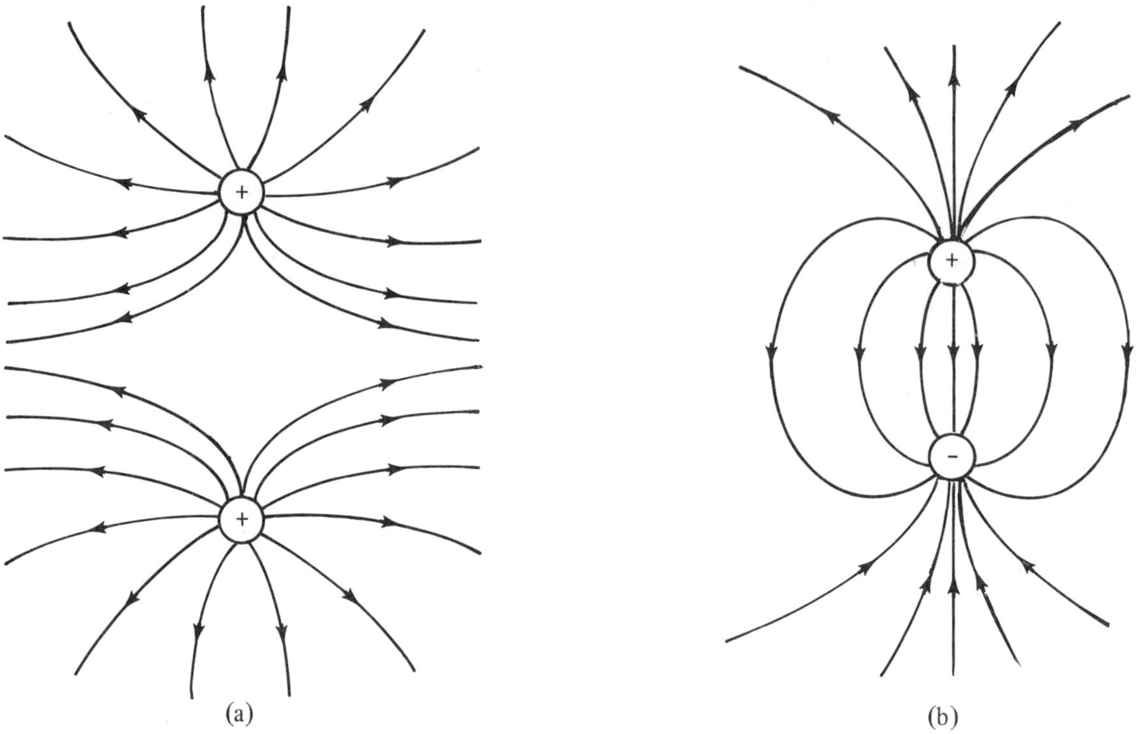

Figure 2–5 (*a*) Lines of electric field for two equal positive charges. (*b*) Lines of electric field for two equal but opposite charges.

2.4 CALCULATION OF ELECTRIC FIELDS

We begin our calculation of electric field intensities by first treating a point charge. If a test charge q' is at a distance r from a second charge q, the Coulomb force on q' is given by

$$\mathbf{F} = k\frac{qq'}{r^2}\hat{r}$$

Therefore, from Equation (2.6) it follows that the electric field intensity at q' *due to the charge q* is given by

$$\mathbf{E} = \frac{\mathbf{F}}{q'} = k\frac{q}{r^2}\hat{r} \qquad (2.7)$$

The direction of **E** is radially outwards from q if q is positive, and radially inwards if q is negative, as in Figure 2–4.

If we wish to calculate the electric field intensity of a group of charges, we first calculate the electric field vectors individually and then add them *vectorially.* In other words, the electric field intensity due to several individual charges is equal to the *vector sum* of the electric field intensities that each charge creates. This is the *superposition principle* applied to electric fields.

Example 2.2

Two charges of 5 μC and -5 μC are placed 0.3 m apart along the x axis, as in Figure 2-6. (a) Find the electric field intensity at the point P whose coordinates are (0, 0.4) m.

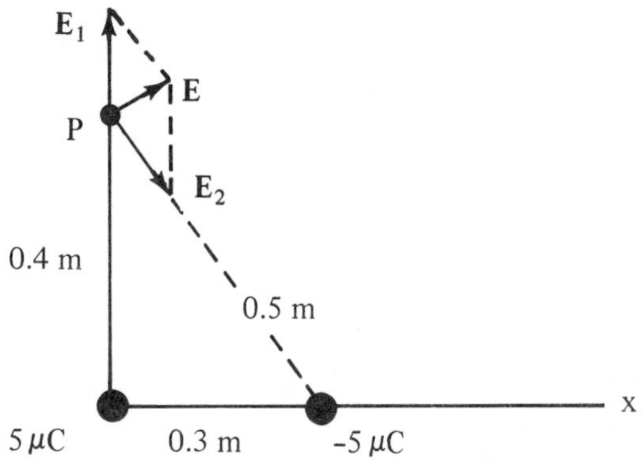

Figure 2-6

The *magnitudes* of the individual electric field intensities are given by

$$E_1 = 9 \times 10^9 \ N\frac{m^2}{C^2} \times \frac{(5 \times 10^{-6} C)}{(0.4 \ m)^2} = 2.8 \times 10^5 \ N/C$$

$$E_2 = 9 \times 10^9 \ N\frac{m^2}{C^2} \frac{(5 \times 10^{-6} C)}{(0.5 \ m)^2} = 1.8 \times 10^5 \ N/C.$$

The x component of $\mathbf{E}_2$ is given by $\frac{3}{5}\mathbf{E}_2$, whereas the y component of $\mathbf{E}_2$ is negative and given by $-\frac{4}{5}\mathbf{E}_2$.

The electric field *vectors* are therefore given by

$$\mathbf{E}_1 = (2.8 \times 10^5 \ \mathbf{j}) \ N/C$$

$$\mathbf{E}_2 = (1.1 \times 10^5 \ \mathbf{i} - 1.4 \times 10^5 \ \mathbf{j}) \ N/C$$

The resultant electric field at the point P is the superposition of fields $\mathbf{E}_1$ and $\mathbf{E}_2$. That is,

$$\mathbf{E} = \mathbf{E}_1 + \mathbf{E}_2 = (1.1 \times 10^5 \ \mathbf{i} + 1.4 \times 10^5 \ \mathbf{j}) \ N/C$$

Therefore, **E** has a magnitude of 1.7×10^5 N/C and is at an angle of 52° with the +x axis.

(b) Find the force on a test charge of 2 μC placed at P.

$$\mathbf{F} = q'\mathbf{E} = (2 \times 10^{-6} \ C)(1.1 \times 10^5 \ \mathbf{i} + 1.4 \times 10^5 \ \mathbf{j}) \ N/C$$

$$\mathbf{F} = (0.11 \ \mathbf{i} + 0.28 \ \mathbf{j}) \ N$$

Now suppose we consider a *continuous* distribution of charge. We can calculate the electric field intensity for such a distribution by breaking the system into *infinitesimal* charge elements dq. Each infinitesimal charge element is treated as a point charge, and Equation (2.7) is used to find the electric field dE for that element. This gives, for the magnitude of dE, the relation

$$dE = k\frac{dq}{r^2} \quad (2.8)$$

The resultant **E** field is obtained by integrating, keeping in mind that dE is a *vector* quantity and must be handled carefully. Therefore, E becomes

$$E = \int dE$$

Example 2.3

Consider a uniformly charged rod, bent into the shape of a semicircular hoop of radius R, as in Figure 2-7. Find the electric field at the center of the semicircle, assuming the total charge on the rod is +q.

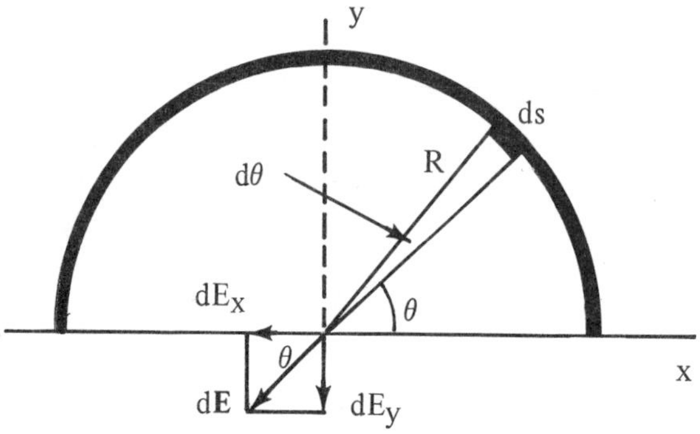

Figure 2-7

The charge per unit length of the rod, λ, is given by

$$\lambda = \frac{q}{\pi R}$$

The unit length ds has a charge dq = λds, where ds = Rdθ, therefore

$$dq = \lambda ds = \frac{q}{\pi R} R d\theta = \frac{q}{\pi} d\theta$$

Now, note that $dE_y = -dE\sin\theta$, and by symmetry, $E_x = 0$, since contributions from the left side of the hoop will cancel those from the right. Therefore, the resultant field is in the −y direction. Since every element is at the same distance from the center, r = R; therefore, we can write the total electric field as

$$E_y = \int dE_y = -\int_0^{\pi} dE\sin\theta$$

$$E_y = -k\int_0^{\pi} \frac{dq}{r^2}\sin\theta = -\frac{kq}{\pi R^2}\int_0^{\pi}\sin\theta d\theta$$

$$E_y = -\frac{kq}{\pi R^2}(-\cos\theta)\Big]_0^{\pi} = -\frac{2kq}{\pi R^2}$$

where $k = 1/4\pi\epsilon_0$ in SI units. Note also that $E_x = 0$ follows from the fact that $dE_x = dE \cos\theta$, and $E_x \sim \int_0^\pi \cos\theta \, d\theta = 0$.

2.5 FLUX OF THE ELECTRIC FIELD AND GAUSS'S LAW

Whenever the term "flux" is used in physics, it refers to a quantity which is a measure of the number of field lines that penetrate through a hypothetical surface. Electric flux, ϕ_E, is therefore a measure of the number of electric field lines that penetrate some surface. For example, if a sphere of radius r surrounds a point charge q, as in Figure 2–4, the electric flux is EA, where A is the surface area of the sphere, namely, $4\pi r^2$. Since E is given by Equation (2.7) for a *point charge,* we get

$$\phi_E = EA = \frac{q}{4\pi\epsilon_0 r^2}(4\pi r^2) = \frac{q}{\epsilon_0} \tag{2.9}$$

A more precise definition of ϕ_E can be obtained by considering an infinitesimal *area* element dA, as in Figure 2–8. We can represent such an area element by a vector **dA**, whose magnitude is dA and whose direction is *normal* to the surface and points away from it.

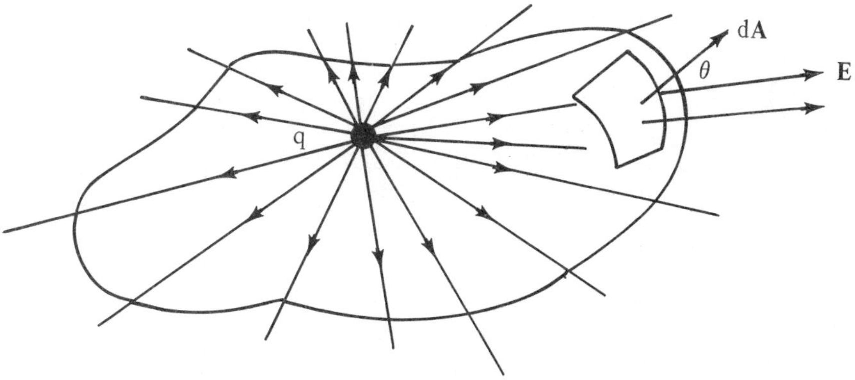

Figure 2–8 A hypothetical surface surrounding a charge q.

The component of **E** normal to the surface element dA is $E\cos\theta$, where θ is the angle between **E** and **dA**, as in Figure 2–8. The electric flux $d\phi_E$ passing through dA is given by

$$d\phi_E = E\cos\theta \, dA \tag{2.10}$$

In vector form, we can write this as

$$d\phi_E = \mathbf{E} \cdot \mathbf{dA} \tag{2.11}$$

To get the *total* flux through some surface, we integrate Equation (2.11) over the entire surface. This gives a *surface integral* of the form

$$\phi_E = \int \mathbf{E} \cdot d\mathbf{A} \tag{2.12}$$

Earlier, in Equation (2.9), the total electric flux through a spherical surface surrounding a point charge q was shown to be

$$\phi_E = \frac{q}{\epsilon_0}$$

Gauss's theorem states that *the total electric flux surrounding some charge q* equals q/ϵ_0 *regardless of the shape of the hypothetical surface.* The mathematical statement of Gauss's law follows from Equation (2.12).

$$\phi_E = \oint \mathbf{E} \cdot d\mathbf{A} = \frac{q}{\epsilon_0} \tag{2.13}$$

The symbol $\oint$ represents an integral over a *closed* surface. The proof of Gauss's law will be left as an exercise. However, it is important to note that the *net charge* q within the hypothetical surface can either be a number of point charges *or any distribution* of charge. We will normally refer to the hypothetical surface as a *Gaussian* surface.

Gauss's law is an extremely powerful tool for determining the electric field intensity for a charge distribution having a *high degree of symmetry.* For example, when spherically symmetric or cylindrically symmetric charge distributions are involved, the surface integrals reduce to a very simple form. The surface integrals can be quite cumbersome for systems having low symmetry. We will not treat such low symmetry charge distributions.

Usually we will assume that a given charged body is either a good insulator or a good conductor. In a good insulator, the free charges are not mobile; therefore, it is physically possible to obtain a uniform (or nonuniform) charge distribution in an insulator. However, free charges in a good conductor are very mobile and will always reside at the *surface* of the conductor under equilibrium conditions. This is due to the Coulomb repulsion between like charges, which favors the surface location of charges as the lowest energy configuration. For example, a charged metallic sphere will have *all of its free charge on the surface under equilibrium conditions.* Consequently, the electric field is *zero* within the metallic sphere. This follows from Gauss's law applied to a hypothetical surface inside the sphere. An equivalent description of the equilibrium state of a charged conductor is the condition that no charges are in motion; hence, there is no current. In addition, it follows that the electric field lines *just outside* the surface of a metal or conducting surface must be *perpendicular* to the surface under equilibrium conditions, since any component of **E** parallel to the surface would cause the motion of charges, and hence a current.

Example 2.4

An insulating sphere of radius R has a uniform charge distribution ρ, and a total positive charge Q [Figure 2-9(*a*)]. (a) What is the electric field intensity at a point *outside* the sphere, that is, for $r > R$?

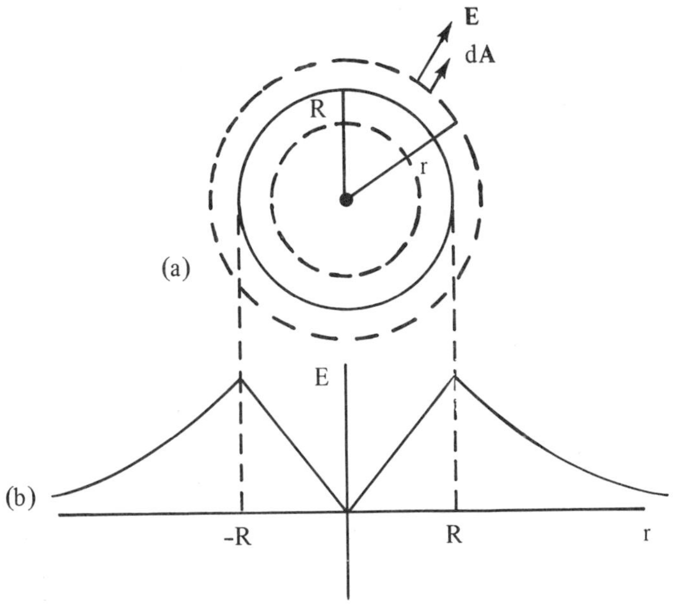

Figure 2-9

We construct a *spherical* Gaussian surface surrounding the sphere and concentric with it, as in Figure 2-9(*a*). Since Q is positive, **E** points *outwards,* and by symmetry **E** is normal to the sphere. Therefore, **E** ∥ d**A**, and $\mathbf{E}\cdot d\mathbf{A} = EdA$. Since E is constant over the whole surface, ϕ_E becomes

$$\phi_E = \oint EdA = E\oint dA = EA$$

Since $A = 4\pi r^2$ for a sphere, and the charge within the surface is Q, Gauss's law gives

$$E4\pi r^2 = \frac{Q}{\epsilon_0}$$

or,

$$E = \frac{Q}{4\pi\epsilon_0 r^2} \qquad (r > R)$$

In vector form, this can be written as

$$\mathbf{E} = \frac{Q}{4\pi\epsilon_0 r^2}\hat{r} = k\frac{Q}{r^2}\hat{r} \qquad (r > R)$$

Note that this result is identical to that obtained for a point charge. Therefore, we conclude that a uniformly charged sphere is *equivalent* to a point charge in the region exterior to the sphere.

(b) What is the electric field intensity at a point *inside* the charged sphere, that is, for $r < R$?

Now we construct a *spherical* Gaussian surface whose radius is $r < R$, concentric with the charge distribution. In this case, the charge q *within* the hypothetical surface is *not* Q, but something less than Q. To calculate this charge, note that $q = \rho V$, where ρ is the charge per unit volume, and V is the volume enclosed by the *inner* Gaussian surface given by $V = \frac{4}{3}\pi r^3$.

Therefore, the charge q enclosed by the inner Gaussian surface is

$$q = \rho V = \rho \frac{4}{3}\pi r^3$$

Application of Gauss's law for $r < R$ gives

$$\phi_E = EA = E\, 4\pi r^2 = \frac{q}{\epsilon_0}$$

$$E = \frac{q}{4\pi\epsilon_0 r^2} = \frac{\rho \frac{4}{3}\pi r^3}{4\pi\epsilon_0 r^2}$$

$$E = \frac{\rho r}{3\epsilon_0} \qquad (r < R)$$

The fact that this result differs from that obtained in part (a) should not be surprising. First, it eliminates the singularity that would exist at $r = 0$ if $E \sim 1/r^2$ (a nonphysical situation). Secondly, $E \rightarrow 0$ as $r \rightarrow 0$, which again one could guess on the basis of the spherically symmetric charge distribution. A plot of E *vs* r is shown in Figure 2-9(*b*).

Example 2.5

A good conductor has the shape of a very long cylinder of radius R. The cylinder has a total charge +Q, and it has a uniform charge density on its *surface* (Figure 2-10). (a) Find the electric field intensity at a point *outside* the cylinder, that is, $r > R$.

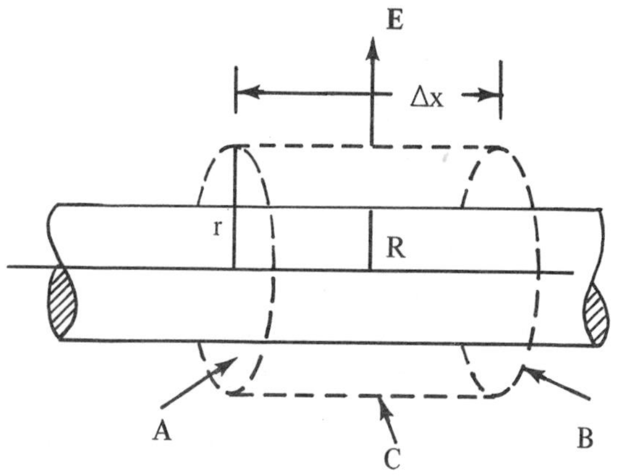

Figure 2-10

Let us call the length of the cylinder l, and let its charge per unit length by λ, where $\lambda = Q/l$. Now, construct a cylindrical Gaussian surface of radius r, concentric with the charged cylinder, whose length is Δx. The charge enclosed by this surface, q, is

$$q = \lambda \Delta x = \frac{Q}{l} \Delta x$$

If the length of the cylinder is large compared to R, that is, $l \gg R$, then end effects can be neglected and the **E** field points radially outwards. Consequently, no electric flux passes through the *ends* A and B of the Gaussian surface (that is, $\mathbf{E} \perp d\mathbf{A}$ for the *ends*, and $\mathbf{E} \cdot d\mathbf{A} = 0$). The electric flux passes only through surface C, where **E** is parallel to d**A** over the entire surface and is constant in magnitude. Therefore, for surface C we get

$$\phi_E = \oint_C \mathbf{E} \cdot d\mathbf{A} = EA = E\, 2\pi r\, \Delta x$$

Applying Gauss's law for this surface gives

$$\phi_E = \frac{q}{\epsilon_0} = \frac{\lambda \Delta x}{\epsilon_0}$$

$$E\, 2\pi r\, \Delta x = \frac{\lambda}{\epsilon_0} \Delta x$$

$$E = \frac{\lambda}{2\pi \epsilon_0 r} \qquad (r > R)$$

(b) What is the electric field intensity at a point inside the cylinder, that is for $r < R$?

Since all of the charge resides at the *surface* of the conductor, the charge enclosed by a Gaussian surface within the conductor will be *zero.* Therefore, from Gauss's law we conclude that **E** = 0 *everywhere inside* the conductor.

2.6 MOTION OF CHARGED PARTICLES IN A UNIFORM ELECTRIC FIELD

When a charged particle, such as an electron or proton, is placed in a *uniform* electric field, the motion will be quite simple, namely, one of constant acceleration. In fact, for nonrelativistic speeds, the kinematics is equivalent to the motion of a mass m in the earth's gravitational field. A charge q moving in a uniform electric field **E** experiences a force q**E**; hence, by Newton's second law,

$$\mathbf{F} = q\mathbf{E} = m\mathbf{a}$$

where m is the mass of the charge. Therefore, if the electric field is along the y direction, as in Figure 2–11, the charge will experience an acceleration in the +y or −y direction, depending on the sign of q. If the charge is negative, such as an electron, the acceleration will *oppose* **E**, and will be downwards in the figure. If the charge is positive, the accelerations will be *along* **E**, or upwards in the figure. The electric field in Figure 2–11 is produced by two oppositely charged metal plates, and in such a configuration the field will be roughly uniform, apart from end effects. Therefore, the trajectory of the charged particle will be a parabola.

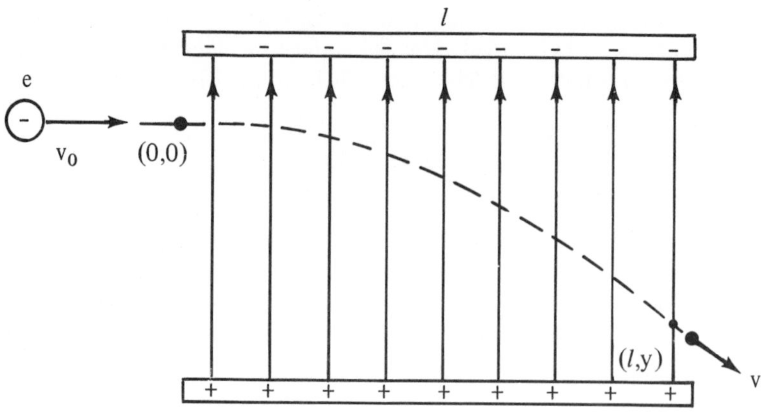

Figure 2-11 The deflection of an electron in a uniform electric field.

Let us assume the charge q has an initial velocity $v_0 \mathbf{i}$ before entering the field region. The acceleration is then given by

$$a = \frac{qE}{m} = \frac{qE}{m}\mathbf{j} \tag{2.14}$$

Therefore, applying our kinematic expressions for constant acceleration, with $v_{yo} = 0$, gives

$$v_x = v_0 = \text{constant} \tag{2.15}$$

$$v_y = at = \frac{eE}{m}t \tag{2.16}$$

Likewise, the coordinates of the particle as a function of time can be predicted from the relations

$$x = v_0 t \tag{2.17}$$

$$v = \frac{1}{2}at^2 = \frac{qE}{2m}t^2 \tag{2.18}$$

After the particle leaves the region of the uniform electric field, it continues to move in a *straight* line with a velocity $v > v_0$.

Note that we have neglected the gravitational force on the charged particle. This is a good approximation when dealing with atomic-sized particles. For a proton, the ratio of the electric force qE to the gravitational force mg is of the order of 10^{11} for an electric field of 10^4 N/C.

Example 2.6

Suppose an electron enters the region of a uniform electric field, with $v_0 = 3 \times 10^6$ m/sec, $l = 0.1$ m, and $E = 2 \times 10^3$ N/C, as in Figure 2–11. (a) What is the acceleration of the electron in the region of the electric field?

Since $q = -e = -1.6 \times 10^{-19}$ C, and $m = 9.1 \times 10^{-31}$ kg, Equation (2.14) gives

$$a = -\frac{eE}{m}\mathbf{j} = -\frac{1.6 \times 10^{-19}\ \text{C} \times 2 \times 10^3\ \mathbf{j}\ \frac{\text{N}}{\text{C}}}{9.1 \times 10^{-31}\ \text{kg}}$$

$$a = -\ 3.5 \times 10^{14}\ \mathbf{j}\ \text{m/sec}^2$$

(b) Find the time it takes the electron to traverse the region of the uniform electric field.

Using Equation (2.17), we set $x = l = 0.1$ m, and $v_0 = 3 \times 10^6$ m/sec. This gives.

$$t = \frac{l}{v_0} = \frac{0.1\ \text{m}}{3 \times 10^6\ \text{m/sec}} = 3.3 \times 10^{-8}\ \text{sec}$$

(c) What distance, y, does the electron fall through in the region of the uniform electric field?

This is calculated by using Equation (2.18) and the results to parts (a) and (b)

$$y = \frac{1}{2}at^2 = -\frac{1}{2}(3.5 \times 10^{14})\frac{\text{m}}{\text{sec}^2} \times (3.3 \times 10^{-8}\ \text{sec})^2$$

$$y = -\ 5.8 \times 10^{-2}\ \text{m}$$

(d) What is the speed of the electron as it emerges from the electric field?

The x component of velocity remains constant, and equals v_0. The y component of velocity as the electron emerges from the field can be calculated from Equation (2.16), with $t = 3.3 \times 10^{-8}$ sec.

$$v_y = at = -\ 3.5 \times 10^{14}\ \frac{\text{m}}{\text{sec}^2} \times 3.3 \times 10^{-8}\ \text{sec}$$

$$v_y = -\ 1.15 \times 10^7\ \text{m/sec}$$

Therefore, the speed of the electron as it emerges from the field is given by

$$v = \sqrt{{v_x}^2 + {v_y}^2} = \sqrt{(3 \times 10^6)^2 + (-1.15 \times 10^7)^2}\ \frac{\text{m}}{\text{sec}} \cong 1.2 \times 10^7\ \text{m/sec}$$

2.7 PROGRAMMED EXERCISES

1 Two charges, q_1 and q_2, are separated by a distance r, as in the figure below.

1.A

If q_1 and q_2 are both positive, sketch vectors representing the directions of the Coulomb force on each particle.

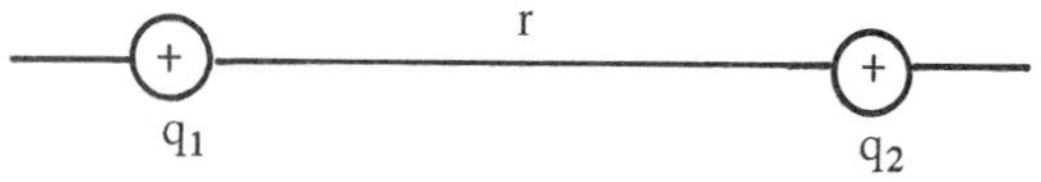

The same results hold if q_1 and q_2 are *negative*, since like charges *repel.*

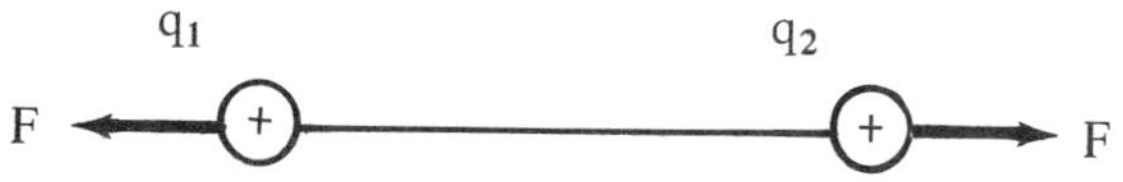

1.B

Sketch vectors representing the directions of the Coulomb force on each particle if q_1 is positive and q_2 is negative.

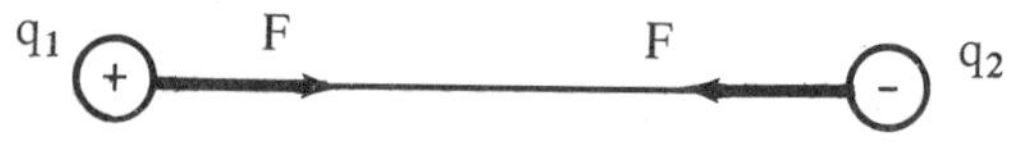

Unlike charges attract.

1.C

What is the magnitude of the Coulomb force?

$$F = k\frac{q_1 q_2}{r^2} \qquad (1)$$

where $k = \dfrac{1}{4\pi\epsilon_0} = 9 \times 10^9\ N\dfrac{m^2}{C^2}$ in SI units.

1.D

Does Coulomb's law hold for any kind of charged pair of bodies? Explain.

No. Coulomb's law is valid only for *point* charges.

1.E

What is the magnitude of the electric field intensity at q_2 due to the charge q_1?

$$E = k\frac{q_1}{r^2} \qquad (2)$$

1.F

Show that the force on q_2 can be obtained from the calculated electric field at this point.

Using (2), we get

$$F = q_2 E = k\frac{q_1 q_2}{r^2}$$

This is consistent with (1).

2 Two stationary, isolated particles have charges of 2q and −q and are separated by a distance d, as in the figure below.

2.A

Is the Coulomb force between the two charges attractive or repulsive? Explain.

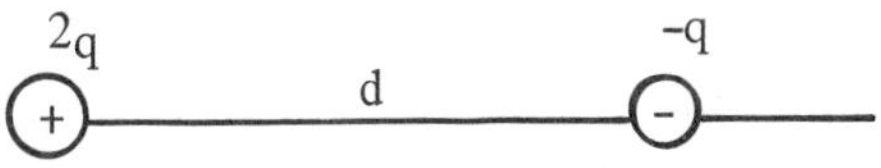

The charges are unlike; therefore, the Coulomb force is *attractive.*

2.B

What is the *magnitude* of the Coulomb force F between the two charges?

$$F = k\frac{q_1 q_2}{r^2} = k\frac{2q^2}{d^2} \qquad (1)$$

2.C

Draw a diagram showing directions of the forces on the individual charges.

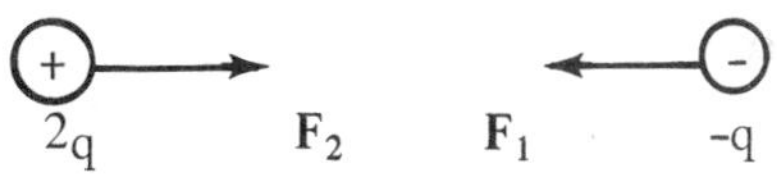

2.D

Why are these forces equal and opposite in direction?

Newton's third law requires that

$$\mathbf{F}_1 = -\,\mathbf{F}_2$$

2.E

If the two charges are released from rest, and have equal masses, m, describe their motion and the motion of the center of mass.

The charge 2q will accelerate to the *right* with an acceleration F/m, while the acceleration of −q will be to the *left,* with a magnitude F/m. Since the center of mass is stationary, it will remain stationary and $a_c = 0$, providing there are *no external* forces on the system.

2.F

What is the *direction* of the electric field at the charge −q due to the charge 2q?

Since 2q is a positive charge, it produces an electric field pointing radially outwards. Hence, **E** is to the *right* at −q.

2.G

Write a vector expression for the electric field $\mathbf{E}_1$ at −q due to the charge 2q.

$$\mathbf{E}_1 = k\frac{2q}{d^2}\,\mathbf{i} \qquad (2)$$

2.H

Write a vector expression for the force $\mathbf{F}_1$ on $-q$ due to its presence in the electric field of 2q.

$$\mathbf{F}_1 = -q\mathbf{E}_1$$

$$\mathbf{F}_1 = -k\frac{2q^2}{d^2}\,\mathbf{i} \qquad (3)$$

2.I

Write a vector expression for the electric field $\mathbf{E}_2$ at 2q due to the charge $-q$.

$$\mathbf{E}_2 = k\frac{q}{d^2}\,\mathbf{i}$$

2.J

Write a vector expression for the force $\mathbf{F}_2$ on 2q due to its presence in the electric field of $-q$.

$$\mathbf{F}_2 = 2q\,\mathbf{E}_2$$

$$\mathbf{F}_2 = k\frac{2q^2}{d^2}\,\mathbf{i} \qquad (4)$$

Note that $\mathbf{F}_1 = -\mathbf{F}_2$, which is consistent with Newton's third law.

3 Two cork balls, each having a mass of 50 g, are suspended by light threads. Each ball is given total charge of $+1\ \mu C$; both are observed to be in equilibrium when separated by a distance $d = 0.1$ m.

3.A

Draw a free body diagram for each ball. Note that the electric force is *repulsive* for like charges.

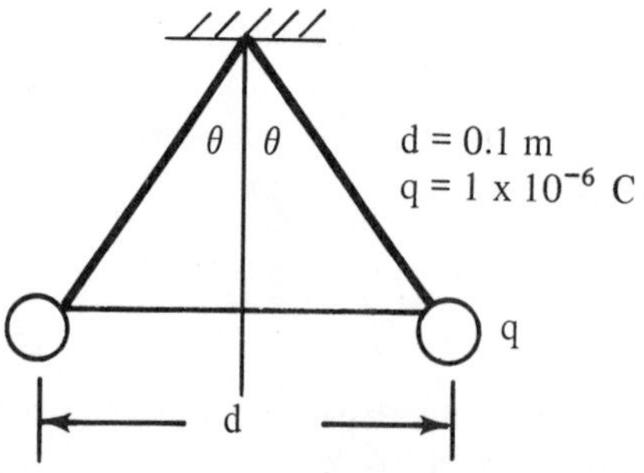

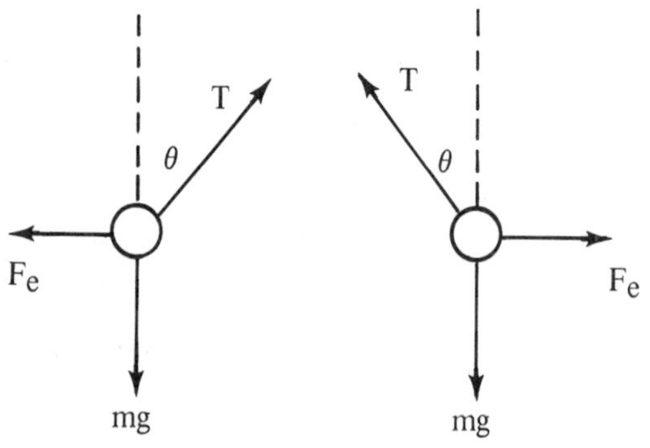

3.B

What are the conditions of equilibrium of the system? That is, for what values of θ and T are the charged masses stationary?

$$\Sigma F_x = T\sin\theta - F_e = 0 \qquad (1)$$

$$\Sigma F_y = T\cos\theta - mg = 0 \qquad (2)$$

3.C

Obtain explicit relations for θ in terms of F_e and mg.

From (1),

$$T = \frac{F_e}{\sin\theta} \qquad (3)$$

Substitute (3) into (2). This gives

$$\tan\theta = \frac{F_e}{mg} \qquad (4)$$

3.D

To obtain a value for θ, we must know F_e. What is the magnitude of F_e?

$$F_e = k\frac{q_1 q_2}{d^2}$$

$$F_e = 9 \times 10^9\ N\frac{m^2}{C^2}\ \frac{(1 \times 10^{-6}\,C)^2}{(0.1\ m)^2}$$

$$F_e = 0.9\ N \qquad (5)$$

3.E

What is the value of θ at equilibrium?

From (4) and (5), we have

$$\tan\theta = \frac{0.9\ N}{(0.05\ kg)\ 9.8\ \frac{m}{sec^2}}$$

$$\tan\theta = 1.8$$

$$\theta \cong 61^\circ$$

3.F

What is the tension in the threads at equilibrium?

Using (3) and (5), with $\theta = 61^\circ$ gives

$$T = \frac{F_e}{\sin\theta} = \frac{0.9\ N}{\sin(61)}$$

$$T \cong 1.0\ N$$

4 A +5 μC charge is located at the point (−0.3, 0) m, while a second charge of −2 μC is located at (0, 0.1) m as shown below.

4.A

Draw a vector representing the electric field E_1 at the origin (0, 0) due to the +5 μC charge. In the same sketch, draw the electric field E_2 at (0, 0) due to the −2 μC charge.

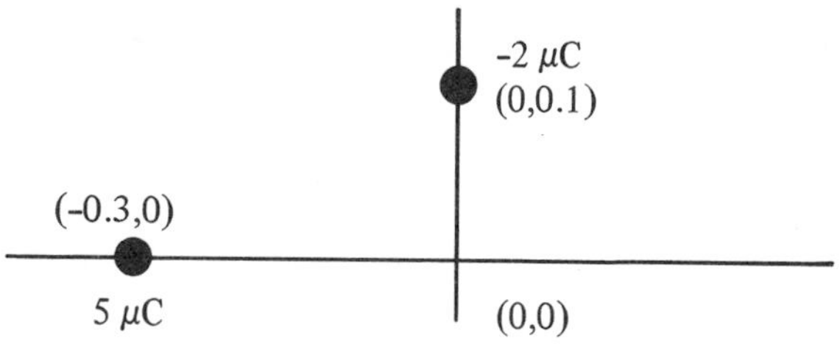

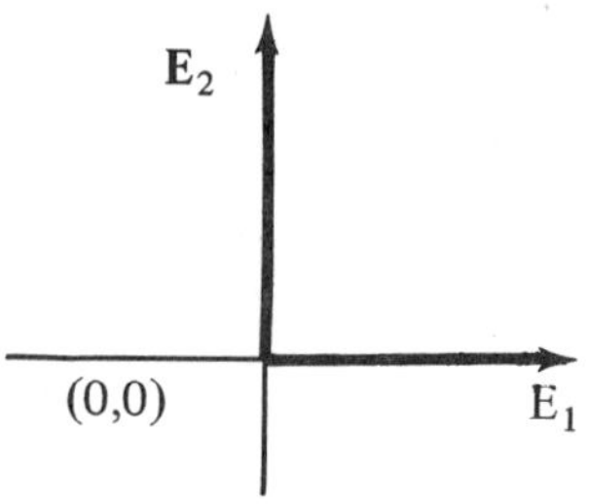

Note: These directions follow from the fact the **E** points *away* from a positive charge and towards a negative charge.

4.B

What are the magnitudes of E_1 and E_2?

$$E_1 = k\frac{q_1}{r^2} = 9 \times 10^9\ N\frac{m^2}{C^2} \times \frac{5 \times 10^{-6}\ C}{(0.3\ m)^2}$$

$$E_1 = 5 \times 10^5\ N/C$$

$$E_2 = 9 \times 10^9\ N\frac{m^2}{C^2}\ \frac{2 \times 10^{-6}\ C}{(0.1\ m)^2}$$

$$E_2 = 18 \times 10^5\ \frac{N}{C}$$

4.C

Write vector expressions for the electric fields E_1 and E_2, and the resultant field **E** at (0, 0).

$$\mathbf{E}_1 = 5 \times 10^5\ \mathbf{i}\ N/C$$

$$\mathbf{E}_2 = 18 \times 10^5\ \mathbf{j}\ N/C$$

$$\mathbf{E} = \mathbf{E}_1 + \mathbf{E}_2$$

$$\mathbf{E} = (5 \times 10^5\ \mathbf{i} + 18 \times 10^5\ \mathbf{j})\frac{N}{C}$$

4.D

What is the magnitude and direction of **E**?

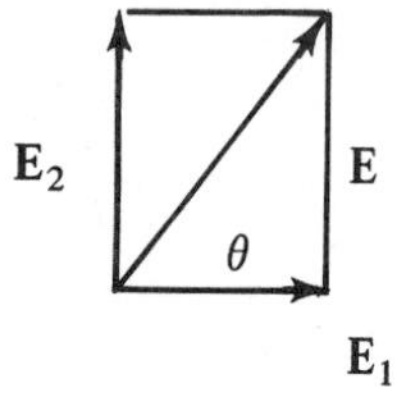

$$E = \sqrt{E_1{}^2 + E_2{}^2} = \sqrt{5^2 + 18^2} \times 10^5\ N/C$$

$$E \cong 1.9 \times 10^6\ N/C$$

$$\tan\theta = E_2/E_1 = 18/5 = 3.6$$

$$\theta \cong 75°$$

4.E

Suppose a third charge of $-3\ \mu C$ is placed at (0, 0). What is the electric force on this charge?

$$F = qE$$

$$F = -3 \times 10^{-6} C\,[5 \times 10^5\, \mathbf{i} + 18 \times 10^5\, \mathbf{j}]\ \frac{N}{C}$$

$$F = (-1.5\, \mathbf{i} - 5.4\, \mathbf{j})\ N$$

5 *The Electric Dipole.* An electric dipole consists of two charges equal in magnitude but opposite in sign, placed a distance 2d apart, as in the figure. The electric field of a dipole is a useful quantity, especially for distances $r \gg d$.

5.A

Draw vectors representing the electric fields due to +q and −q *at the point P.* Show the resultant field $\mathbf{E_P}$ in the same diagram.

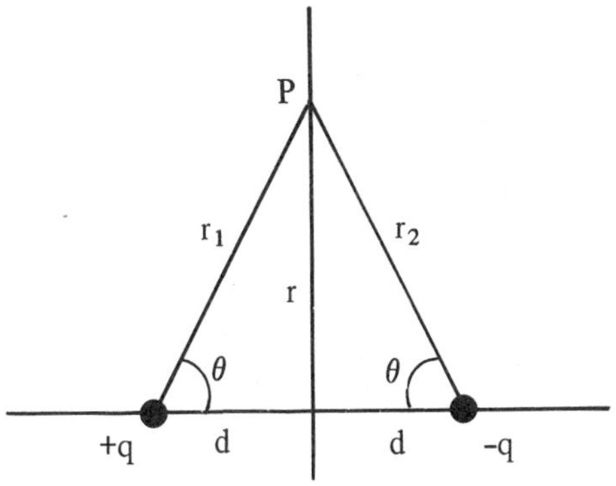

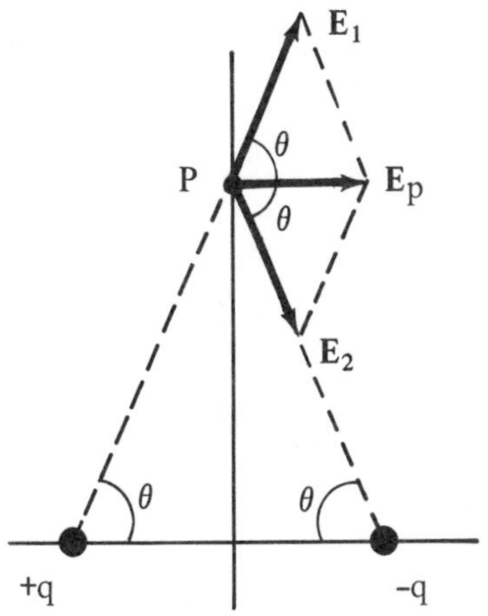

Note: $E_1 = E_2$

5.B

What are the magnitudes of $\mathbf{E_1}$ and $\mathbf{E_2}$?

$$E_1 = k\frac{q}{r_1^2} = k\frac{q}{(r^2 + d^2)}$$

$$E_2 = k\frac{q}{r_2^2} = k\frac{q}{(r^2 + d^2)}$$

5.C

Write a vector expression for the *resultant* electric field $\mathbf{E_p}$. To make the calculation simple, refer to the vector diagram in frame 5.A, and use the *superposition* principle.

From the vector diagram in frame 5.A, we see that $E_y = 0$. But $E_x = E_{x1} + E_{x2}$

$$E_p = E_x = 2E_1 \cos\theta = \frac{2kq}{(r^2 + d^2)} \cos\theta$$

Since $\cos\theta = \dfrac{d}{r_1} = \dfrac{d}{(r^2 + d^2)^{1/2}}$

$$E_p = \frac{2kqd}{(r^2 + d^2)^{3/2}}\ \mathbf{i} \qquad (1)$$

5.D

For large distances from the dipole, that is, $r \gg d$, what is an approximate value for $\mathbf{E}_p$?

For $r \gg d$, we can neglect d in the denominator of (1); therefore,

$$\mathbf{E}_p \cong \frac{2kqd}{r^3}\,\mathbf{i} \qquad (2)$$

5.E

The quantity 2qd is called the *electric dipole moment p.* Write **E** in terms of p for $r \gg d$. (Note that $E \sim 1/r^3$ for a dipole, whereas $E \sim 1/r^2$ for a point charge. That is, the field drops off much faster for a dipole than a point charge.)

$$\mathbf{E}_p \cong k\frac{p}{r^3}\,\mathbf{i} \qquad (3)$$

for $r \gg d$.

†6 A circular loop of radius R has a uniform positive charge per unit length λ, and a total charge Q. We wish to calculate the electric field intensity at points along the axis of the loop. (The calculation of E for arbitrary points is very cumbersome.)

6.A

Consider a segment of the loop whose length is ds. This segment carries a charge dq. What is dq in terms of ds?

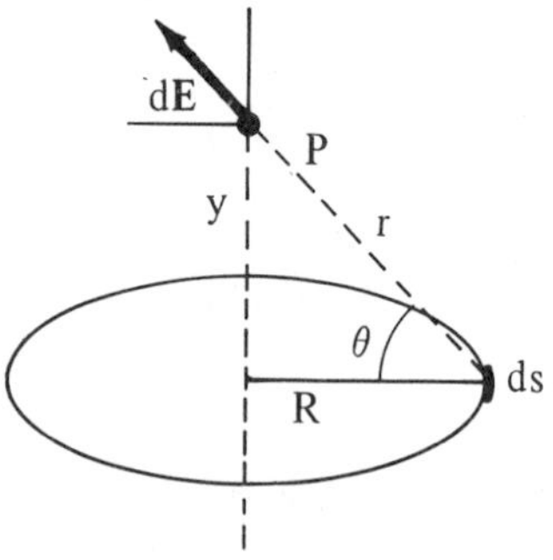

$$dq = \lambda ds \qquad (1)$$

6.B

The segment ds gives rise to an electric field dE at P, as shown in frame 6.A. What is the magnitude of dE at P?

$$dE_p = k\frac{dq}{r^2} = k\frac{\lambda ds}{r^2} \qquad (2)$$

where $k = \dfrac{1}{4\pi\epsilon_0}$.

6.C

To get the *total* field at P, we must sum up contributions to all segments over the loop. But this is a *vector* sum. Show in a *vector* diagram that **E** at P will be *only* in the y direction.

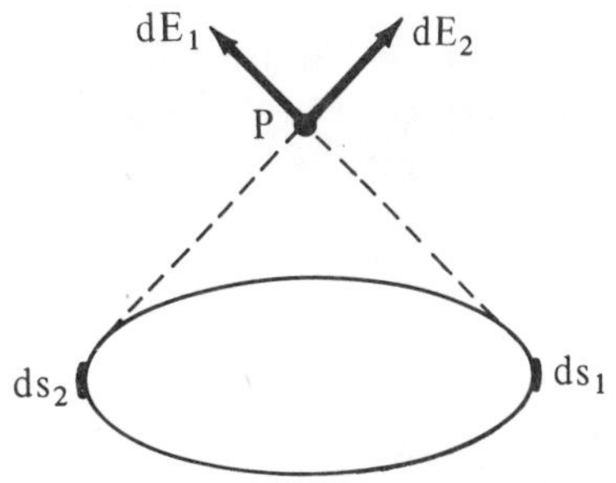

Consider two segments diametrically opposite one another as shown above. Obviously, the x components of dE *cancel* and the y components reinforce each other, so the total field is in the +y direction.

6.D

What is the y component of the electric field at P due to the segment ds?

$$dE_y = dE \sin\theta$$

$$dE_y = k\frac{\lambda \sin\theta \, ds}{r^2} \qquad (3)$$

6.E

We now must integrate (3) to get the total field at P. Reduce this integral to a simple form by noting that $\sin\theta = \frac{y}{r}$, and $r^2 = y^2 + R^2$.

$$E_p = \int dE_y = \int \frac{k\lambda \sin\theta}{r^2} ds$$

$$E_p = k\lambda \int \frac{y}{r^3} ds \qquad (4)$$

6.F

Both y and r are constants for a given point P. Therefore, we can extract them from the integrand. What is the final result for E_p?

We can write (4) as

$$E_p = k\lambda \frac{y}{(y^2 + R^2)^{3/2}} \int ds$$

But $\int ds = s = 2\pi R$, the total circumference of the loop. Since $\lambda = Q/2\pi R$, E_p becomes

$$E_p = k\frac{Qy}{(y^2 + R^2)^{3/2}} \qquad (5)$$

6.G

What is E at the center of the loop, that is, for y = 0? Explain this result.

From (5), we see that $E_p = 0$ at $y = 0$. At this special point, segments diametrically opposite one another give fields which *exactly* cancel each other.

6.H

Consider a point far from the loop along the axis, that is, $y \gg R$. What is an approximate form for E_p? Explain.

For $y \gg R$, we can neglect R in (5). Therefore, (5) reduces to

$$E_p \cong k\frac{Q}{y^2}$$

This is equivalent to the field of a point charge Q. That is, the loop "appears" to be a point charge at large distances.

7 *Electric Flux.* A uniform electric field E exists in some region of space as shown in the figure below. A plane surface of area A makes an angle θ with a plane perpendicular to the electric field lines (the dotted plane).

7.A

What is the qualitative meaning of electric flux, ϕ_E?

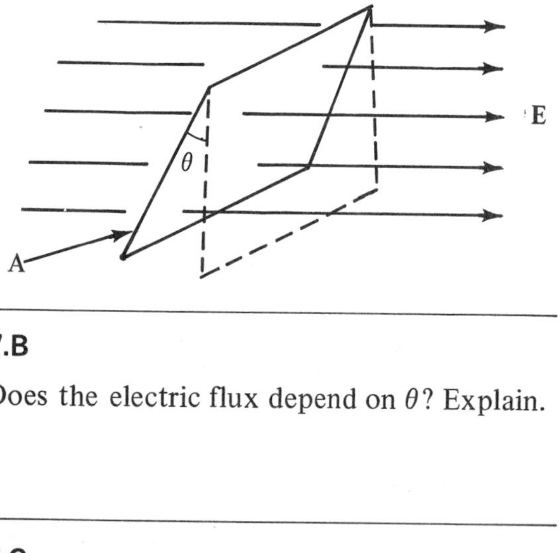

The electric flux is the number of electric field lines that pass through the surface whose area is A.

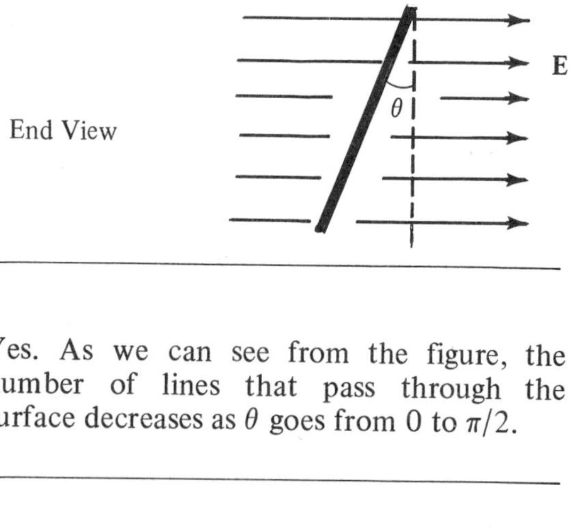

7.B

Does the electric flux depend on θ? Explain.

Yes. As we can see from the figure, the number of lines that pass through the surface decreases as θ goes from 0 to $\pi/2$.

7.C

What is the relation for electric flux in this case? Note that **E** is *constant* over the surface in this case.

$$\phi_E = \int \mathbf{E}\cdot d\mathbf{A} = \mathbf{E}\cdot\int d\mathbf{A} = \mathbf{E}\cdot\mathbf{A}$$

$$\phi_E = EA\cos\theta \qquad (1)$$

7.D

When is the electric flux a maximum? When is it zero?

From (1), we see that ϕ_E is a maximum for $\theta = 0$ [$\cos 0 = 1$], while ϕ_E is zero for $\theta = \pi/2$ [$\cos \pi/2 = 0$].

7.E

What are the units of electric flux in SI units?

$$[\phi_E] \sim [EA] \sim \frac{N}{C} m^2$$

7.F

Suppose $A = 3\ m^2$ and $E = 2 \times 10^4$ N/C. Find ϕ_E if the plane surface makes an angle of 37° with **E**.

$$\phi_E = EA \cos\theta$$

$$\phi_E = 2 \times 10^4 \frac{N}{C} 3\ m^2 \cos(37)$$

$$\phi_E = 4.8 \times 10^4 \ \frac{Nm^2}{C}$$

8 A closed surface in the shape of a cube, whose side is a, is situated in a uniform electric field E directed along the +x axis, as shown below. There are *no* charges within the cube.

8.A

What is the electric flux through surface I, the *right* face of the cube?

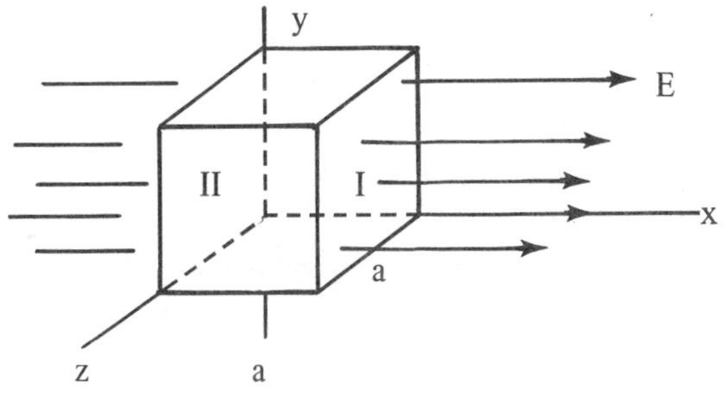

On this face, d**A** points out of the surface, hence to the right. Therefore, $\theta = 0$ in the expression

$$\phi_E = EA \cos\theta \qquad (1)$$

$$\therefore \qquad \phi_E(I) = Ea^2 \qquad (2)$$

Note: (1) applies only if **E** is uniform over the surface.

8.B

What is the electric flux through surface II, the *left* face of the cube coincident with the yz plane?

In this case, d**A** points to the *left,* that is, in the negative x direction. Therefore, $\theta = \pi$ in the expression for ϕ_E.

$$\phi_E(II) = EA\cos(\pi) = -Ea^2 \qquad (3)$$

8.C

What is the electric flux through the remaining four surfaces of the cube? Explain your answers.

The electric flux through the remaining four faces is *zero,* since **E** is parallel to each of these planes; hence, no electric field lines pass through them. Or, we can say $\mathbf{E} \perp d\mathbf{A}$ for each face, and since $\cos(\pi/2) = 0$, $\phi_E = 0$ for each such face.

8.D

What is the net electric flux through the cube?

$\phi_E = \phi_E(I) + \phi_E(II) = 0$

8.E

How can you interpret the result that the net electric flux is zero?

ϕ_E is zero because the number of lines that enter the cube from the left leave the cube at the right face.

8.F

Is it possible to have some *net charge* within the cube and still maintain a net flux of zero? Explain.

No. Gauss's law states that ϕ_E through a *closed* surface must equal q/ϵ_0, where q is the *net* charge within an imaginary surface. Therefore, if $\phi_E = 0$, it follows that $q = 0$.

9 *Gauss's law.* A plane sheet of metal has a uniform positive charge distributed on both surfaces, a cross-section of which is shown in the figure below. The charge *per unit area* is σ, and the sheet is long in all directions.

9.A

What are the directions of the **E** fields in regions I and III? Show these in a diagram.

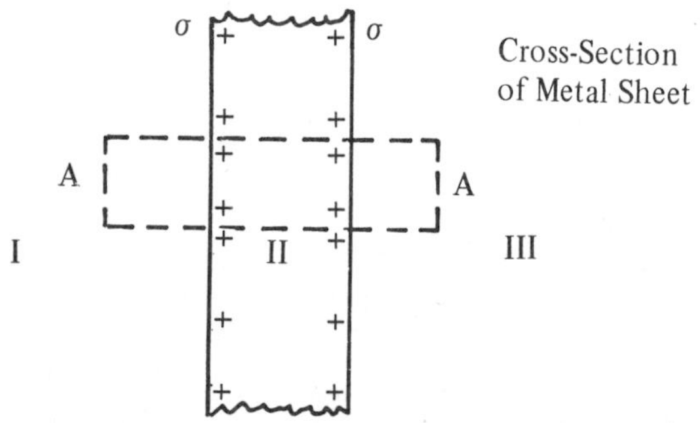

The electric field *must* be normal to the metal surface. In region I, **E** points to the left, while in region III, **E** points to the right.

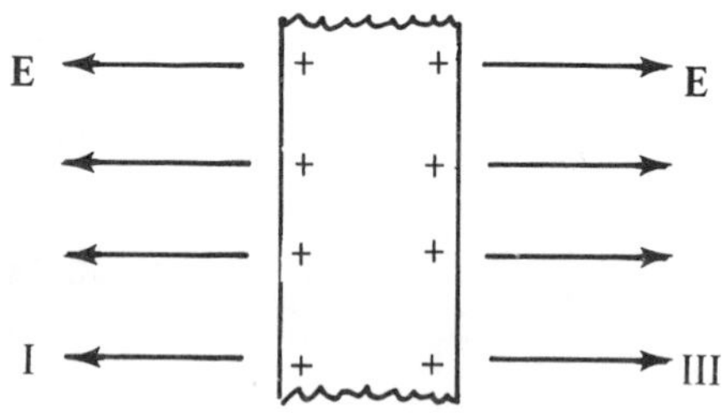

9.B

Why must **E** just outside the metal be *normal* to the surface?

If **E** has a component parallel to the surface, then there would be such a component just inside the surface. Such a field in the metal would move charges in the metal and create currents. No currents can exist at equilibrium.

9.C

What is the electric field in region II?

E = 0 in region II. This has been explained in 9.B.

9.D

To calculate **E** in regions I and III, we construct a Gaussian surface in the shape of a pillbox, whose ends are parallel with the surface (see the dotted lines in the figure of frame 9.A). The ends have an area A. What is the total charge enclosed by this pillbox?

The charge at the right face of the metal is σA. The charge at the left face is σA. Therefore, the *total* charge enclosed is $2\sigma A$.

$$q = 2\sigma A \qquad (1)$$

9.E

What is the electric flux through the right end of the pillbox?

dA and **E** both point to the right for this surface, and **E** is constant over this surface. $\therefore \phi_E = EA$.

9.F

What is the electric flux through the left end of the pillbox?

dA and **E** both point to the *left* for this surface, and **E** is again constant. $\therefore \phi_E = EA$.

9.G

What is the electric flux through the remaining surface of the pillbox?

$\phi_E = 0$ over the remaining surface, since **E** is parallel everywhere to this surface.

9.H

What is the *total* electric flux through the pillbox?

We add contributions from the left and right faces to get

$$\phi_E = 2EA \qquad (2)$$

9.I

Use Gauss's theorem and (1) and (2) to find the **E** field in regions I and III. Recall that q in Gauss's law is the *net* charge enclosed by the Gaussian surface.

$$\phi_E = \oint \mathbf{E} \cdot d\mathbf{S} = q/\epsilon_0$$

$$2EA = \frac{2\sigma A}{\epsilon_0}$$

$$\therefore \qquad E = \frac{\sigma}{\epsilon_0} \qquad (3)$$

or,

$$\mathbf{E} = \frac{\sigma}{\epsilon_0}\mathbf{i}, \quad \text{in region III}$$

$$\mathbf{E} = -\frac{\sigma}{\epsilon_0}\mathbf{i}, \quad \text{in region I}$$

†10 An infinitely long insulating hollow cylinder has a uniform *negative* charge density ρ. The radius of the hollow section is a, and the radius of the outer surface is b, as in the figure below. We wish to find the electric fields in the regions $r > b$, $b > r > a$, and $r < a$.

10.A

Construct a Gaussian surface that will lead to a direct calculation of **E** for $r > b$.

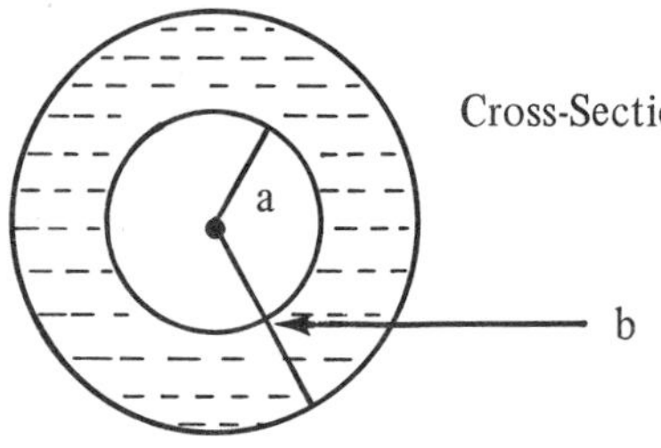

Cross-Section of Cylinder

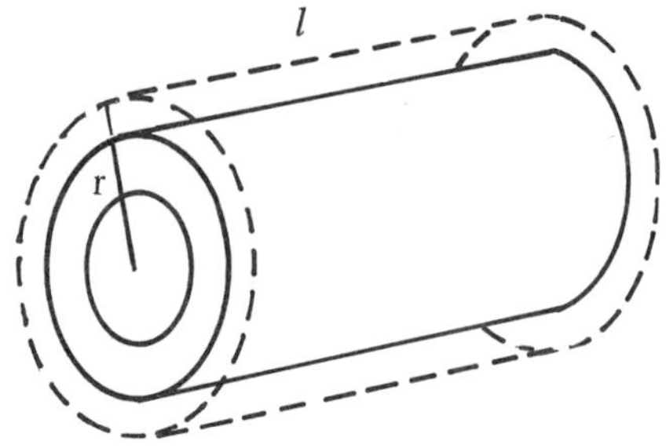

Dotted-line is of Gaussian cylinder of length *l*, radius r.

10.B

What is the direction of the electric field for $r > b$?

By symmetry, the **E** field is radial. Since the cylinder is *negatively* charged, the field lines point radially *inwards.*

10.C

Show the direction of **E** in a diagram for $r > b$, and the direction of the *area* element **dA** at the Gaussian surface.

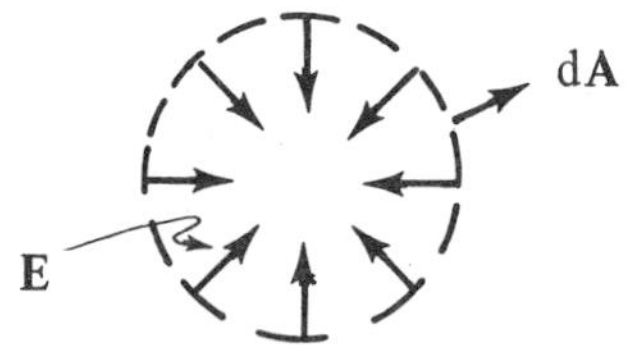

10.D

Note that **E** is constant in magnitude over this Gaussian surface, and **E** is *antiparallel* to **dA**. Call the length of the Gaussian cylinder *l*, and its radius r. Find an expression for the electric flux ϕ_E through the cylinder.

Since **E** is *opposite* in direction to **dA**, $\mathbf{E}\cdot d\mathbf{A} = EdA \cos(\pi) = -EdA$; so ϕ_E becomes

$$\phi_E = \oint \mathbf{E}\cdot d\mathbf{A} = -E\int dA$$

$$\phi_E = -EA = -E\,(2\pi r l) \qquad (1)$$

Note that the *end* walls do not contribute to ϕ_E, since no flux lines pass through them; that is, **E** is parallel to these surfaces.

10.E

What is the *net* charge enclosed by the Gaussian cylinder at $r > b$ in terms of the charge density ρ and the radii of the cylinder? Note that you must subtract out the volume of the hollow section of the cylinder.

$q = \rho V$, where V is the volume of the cylinder enclosed by the Gaussian surface.

$$V = \pi b^2 l - \pi a^2 l$$

$$V = \pi l(b^2 - a^2)$$

so

$$q = \rho \pi l(b^2 - a^2) \qquad (2)$$

10.F

Apply Gauss's law and use (1) and (2) to find E in the region $r > b$.

$$\phi_E = \oint \mathbf{E} \cdot d\mathbf{A} = \frac{q}{\epsilon_0}$$

$$-E\,(2\pi r l) = \frac{\rho \pi l\,(b^2 - a^2)}{\epsilon_0}$$

$$E = -\frac{\rho(b^2 - a^2)}{2\epsilon_0 r} \quad \text{for } r > b \qquad (3)$$

10.G

Now construct a Gaussian cylinder of radius $a < r < b$ to find the electric field in this region. What is the electric flux through this surface?

ϕ_E is still given by (1), since there is cylindrical symmetry, and **E** is radial inward.

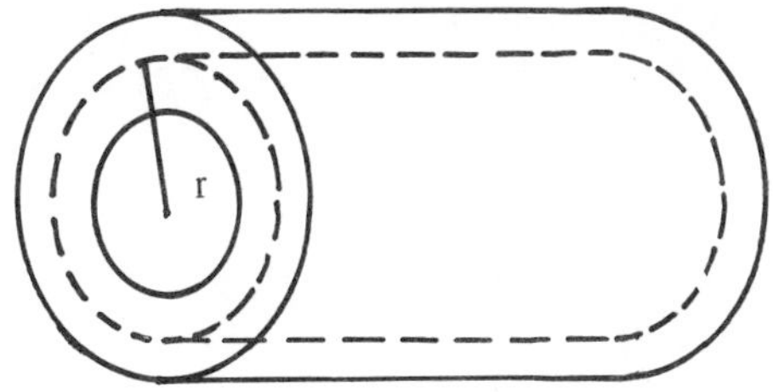

10.H

What is the total charge q' enclosed by the Gaussian cylinder whose radius is $a < r < b$, as shown to the right of frame 10.G?

$$q' = \rho V'$$

where V' is the volume of the cylinder of radius r.

$$V' = \pi l(r^2 - a^2)$$

so

$$q' = \rho \pi l(r^2 - a^2) \qquad (4)$$

10.I

Apply Gauss's law, together with (1) and (4), to find E in the region $a < r < b$.

$$\oint \mathbf{E} \cdot d\mathbf{A} = \frac{q'}{\epsilon_0}$$

$$-E\,(2\pi r l) = \frac{\rho \pi l (r^2 - a^2)}{\epsilon_0},$$

$$E = -\frac{\rho}{2\epsilon_0}\left(\frac{r^2 - a^2}{r}\right) \quad \text{for } a < r < b \qquad (5)$$

10.J

What is E in the region $r \leqslant a$? Explain.

There is no charge in this region; hence, by Gauss's law, $\oint \mathbf{E} \cdot d\mathbf{A} = 0$ for a Gaussian surface within this region. ∴ E = 0.

10.K

Show that (3) and (4) are equivalent solutions at r = b.

By inspection, (3) and (5) reduce to

$$E = -\frac{\rho}{2\epsilon_0}\left(\frac{b^2 - a^2}{b}\right) \quad \text{at } r = b$$

11 A solid conducting sphere of radius a has a net positive charge of 3Q. Concentric with this sphere is a conducting spherical shell of radius b ($b > a$), which has a net negative charge of $-Q$.

11.A

Construct a spherical Gaussian surface with $r > b$. What is the net charge q_1 enclosed by this surface?

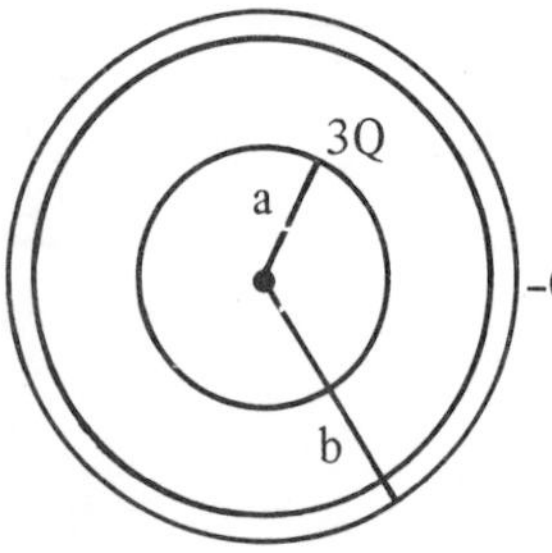

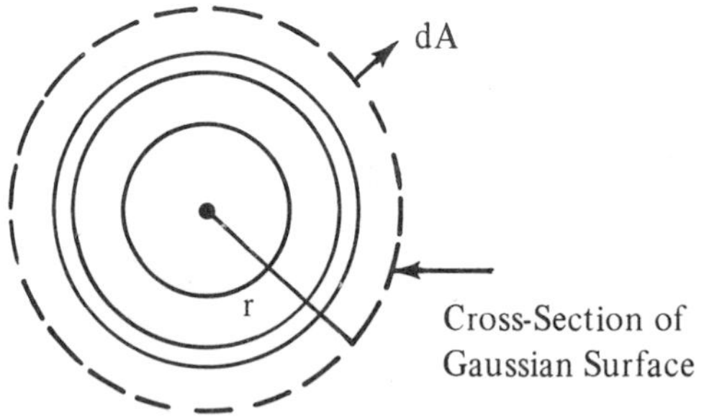

$$q_1 = 3Q + (-Q)$$

$$q_1 = 2Q \qquad (1)$$

11.B

What is the direction of the **E** field at points $r > b$?

Since the net charge enclosed by the Gaussian surface is *positive*, **E** must point *radially outwards* for $r > b$.

$$\therefore \quad \mathbf{E} \cdot d\mathbf{A} = E\,dA$$

11.C

By symmetry, we see that **E** is constant in magnitude over the spherical Gaussian surface concentric with the spheres. Use this fact and Gauss's law to find **E** at $r \geqslant b$.

$$\oint \mathbf{E} \cdot d\mathbf{A} = \frac{q_1}{\epsilon_0}$$

$$EA = E\, 4\pi r^2 = \frac{2Q}{\epsilon_0}$$

$$E = \frac{Q}{2\pi\epsilon_0 r^2} \quad \text{for } r \geqslant b \qquad (2)$$

11.D

Now construct a spherical Gaussian surface with $a < r < b$. What is the charge q_2 enclosed by this surface?

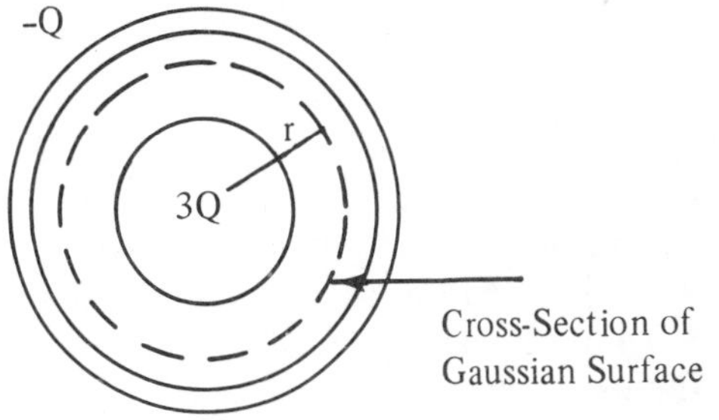

$$q_2 = 3Q \qquad (3)$$

11.E

Use (3) and Gauss's law to find the electric field in the region $a < r < b$. Again, note that **E** is constant in magnitude over the Gaussian surface.

$$\oint \mathbf{E} \cdot d\mathbf{A} = \frac{q_2}{\epsilon_0}$$

$$EA = E 4\pi r^2 = \frac{3Q}{\epsilon_0}$$

$$E = \frac{3Q}{4\pi\epsilon_0 r^2} \quad \text{for } a < r < b \qquad (4)$$

11.F

What is the electric field intensity in the region $r < a$ and in the region of $r = b$ (that is, within the walls of the shell)?

$E = 0$ in both cases, since they are both conducting bodies, and no electric field can exist within the conductor at equilibrium.

11.G

A spherical Gaussian surface can be constructed just within the walls of the spherical shell. Noting that **E** = 0 in this region, determine the *total* charge q_3 on the *inner* wall of the shell.

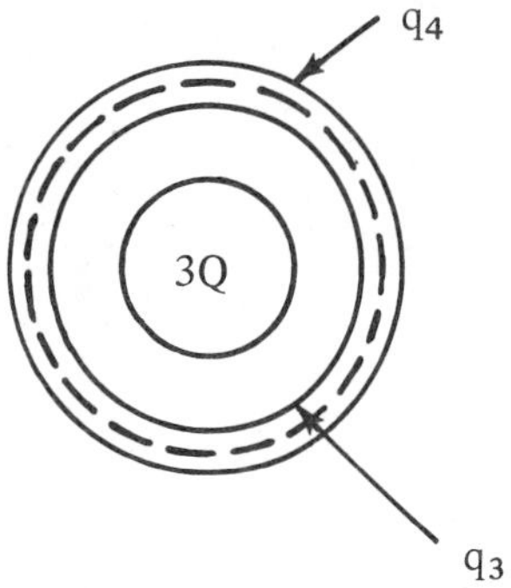

From Gauss's law,

$$\oint \mathbf{E}\cdot d\mathbf{A} = \frac{q}{\epsilon_0} = q_3 + 3Q$$

But **E** = 0 in the region of the Gaussian surface, ∴

$$q_3 + 3Q = 0$$

$$q_3 = -3Q \qquad (5)$$

11.H

Since the charge on the *inner* surface of the shell is −3Q, and the *total* charge on the shell is −Q, what necessarily is the total charge q_4 on the *outer* surface of the shell?

We require that

$$q_3 + q_4 = -Q$$

$$-3Q + q_4 = -Q$$

$$\therefore \quad q_4 = 2Q \qquad (6)$$

†12 An insulating solid sphere of radius R has a *nonuniform* charge density given by $\rho = \rho_0 r$ C/m^3. In other words, $\rho = 0$ at r = 0, and increases linearly to a maximum at r = R. The total charge is positive and is taken to be Q. We wish to find the electric field intensity at points inside and outside the sphere.

12.A

If we construct a concentric spherical Gaussian surface, with r > R, the total charge enclosed is obviously Q. What is the electric field intensity outside the sphere?

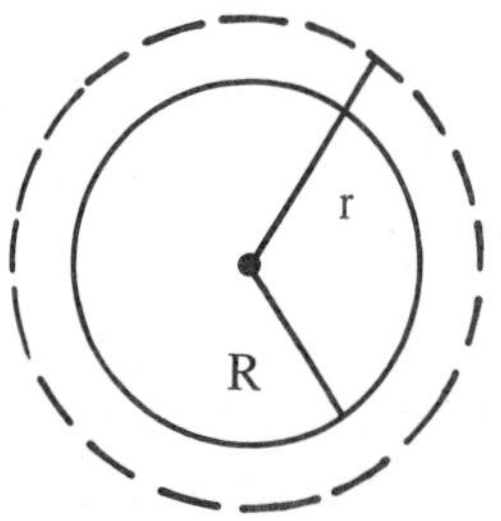

Again, we have spherical symmetry. **E** is radial outwards and is constant in magnitude everywhere on this Gaussian surface. ∴

$$\oint \mathbf{E}\cdot d\mathbf{A} = EA = E(4\pi r^2) = \frac{Q}{\epsilon_0}$$

$$E = \frac{Q}{4\pi\epsilon_0 r^2} \quad \text{for } r > R \qquad (1)$$

12.B

Charge density in this problem means charge per unit volume. Therefore, $\rho = q/V$. Hence, we can say that a small chunk of the charged sphere of volume dV has a charge dq, or $dq = \rho dV$. From this, find a relation between the total charge Q, and the constants ρ_0 and R. *Hint:* Since there is spherical symmetry, let dV be the volume of a spherical shell and reduce the integral to one over r.

$$dq = \rho dV$$

$$Q = \int dq = \int \rho dV$$

Since $\rho = \rho(r)$, we can write dV as the volume of a spherical shell of thickness dr. This shell has a volume $dV = 4\pi r^2 dr$.

$$Q = \int_0^R \rho_0 r (4\pi r^2 dr)$$

$$Q = \pi \rho_0 R^4 \qquad (2)$$

12.C

To find E *within* the charged sphere we construct another spherical Gaussian surface of radius $r < R$. Note that the charge q enclosed by this surface is less than Q. Calculate q in terms of ρ_0 and r. (Remember that *only* the charge *within* the Gaussian surface contributes to E at r.)

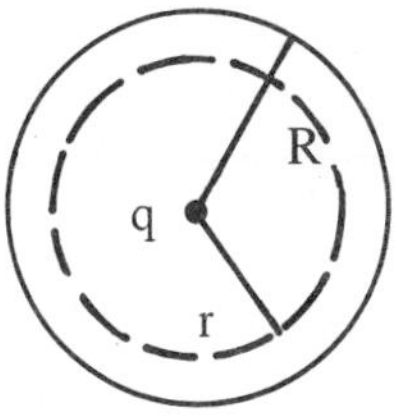

$$q = \int_0^r \rho dV$$

$$q = \int_0^r \rho_0 r (4\pi r^2 dr)$$

$$q = \pi \rho_0 r^4 \qquad (3)$$

Note that (3) agrees with (2) when r = R.

12.D

Apply Gauss's law and (3) to find E in the region $r < R$.

$$\oint \mathbf{E} \cdot d\mathbf{A} = \frac{q}{\epsilon_0}$$

$$EA = E(4\pi r^2) = \frac{\pi \rho_0 r^4}{\epsilon_0}$$

$$E = \frac{\rho_0 r^2}{4\epsilon_0} \quad \text{for } r < R \qquad (4)$$

12.E

Make a rough plot of E *vs* r in the regions $r < R$ and $r \geqslant R$. [Note that (4) is consistent with (1) at $r = R$, since $Q = \rho_0 R^4$ is the total charge on the sphere.]

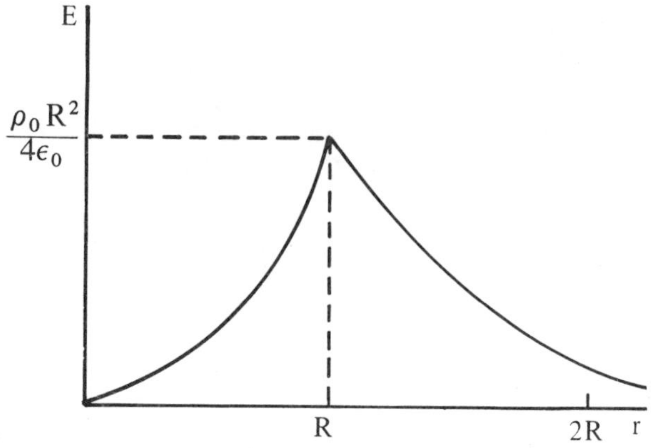

13 *Motion of a Charged Particle in a Uniform Electric Field.* A proton enters a region of a uniform electric field of intensity 2×10^4 N/C in the negative y direction. The proton has an initial speed of 5×10^6 m/sec, and its velocity vector makes an angle of 37° with the horizontal as shown below.

13.A

Write a vector expression for $\mathbf{v}_0$, from the information above.

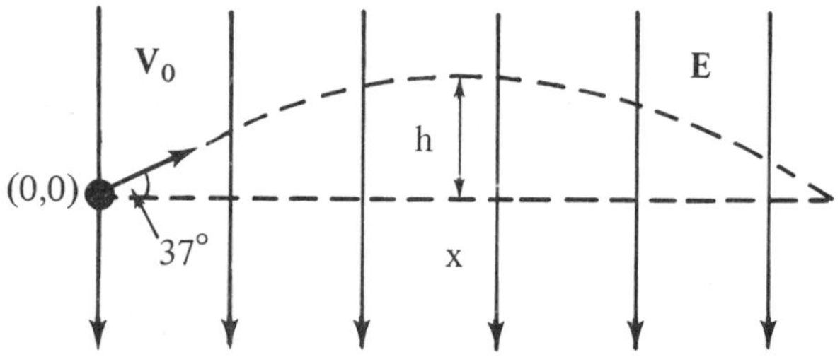

$$v_{x0} = v_0 \cos \theta_0 = 5 \times 10^6 \cos(37) \text{ m/sec}$$

$$v_{x0} = 4 \times 10^6 \text{ m/sec}$$

$$v_{y0} = v_0 \sin \theta_0 = 5 \times 10^6 \sin(37) \text{ m/sec}$$

$$v_{y0} = 3 \times 10^6 \text{ m/sec}$$

$$\therefore \quad \mathbf{v}_0 = (4 \times 10^6 \, \mathbf{i} + 3 \times 10^6 \, \mathbf{j}) \text{ m/sec} \qquad (1)$$

13.B

Determine the acceleration of the proton due to the *electric* force (that is, neglect gravity, since $mg \ll eE$). Use the fact that the mass of a proton is $m = 1.67 \times 10^{-27}$ kg, and its charge is $e = 1.6 \times 10^{-19}$ C.

$$\mathbf{F} = e\mathbf{E} = -eE\mathbf{j} = m\mathbf{a}$$

$$\mathbf{a} = -\frac{eE}{m}\mathbf{j}$$

$$\mathbf{a} = -\frac{1.6 \times 10^{-19} \text{ C} \left(2 \times 10^4 \frac{\text{N}}{\text{C}}\right)}{1.67 \times 10^{-27} \text{ kg}} \mathbf{j}$$

$$\mathbf{a} = -1.9 \times 10^{12} \, \mathbf{j} \text{ m/sec}^2 \qquad (2)$$

13.C

Write an expression for the x component of velocity as a function of time.

Since $a_x = 0$, $v_x = v_{x0}$, that is, the x component of velocity remains *constant.*

$$v_x = v_{x0} = 4 \times 10^6 \text{ m/sec}$$

13.D

Write an expression for the y component of velocity as a function of time.

Since a_y is constant, we can write

$$v_y = v_{yo} + a_y t$$

Using (1) and (2) gives

$$v_y = (3 \times 10^6 - 1.9 \times 10^{12}\ t)\ \text{m/sec}$$

13.E

Since the acceleration of the proton is in the *negative* y direction, it behaves like a projectile in the earth's gravitational field. Find the time t_1 it takes the proton to reach its peak altitude. Take $t = 0$ to be the time it first enters the **E** field.

When the proton reaches its peak, $v_y = 0$. ∴, from (3) we get

$$t_1 = \frac{3 \times 10^6}{1.9 \times 10^{12}}\ \text{sec}$$

$$t_1 \cong 1.6 \times 10^{-6}\ \text{sec} \qquad (4)$$

13.F

What is the maximum height h reached by the proton? Take the origin of coordinates at the position of the proton when it first enters the **E** field, as in frame 13.A.

For constant acceleration, we have

$$y = v_{yo} t + \frac{1}{2} a_y t^2$$

When the proton is at its peak, $t = t_1$ so

$$h = \left(3 \times 10^6\ \frac{\text{m}}{\text{sec}}\right)(1.6 \times 10^{-6}\ \text{sec})$$

$$-\frac{1}{2}\left(1.9 \times 10^{12}\ \frac{\text{m}}{\text{sec}^2}\right)(1.6 \times 10^{-6}\ \text{sec})^2$$

$$h = 2.4\ \text{m} \qquad (5)$$

13.G

What is the *horizontal* distance covered by the proton when it once again has a zero y coordinate?

This distance is covered in a time $2t_1$, that is, twice the time it takes to reach its peak. Therefore,

$$x = v_{xo}(2t_1)$$

$$x = 4 \times 10^6\ \frac{\text{m}}{\text{sec}}(3.2 \times 10^{-6}\ \text{sec})$$

$$x = 12.8\ \text{m} \qquad (6)$$

13.H

Calculate the ratio of eE to mg for the proton, and verify that eE ≫ mg. Use SI units.

$$\frac{eE}{mg} = \frac{1.6 \times 10^{-19}\ (2 \times 10^{4})\ N}{1.67 \times 10^{-27}\ (9.8)\ N}$$

$$\frac{eE}{mg} \cong 2 \times 10^{11} \qquad (7)$$

2.8 SUMMARY

The electric force between two point charges q_1 and q_2 is given by *Coulomb's law:*

$$\mathbf{F}_{12} = k\frac{q_1 q_2}{r^2}\,\hat{r} \tag{2.5}$$

where r is the separation of the charges and $k = 1/4\pi\epsilon_0 = 9 \times 10^9\ \mathrm{Nm^2/C^2}$. The force is attractive for unlike charges and repulsive for like charges.

The electric field intensity **E** at a distance r from a point charge q is given by

$$\mathbf{E} = k\frac{q}{r^2}\,\hat{r} \tag{2.7}$$

Gauss's theorem states that the total electric flux through a closed surface surrounding a charge q is given by

$$\phi_E = \oint \mathbf{E}\cdot d\mathbf{A} = \frac{q}{\epsilon_0} \tag{2.13}$$

The force experienced by a charge q in an electric field E is given by

$$\mathbf{F} = q\mathbf{E}$$

2.9 PROBLEMS

1. Two point charges of 3 μC and 8 μC are separated by 0.2 m. (a) Find the force of repulsion between the two charges. (b) Calculate the electric field intensity at the position of the 3 μC charge due to the 8 μC charge.

2. The electron and proton in the hydrogen atom as a "mean" separation of 0.52 A (where 1A = 1 Angstrom = 10^{-10} m). (a) Determine the electric field due to the proton at the "mean" position of the electron. (b) Find the force of attraction between the two particles for this separation. (Use the data in Table 2-1.)

3. Three point charges of 2 μC, -5 μC and 3 μC are located at the corners of an *equilateral* triangle of side 0.1 m, as in Figure 2-12. (a) Determine a *vector* expression for the electric force on the 3 μC charge. (b) Determine a *vector* relation for the electric field at the position of the 3 μC charge.

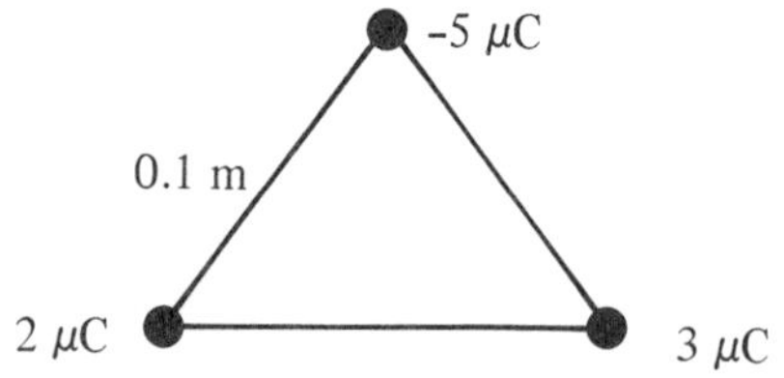

Figure 2-12

4. Four charges are positioned at the corners of a rectangle, as in Figure 2-13. Find the x and y components of the electric field at the *center* of the rectangle.

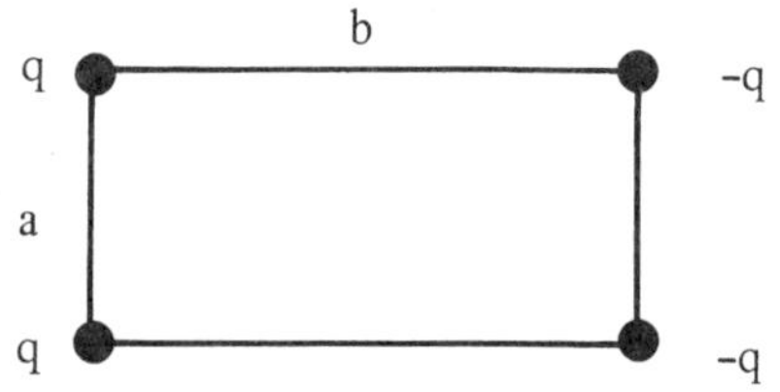

Figure 2-13

5. Charges of 5 μC, -5 μC and 16 μC are positioned at three corners of a square of side 0.1 m, as in Figure 2-14. (a) Obtain a vector expression electric field intensity at P. (b) Determine the force which would act on a 2 μC charge placed at P.

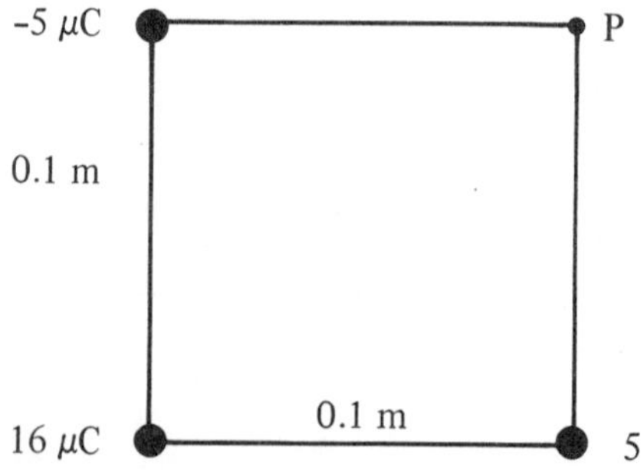

Figure 2-14

6. In a particular experiment, a "free" electron is present in a vacuum. (a) What would be the magnitude and direction of the electric field necessary to maintain uniform motion of the electron in a direction parallel to the earth? (b) Repeat the calculation for a proton. (*Hint:* In this situation, uniform motion implies *constant* velocity. Consider the combined action of the electric force and the gravitational force, and use Table 2-1.)

7. A charged cork ball, whose mass is 1 g, is suspended on a light string in the presence of a uniform electric field, as in Figure 2-15. When $\mathbf{E} = (3\,\mathbf{i} + 5\,\mathbf{j}) \times 10^5$ N/C, the ball is in equilibrium at $\theta = 37°$. Find (a) the charge on the ball and (b) the tension in the string.

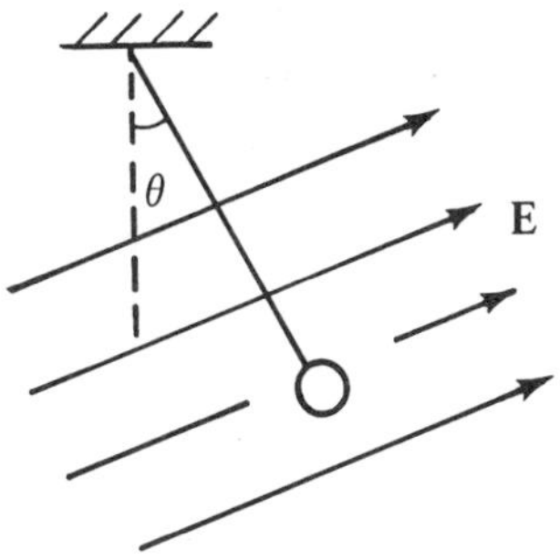

Figure 2-15

8. Two small spheres of equal mass 2 g are suspended on light strings 10 cm in length in a uniform electric field produced by two charged plates, as in Figure 2-16. The spheres have charges of -5×10^{-8} C and $+5 \times 10^{-8}$ C. (a) Determine the electric field intensity which enables the spheres to be in equilibrium at $\theta = 10°$. (b) Find the surface charge density on the plates for this value of E.

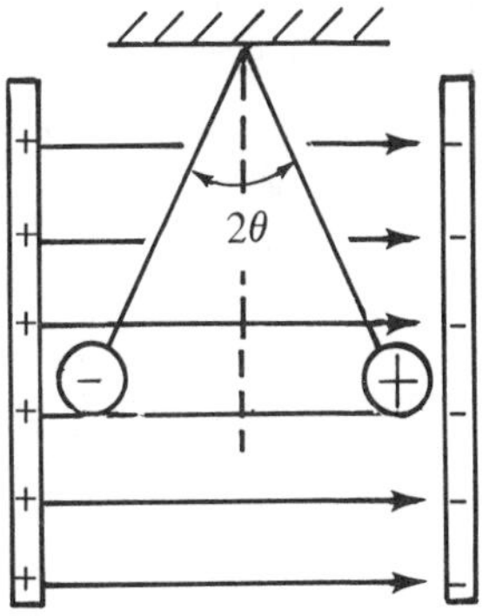

Figure 2-16

9. A charge of $-3\ \mu$C is located at the origin, while a second charge of $-7\ \mu$C is located at the position (0, 0.2) m. At what point along the axis connecting the two charges is the electric field zero? Are there any other points for which $\mathbf{E} = 0$?

10. Three charges of q, -2q, and q are located along the x axis, as in Figure 2-17. Show that the electric field intensity at P for $r \gg a$ is given by

$$\mathbf{E} = -\,k\frac{2qa^2}{r^4}\,\mathbf{j}$$

Such a charge distribution, which is essentially two electric dipoles, is called an *electric quadrupole,* and the *quadrupole moment* is defined as $Q = 2qa^2$. Note that $E \sim 1/r^4$ for a quadrupole, $E \sim 1/r^3$ for a dipole and $E \sim 1/r^2$ for a monopole (a single charge).

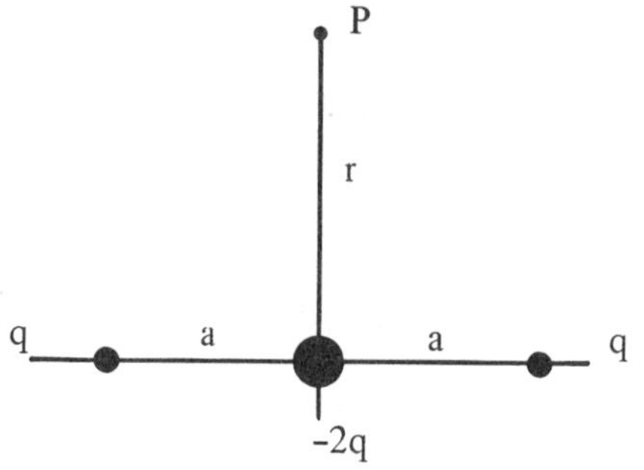

Figure 2-17

11. A uniformly positive charged filament is bent into the shape of a circular arc of radius R, as in Figure 2-18. The arc subtends an angle of 90° at the center, and the charge per unit length is λ. Calculate the electric field intensity at the center of curvature, 0.

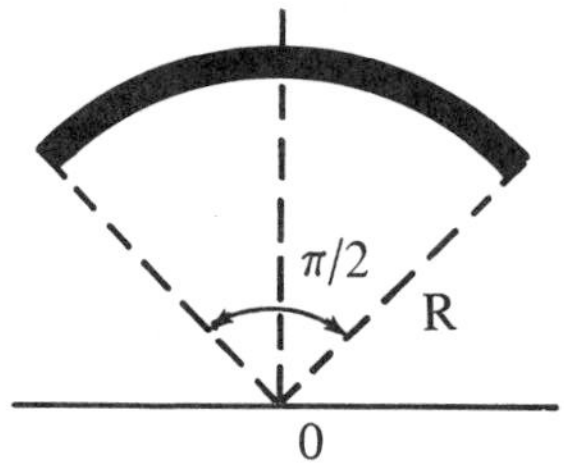

Figure 2-18

12. A flat disc of radius R has a *uniform* positive charge per unit area σ, as in Figure 2-19. Calculate the electric field intensity at the point P, along the axis of the disc. [*Hint:* Think of the disc as a composition of rings of charge (as in Exercise 6) where the charge on each ring is $dq = \sigma dA$, where $dA = 2\pi r dr$ is a unit of area for the ring.]

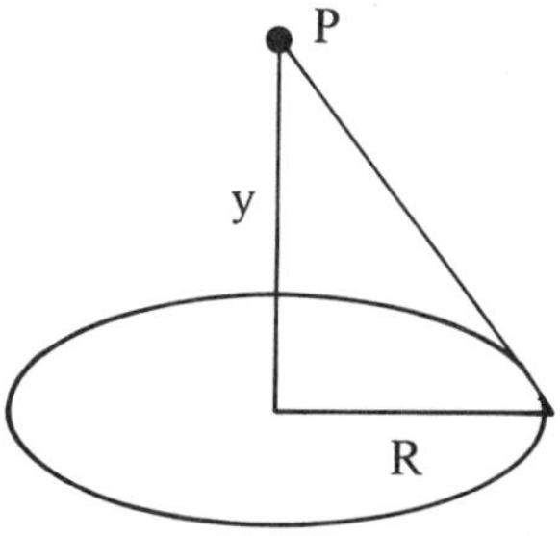

Figure 2-19

13. An insulating filament of length l has a *nonuniform* charge per unit length, which varies as $\lambda = \lambda_0(x - d)$. The filament is oriented along the x axis, as in Figure 2-20. Find the electric field intensity at the origin. (*Hint:* An infinitesimal element of length dx has a charge $dq = \lambda dx$.)

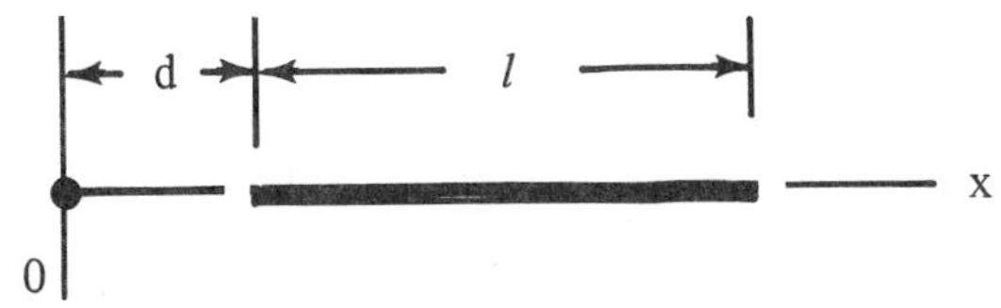

Figure 2-20

14. A uniform electric field of 1500 N/C is directed along the x axis. An uncharged circular disc of radius 1 m is placed in the region of the electric field. What is the electric flux through the disc if: (a) the disc is located in the yz plane? (b) the disc is located in the xy plane? (c) the disc makes an angle of 45° with the x and y axes?

15. A uniformly charged straight filament 10 m in length has a total positive charge of 5×10^{-6} C. An uncharged cardboard cylinder 2 cm in length, 10 cm in radius, surrounds the filament at its center and is concentric with the filament. Find (a) the electric field intensity at the surface of the cylinder and (b) the electric flux through the cardboard cylinder.

16. In Example 2.5, by using Gauss's law we found that the electric field of an infinitely long, uniformly charged filament is given by

$$\mathbf{E} = k\frac{2\lambda}{r}\hat{r}$$

where λ is the charge per unit length. Show that the same result is obtained by integration, choosing the differential element of charge to be $dq = \lambda dx$, as in Figure 2-21. (*Hint:* a convenient variable to use is θ, defined in Figure 2-21. Note that $x = r\cot\theta dx = -r\csc^2\theta d\theta$.)

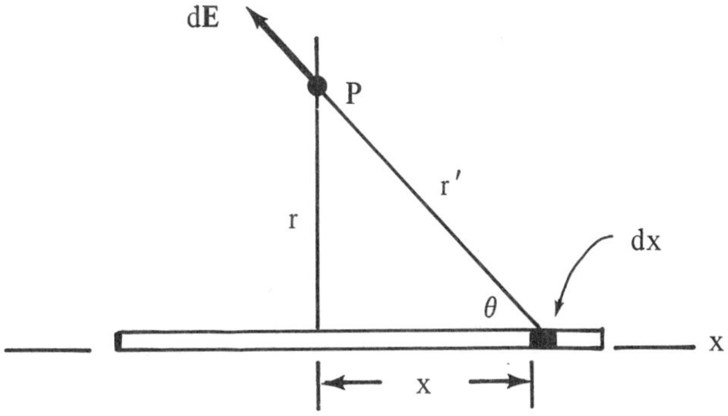

Figure 2-21

17. Derive Gauss's law by constructing an arbitrarily shaped imaginary surface surrounding a point charge q, as in Figure 2-8. Note that the *solid angle* $d\Omega$ subtended by the surface element dA is defined as $d\Omega = dA \cos\theta/r^2$, as in Figure 2-22. The *total* solid angle subtended at a point is 4π steradians, where the unit solid angle is called one steradian. Therefore, regardless of the shape of the surface surrounding q, $\oint d\Omega = 4\pi$.

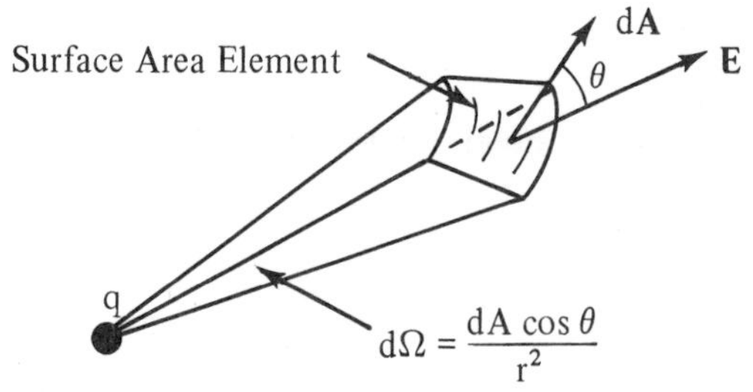

Figure 2-22

18. Three large conducting plates are placed parallel to one another. A cross-section is shown in Figure 2-23. The outer plates have a uniform charge per unit area of 3σ and $-\sigma$, respectively, while the middle plane is *grounded.* Use Gauss's law to find (a) the surface charge densities of the left and right surfaces of the *center plate* and (b) the electric field intensity in regions I, II, and III.

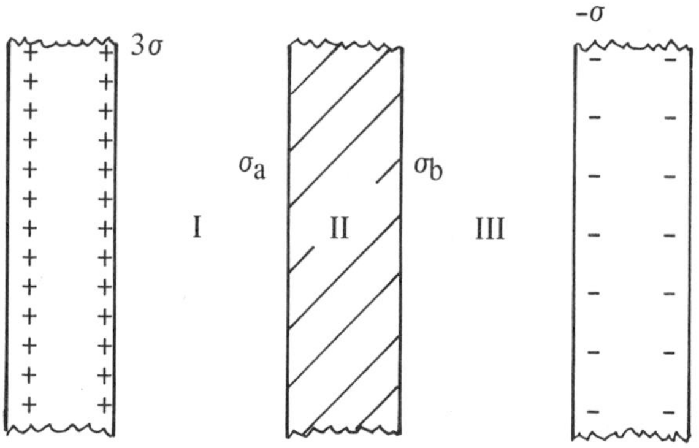

Figure 2-23

19. A solid insulating sphere of radius a has a uniform charge density, ρ, and total charge Q. Concentric with this sphere is an uncharged conducting hollow sphere, whose inner and outer radii are b and c, as in Figure 2-24. (a) Find the electric field intensity in the regions $r < a$ and $b > r > a$. (b) Determine the induced charge per unit area on the inner and outer surface of the spherical shell.

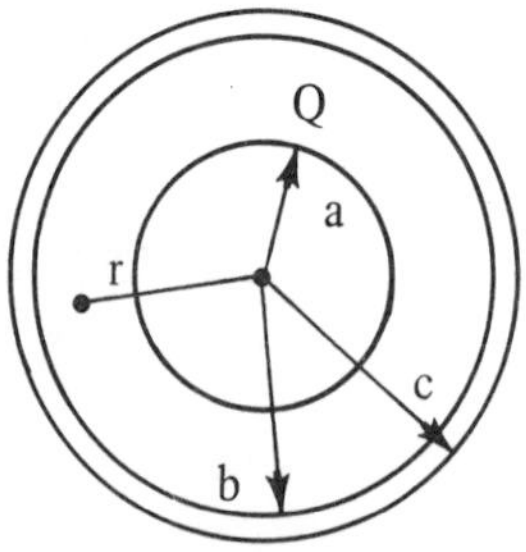

Figure 2-24

20. Suppose Figure 2-24 represents an insulating sphere surrounded by a concentric, metal hollow sphere, with a = 5 cm, b = 20 cm, and c = 25 cm. The electric field intensity at a point 10 cm from the center is known to be 3.6×10^3 N/C radially *inwards,* while at a point 50 cm from the center, the electric field is 2×10^2 N/C radially *outwards.* From this information, find (a) the charge on the insulating sphere, Q, (b) the total charge on the inner and outer surfaces of the hollow sphere and (c) the charges per unit area on the inner and outer surfaces of the hollow sphere. Note that in this case, $Q_b \neq Q_c$.

21. A very long cylinder 5 cm in radius has a *negative* charge per unit length of -5×10^{-6} C/m. A point charge of 3×10^{-8} C is placed 20 cm from the cylinder. Find (a) the electric field intensity at the position of the point charge and (b) the electric force on the point charge.

22. A long metal cylinder of radius a has a uniform charge per unit length λ and is surrounded by a concentric hollow metal cylinder whose *net* charge per unit length is -3λ, as in Figure 2-25. The hollow cylinder has inner and outer radii of b and c, respectively. (a) Find the charge per unit length on the inner and outer surfaces of the hollow cylinder. (b) Determine the electric field intensity in the regions $r < a$, $a < r < b$, $b < r < c$, and $r > c$.

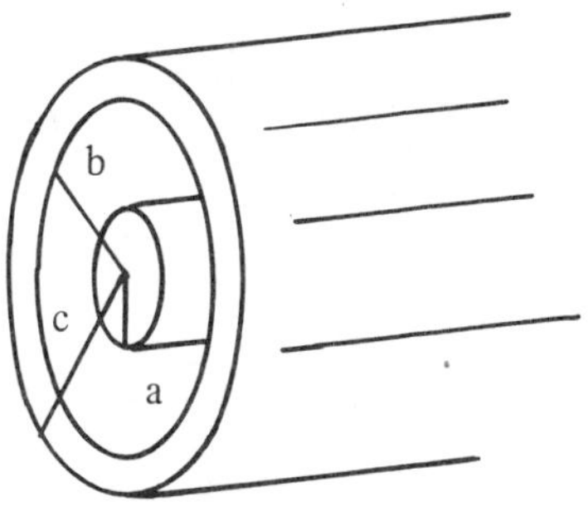

Figure 2-25

23. With reference to Figure 2-25, suppose the inner cylinder has a radius of a = 2 cm, and the hollow cylinder has radii of b = 10 cm and c = 12 cm. The *net* charge per unit length on the hollow cylinder is known to be -8×10^{-6} C/m, while the electric field intensity at r = 6 cm is known to be 6×10^5 N/C directed radially *outwards.* Find (a) the charge per unit length on the inner cylinder, (b) the charge per unit length on the inner and outer surfaces of the hollow cylinder and (c) the magnitude and direction of the electric field intensity at r = 20 cm.

24. A solid insulating sphere of radius R has a *nonuniform* charge density which varies with r according to the expression $\rho = \rho_0 r^2$. Find expressions for the electric field intensity for points inside and outside the sphere, that is, for $r < R$ and $r > R$. (*Hint:* The total charge Q of the sphere is given by $\int_0^R \rho dV$, while the charge q within a radius $r < R$ is given by $\int_0^r \rho dV$. Use $4\pi r^2\, dr$ as the volume element dV, which corresponds to the volume of a spherical shell of radius r, thickness dr.)

25. A hollow insulating sphere has inner and outer radii of a and b, respectively. The sphere has a *uniform* charge density ρ. Find expressions for the electric field intensity in the regions (a) $r < a$, (b) $a < r < b$ and (c) $r > b$.

26. A *positively* charged solid insulating sphere of radius R has a uniform charge density, ρ, as in Example 2.4. A *small* diameter hole is bored through the sphere, passing through its center, as in Figure 2-26. A *negative* point charge $-q$, whose mass is m, is placed in the hole and is released from its surface. (a) What is the force on the charge $-q$ when it is a distance r from the center of the insulating sphere? (b) Neglecting friction, show that the charge $-q$ exhibits simple harmonic motion along the length of the bored hole, with an amplitude R and period T given by

$$T = \frac{1}{2\pi}\sqrt{\frac{3m\epsilon_0}{\rho q}}$$

(c) Use the results to (b) to find the period of an electron moving through a uniformly charged sphere of radius 10^{-8} m, and total charge 4.8×10^{-19} C. (This represents Thomson's model of the atom.)

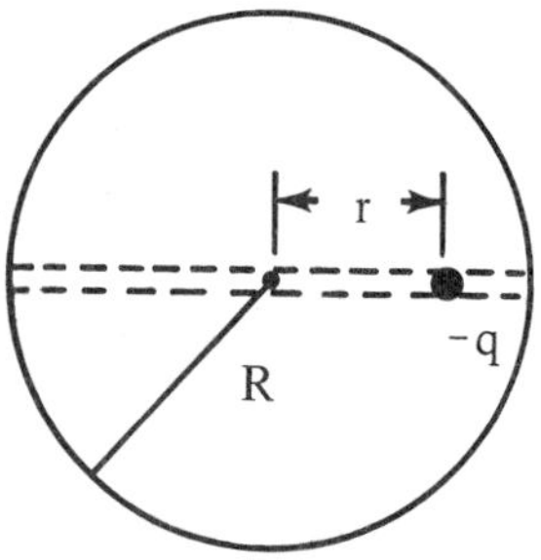

Figure 2-26

27. *Electric Dipole.* Two charges of q and $-q$ are located on the y axis at the positions $y = a$ and $y = -a$, respectively. Find the electric field intensities for the points (x, 0) and (0, y) when x and y are *large* compared to a. Write your expressions in terms of the electric dipole moment $p = 2qa$.

28. A proton is projected into a region of uniform electric field of intensity 3×10^5 N/C, as in Figure 2-27. The proton travels a distance of 4 cm before stopping. Determine (a) the acceleration of the proton, (b) the initial speed of the proton and (c) the time it takes the proton to come to rest.

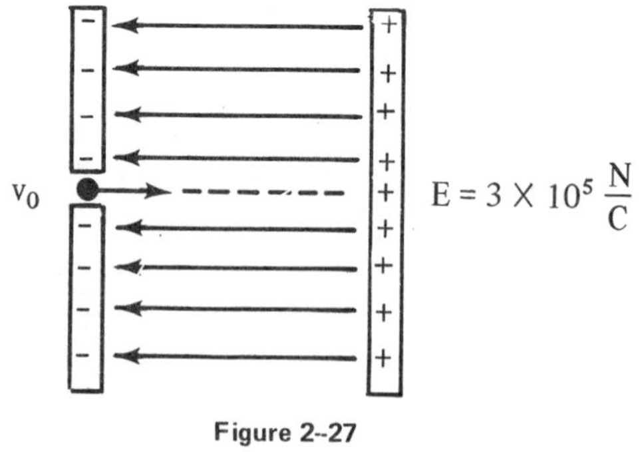

Figure 2-27

29. A positive charge q has an initial velocity $v_0\mathbf{i}$ as it enters the region of a uniform electron field in the y direction, as in Figure 2-28. A screen is placed a distance L from the horizontal charged plates, as shown. Ignore the effects of gravity. (a) Show that the equation of the path followed by the charge is given by

$$y = \left(\frac{qE}{2mv_0^2}\right) x^2$$

(b) If $L \gg l$, show that the ratio q/m is given by

$$\frac{q}{m} \cong \frac{hv_0^2}{ELl}$$

This suggests a means of measuring q/m for charged particles, such as protons or electrons. The technique can also be used to "sort out" the energies of identical particles and is the principle behind the cathode ray tube.

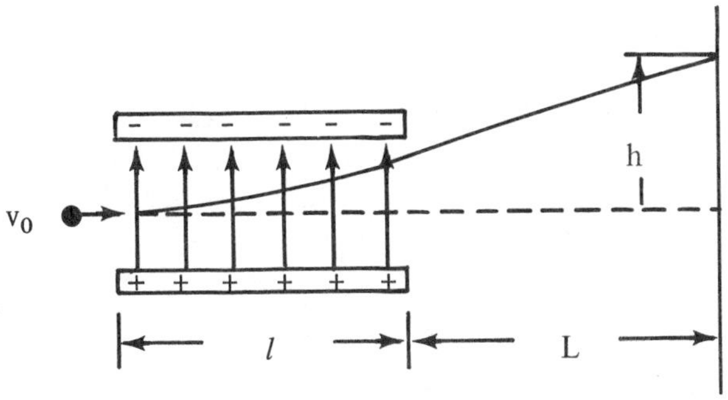

Figure 2-28

30. An electron is projected into the region of a uniform electric field of 2×10^4 N/C in such a way that it *just clears* the bottom positive plate and *just reaches* the negative plate, as in Figure 2-29. (a) Find the initial speed of the electron, v_0. (b) How long does it take the electron to reach the negative plate? (c) What is the horizontal distance x?

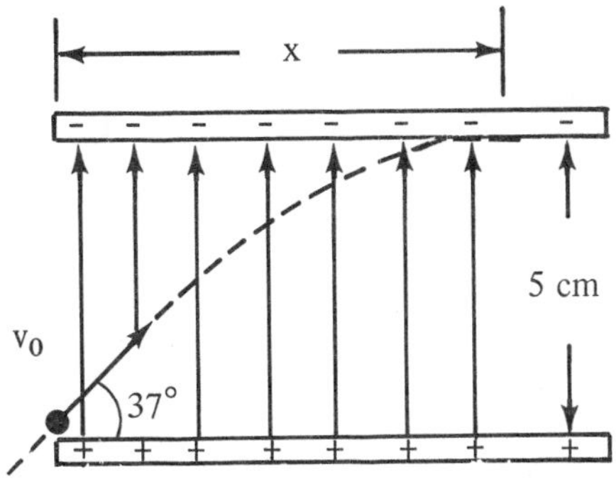

Figure 2-29

31. *Millikan's Oil Drop Experiment.* An ingenious technique for determining the charge on an electron was devised by Millikan. The basic apparatus is sketched in Figure 2-30. Oil is sprayed as fine droplets between two oppositely charged plates, and can become negatively charged by friction with the nozzle of the spraying device. Three forces act on the charged oil droplets: the gravitational force mg, the electric force q**E**, and the buoyancy force, $\mathbf{F_b}$. (Archimedes' principle says that F_b equals the weight of the medium displaced by the oil droplet.) The electric field is varied until the combination of the electric force and buoyancy force equals the weight, so the droplet is stationary between the plates.

Under this condition, show that

$$q = \frac{(\rho_o - \rho_a)\,gV}{E}$$

where ρ_o and ρ_a are the densities of oil and air, respectively, V is the volume of a droplet and E is the electric field intensity.

In practice, q is some integral multiple of the charge on the electron, and *many* measurements have to be made to get e.

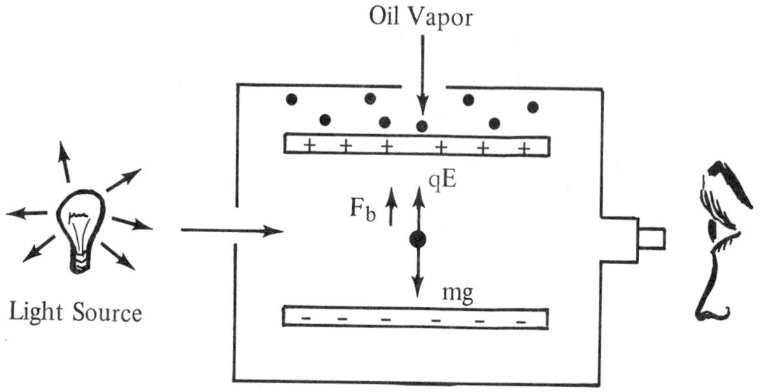

Figure 2-30

3

ELECTRIC POTENTIAL AND CAPACITANCE

The concept of gravitational potential energy, and an associated potential function, was shown to be a useful tool in various mechanics problems. Likewise, the concept of electric potential is introduced in this chapter because of its usefulness in many areas of electricity, such as the characteristics of a battery and the properties of circuit components.

3.1 ELECTRIC POTENTIAL

When a charged particle is subjected to an electric field **E**, we have seen that an electric force equal to q**E** acts on the particle. Therefore, if the particle is allowed to move freely, it will experience an acceleration and its kinetic energy will change in time. Consequently, the work-energy theorem tells us that work is being done on the particle by the electric force, where the work provides the change in kinetic energy. The work done by the electric force, dW, for a infinitesimal displacement ds, is given by

$$dW = \mathbf{F}\cdot d\mathbf{s} = q\mathbf{E}\cdot d\mathbf{s} \tag{3.1}$$

Therefore, the work done on the charged particle for a finite displacement from a to b is given by

$$W = q\int_a^b \mathbf{E}\cdot d\mathbf{s} \tag{3.2}$$

The integral in Equation (3.2) is called a *line integral* and is evaluated along the path of the particle. The lower and upper limits represent the initial and final vector positions, $\mathbf{r}_a$ and $\mathbf{r}_b$, respectively. If **E** is a *static* electric field, the integral given by Equation (3.2) is independent of the path taken between $\mathbf{r}_a$ and $\mathbf{r}_b$. It only depends on the end points. In other words, the electric force q**E** is a *conservative* force if **E** is a static field; therefore, we can associate a potential energy function U with this force. In mechanics, we have seen that the work done by a conservative force equals the negative of the difference in potential energy between the final position and the initial position. That is,

$$W = -\Delta U = -[U_b - U_a] \tag{3.3}$$

Substituting Equation (3.3) into Equation (3.2) gives

$$U_b - U_a = -q\int_a^b \mathbf{E}\cdot d\mathbf{s} \tag{3.4}$$

Now it is convenient to define the *electric potential* V as the potential energy per unit charge.

$$V = \frac{U}{q}$$

Applying this definition to Equation (3.4), we have

$$V_b - V_a = -\int_a^b \mathbf{E}\cdot d\mathbf{s} \tag{3.5}$$

Therefore, when one evaluates the line integral, this is equivalent to a calculation of the *potential difference* $\Delta V = V_b - V_a$. In practice, what one measures is, in fact, a potential difference. For example, the "voltage" of a battery means a potential difference of a specified number of volts between the terminals. Absolute electric potential has no physical significance. A value can be assigned to V at some point, once an arbitrary reference point is selected as the zero of potential.

Since energy is a scalar quantity, electric potential is also a scalar quantity. The unit of potential in SI units is the volt, V, and from the definition of potential we can write

$$1\ \text{V} = 1\ \text{J/C} \tag{3.6}$$

Therefore, it follows that we can express *electric fields* in SI units as either N/C *or* V/m. The proof of this is left as an exercise.

When atomic constituents, such as electrons or protons, are accelerated in electric fields, it is common to speak of the electron volt, eV, as a unit of energy. An electron volt is the equivalent to the increase in kinetic energy of an electron (or proton) when accelerated through a potential difference of one volt. Since $e = 1.6 \times 10^{-19}$C, we have the following conversion factor:

$$1\ \text{eV} = 1.6 \times 10^{-19}\ \text{C}\frac{\text{J}}{\text{C}} = 1.6 \times 10^{-19}\ \text{J}$$

Consider a battery, whose terminal voltage is V, connected to two parallel plates, as in Figure 3-1. The separation of the plates is d, and the electric field is assumed to be uniform. Let us use Equation (3.5) to find a relation for the electric field between the metal plates in terms of ΔV and d.

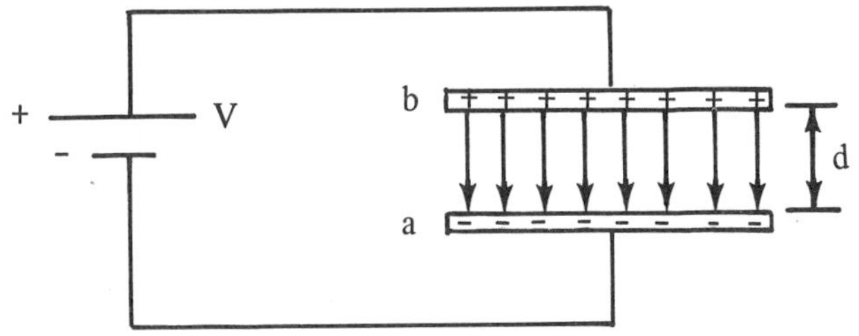

Figure 3–1

In Figure 3–1, the upper plate b is at a *higher* potential than the lower plate a. Thus, $\Delta V = V_b - V_a$. Since **E** is downwards, and the path of the integral in Equation (3.5) can be taken to be a straight vertical line from a to b, then $d\mathbf{s} = dy\,\mathbf{j}$. In this case, $\mathbf{E}\cdot d\mathbf{s}$ becomes

$$\mathbf{E}\cdot d\mathbf{s} = -E\mathbf{j}\cdot dy\,\mathbf{j} = -E dy$$

Since **E** is constant over the path of integration, the line integral reduces to

$$\Delta V = -\int_0^d \mathbf{E}\cdot d\mathbf{s} = E\int_0^d dy = Ed$$

or,

$$E = \frac{\Delta V}{d} \tag{3.7}$$

Therefore, if we connect a 12 V battery to a pair of parallel metal plates separated by 0.1 cm, the electric field between the plates is given by

$$E = \frac{12\ \text{V}}{0.1 \times 10^{-2}\ \text{m}} = 1.2 \times 10^4\ \frac{\text{V}}{\text{m}}$$

We will make further use of Equation (3.7) later, when we treat a device called a capacitor. Keep in mind that Equation (3.7) applies *only* to a pair of charged parallel plates.

3.2 POTENTIAL IN A UNIFORM ELECTRIC FIELD

It was stated earlier that a *static* electric field acting on a charge q gives rise to a *conservative* electric force qE. One of the properties of a conservative force field is that the potential difference between any two points is *independent* of the path between these two points. Let us show that a uniform electric field has this property. Suppose a particle moves from point a to point b in a uniform electric field, as in Figure 3–2. The displacement from a to b can be made by a succession of small horizontal and vertical steps. Since E is in the x direction, and $\mathbf{ds} = \mathbf{i}\,dx + \mathbf{j}\,dy$, the integrand in Equation (3.5) can be written as

$$\mathbf{E}\cdot\mathbf{ds} = E\mathbf{i}\cdot(\mathbf{i}\,dx + \mathbf{j}\,dy) = E\,dx$$

Since E is constant, the potential difference between a and b reduces to

$$V_b - V_a = -\int \mathbf{E}\cdot\mathbf{ds} = -\int_0^d E\,dx$$

$$V_b - V_a = -Ed \tag{3.8}$$

In other words, the potential difference between a and b in a *uniform* electric field depends only on the displacement in the direction *parallel* to E. This is equivalent to the statement that no work is done on the charged particle in the *vertical* steps of the successive displacements. Work is done only along the horizontal steps. The *change* in potential energy of the charge q for the horizontal displacement d is given by

$$\Delta U = q\,(V_b - V_a) = -qEd \tag{3.9}$$

Hence, if q is positive, $V_b < V_a$, so the charge loses potential energy, but *gains* kinetic energy. It gains kinetic energy because the force on a positive charge is to the right in Figure 3–2; therefore, the particle accelerates to the *right.* On the other hand, if q is *negative,* $V_b > V_a$. Consequently, the increase in potential energy is followed by a decrease in kinetic energy. In other words, a negative particle decelerates to the *left* in Figure 3–2, and loses kinetic energy.

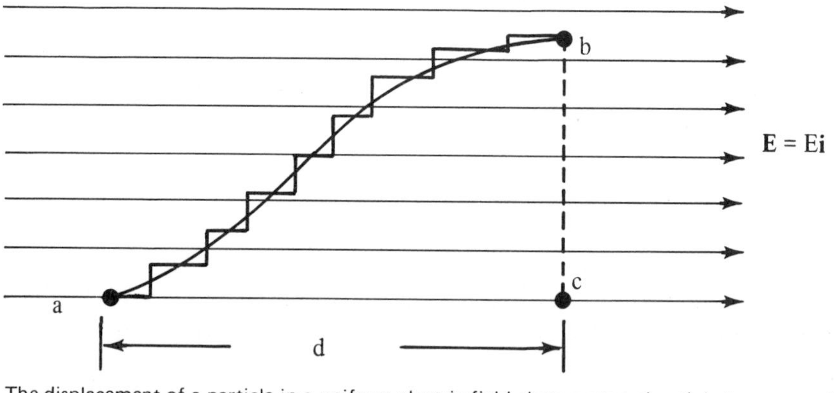

Figure 3–2 The displacement of a particle in a uniform electric field along a general path between a and b.

Our results also show that all points in a line *perpendicular* to the E field are at the same potential. For example, in Figure 3–2, $V_b - V_a = V_c - V_a$; therefore, $V_b = V_c$. We call a surface containing such a set of equal potential points an *equipotential surface.* The equipotential surfaces of a uniform **E** field are a family of planes perpendicular to the **E** field. Therefore, from Equation (3.3), we see that *no work* is done in moving a test charge between any two points on an equipotential surface. Later, we shall see that the equipotential surfaces of an isolated point charge are a family of spheres concentric with the charge.

Example 3.1

A proton is released from rest in a uniform electric field, $\mathbf{E} = (3 \times 10^5\ \mathbf{i})$ V/m. The proton undergoes a displacement of 0.2 m in the direction of **E**. (a) What is the *change* in electric potential of the proton for this displacement?

Using Equation (3.8), we have

$$\Delta V = -Ed = -(3 \times 10^5)\frac{V}{m}(0.2\ m)$$

$$\Delta V = -6 \times 10^4\ V$$

(b) What is the change in potential energy of the proton for this displacement?

$$\Delta U = q\Delta V = 1.6 \times 10^{-19}\ C\ (-6 \times 10^4\ V)$$

$$\Delta U = -9.6 \times 10^{-15}\ J$$

Since

$$1\ eV = 1.6 \times 10^{-19}\ J$$

$$\Delta U = -\frac{9.6 \times 10^{-15}\ J}{1.6 \times 10^{-19}\ J/eV} = -6 \times 10^4\ eV$$

(c) If there are no forces acting on the proton other than the conservative electric force, the work-energy theorem reveals that $\Delta K + \Delta U = 0$. Use this to find the *speed* of the proton after it has been displaced 0.2 m. ($m_p = 1.67 \times 10^{-27}$ kg).

Since $v_0 = 0$, we have

$$\Delta K + \Delta U = \frac{1}{2}m_p v^2 - 9.6 \times 10^{-15}\ J = 0$$

$$v^2 = \frac{19.2 \times 10^{-15}}{1.67 \times 10^{-27}}\frac{m^2}{sec^2} = 11.5 \times 10^{12}\ m^2/sec^2$$

$$v = 3.4 \times 10^6\ m/sec$$

3.3 POTENTIAL DUE TO POINT CHARGES

An isolated point charge q produces an electric field **E** that points in the radial direction. If the charge is positive, the electric field points radially *outwards* from the charge, as in Figure 3-3. Therefore, we can write **E** as some arbitrary point along the path as

$$\mathbf{E} = E\hat{r} = k\frac{q}{r^2}\hat{r}$$

where $k = 1/4\pi\epsilon_0$. Since ds has a component along the radial direction, namely **dr**, only this component contributes to **E**·ds. That is, **E**·ds = **E**·dr = Edr. Therefore, Equation (3.5) reduces to

$$V_b - V_a = -\int \mathbf{E}\cdot d\mathbf{s} = -\int E dr = -kq\int_{r_a}^{r_b} \frac{dr}{r^2}$$

$$V_b - V_a = kq\left[\frac{1}{r_b} - \frac{1}{r_a}\right] \tag{3.10}$$

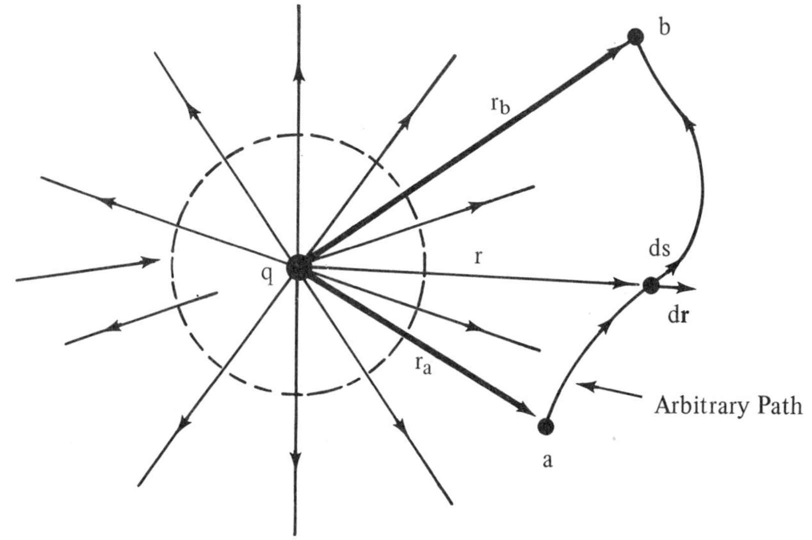

Figure 3–3 The displacement of a test charge in the vicinity of a positive charge q. The test charge moves from a to b along an arbitrary path.

Equation (3.10) is a very important result. The potential difference between points a and b depends *only* on the initial and final coordinates. The result is *independent* of the path, since we are dealing with a conservative force field. It is convenient to choose the point at r = ∞ as the zero of potential. With this choice, we can write the potential of a point charge q at a distance r from the charge as

$$V(r) = k\frac{q}{r} \quad [\text{where } V(\infty) = 0] \tag{3.11}$$

Since V is constant on a spherical surface of radius r, we conclude that the equipotential surfaces for an isolated charge are a family of spheres concentric with the charge, as shown in Figure 3–3.

The potential energy of two or more charges is obtained by applying the law of superposition. Since potential is a scalar quantity, we simply take the potential of each charge separately at some point, and then sum the terms to get the total potential. Thus, for N charges, we have

$$V(r) = k\sum_{i=1}^{N} \frac{q_i}{r_i} \tag{3.12}$$

It is also useful to define the potential energy of a system of charged particles as the work required to bring the particles from an infinite separation, where they are at rest, to some finite separation. If the electric potential at a given point is V_1 due to charge q_1, then the work required to bring a charge q_2 from ∞ to the point where the potential is V_1 is given by

$$W = q_2 V_1$$

But the work done equals the potential energy of the system, U, and $V_1 = k\,q_1/r$, so for a system of two particles,

$$U = k\frac{q_1 q_2}{r} \qquad (3.13)$$

If there are *more* than two particles, the potential energy of the system is calculated by calculating U for every *pair* of charges, and summing the terms algebraically. For example, for *four* charges, the total potential energy would contain six terms:

$$U = U_{12} + U_{13} + U_{14} + U_{23} + U_{24} + U_{34}$$

In general, we can write the total potential energy as

$$U = k \sum_i \sum_j \frac{q_i q_j}{r_{ij}} \quad (i \neq j) \qquad (3.14)$$

Example 3.2

Two charges of 2 μC and -6 μC are located at the positions (0, 0) m and (0, 3) m, respectively (Figure 3-4). (a) Find the total electric potential due to these charges at the point (4, 0) m.

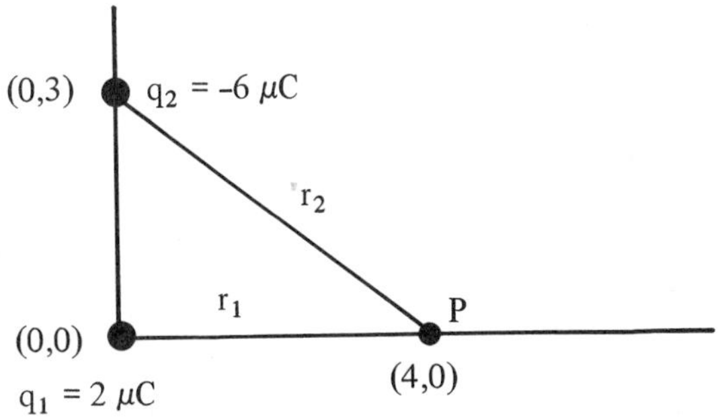

Figure 3-4

For two charges, the sum in Equation (3.12) gives

$$V = k\left[\frac{q_1}{r_1} + \frac{q_2}{r_2}\right]$$

In this example, $q_1 = 2 \times 10^{-6}$ C, $r_1 = 4$ m, $q_2 = -6 \times 10^{-6}$ C, and $r_2 = 5$ m. Therefore, V at P reduces to

$$V_P = 9 \times 10^9 \ N\frac{m^2}{C^2}\left[\frac{2 \times 10^{-6}\ C}{4\ m} - \frac{6 \times 10^{-6}\ C}{5\ m}\right]$$

$$V_P = -6.3 \times 10^3\ V$$

(b) How much work is required to bring a 3 μC charge from ∞ to the point P?

$$W = q_3 V_P = 3 \times 10^{-6}\ C\ (-6.3 \times 10^3\ V)$$

But 1 V = 1 J/C, so W becomes

$$W = -\ 18.9 \times 10^{-3}\ J$$

The negative sign means that work is done on the charge for this displacement from ∞ to P. Therefore, positive work would have to be done to remove the charge from P back to ∞.

(c) What is the *potential energy* of the system of *three* charges?

For three charges, the total potential energy summed over all pairs is given by

$$U = U_{12} + U_{13} + U_{23}$$

$$U = k\left[\frac{(2 \times 10^{-6})(-6 \times 10^{-6})}{3} + \frac{(2 \times 10^{-6})(3 \times 10^{-6})}{4} + \frac{(-6 \times 10^{-6})(3 \times 10^{-6})}{5}\right]$$

$$U = -5.5 \times 10^{-2}\ J$$

3.4 POTENTIAL OF A CONTINUOUS DISTRIBUTION OF CHARGE

The potential of a continuous distribution of charge can be calculated by first calculating the potential dV of a differential element dq of the charge distribution. This is given by

$$dV = k\frac{dq}{r}$$

where r is the distance from dq to the point in question. To get the total potential, we integrate over all charge elements. Since each element is in general at a different distance from the point in question, V becomes

$$V = k \int \frac{dq}{r} \tag{3.15}$$

In effect, we have replaced the sum in Equation (3.12) by an integral.

Example 3.3

Find the electric potential at a point P located along the axis of a uniformly charged ring of radius R (Figure 3-5). Let the charge per unit length be λ, and assume the point is a distance d from the center of the ring.

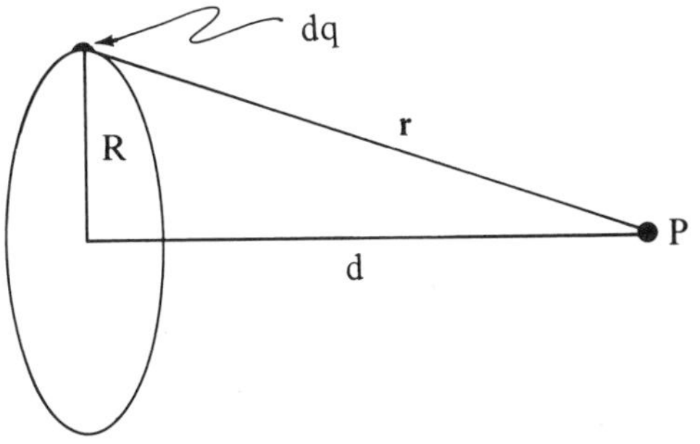

Figure 3-5 Geometry for calculating the electric potential along the axis of a uniformly charged ring of radius R.

The differential element of charge dq, which is a distance r from the point P, can be written as

$$dq = \lambda ds$$

where ds is an element of length measured along the ring. Therefore, Equation (3.15) gives

$$V_P = k \int \frac{\lambda ds}{r}$$

However, all points on the ring are at the *same* distance r from P, and λ is constant, so V_P can be written as

$$V_P = k \frac{\lambda}{r} \int ds$$

Since $\int ds = 2\pi R$, the total circumference of the ring, and $r = \sqrt{R^2 + d^2}$, V_P reduces to

$$V_P = \frac{\lambda R}{2\epsilon_0 \sqrt{R^2 + d^2}}$$

A second method of calculating the potential of a continuous distribution of charge when the **E** field is known is to make use of Equation (3.5), directly. This technique is useful when the charge distribution is highly symmetric, as in the case of charged spheres or cylinders. In such cases, we first evaluate **E** using Gauss's law, and substitute this result into Equation (3.5) to get the potential difference between any two points. We will illustrate this method with an example.

Example 3.4

An insulating solid sphere of radius R has a *uniform* positive charge density ρ and total charge Q (Figure 3-6). (a) Find the electric potential at a point outside the spheres, that is, $r > R$. Take the point at ∞ to be at zero potential.

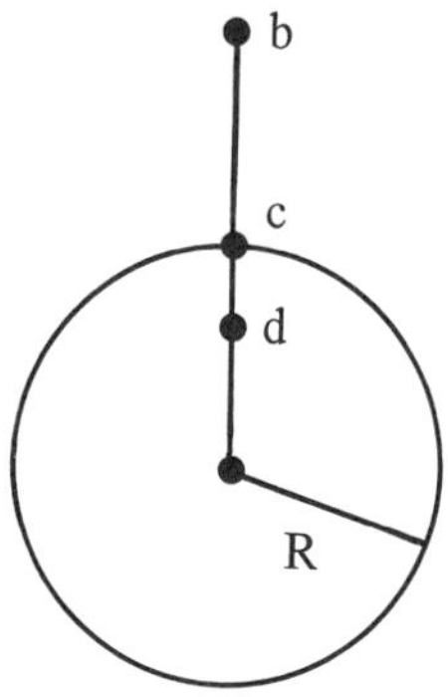

Figure 3-6 A uniformly charged solid sphere of radius R, charge density ρ and total charge Q.

In Example 2.4 of Chapter 2, we found that the electric field *outside* such a charged sphere is given by

$$\mathbf{E}_1 = k\frac{Q}{r^2}\hat{r} \quad (r > R)$$

Substituting this into Equation (3.5) gives

$$V_b - V_\infty = -\int_\infty^r k\frac{Q}{r^2}dr$$

Note that this result follows from the fact that ds = dr for a radial path, and $\hat{r}\cdot\mathbf{dr} = dr$. The potential at point b (see Figure 3-6) with $V_\infty = 0$ reduces to

$$V_b = k\frac{Q}{r} = \frac{\rho R^3}{3\epsilon_0 r} \quad (r \geqslant R) \qquad (1)$$

where we have used the fact that $k = \frac{1}{4\pi\epsilon_0}$ and $Q = \rho V = \rho\frac{4}{3}\pi R^3$. Note that the result is *identical* to that obtained for a *point charge.*

(b) Find the electric potential at a point *inside* the charged sphere, that is, $r < R$.

Referring again to Example 2.4 in Chapter 2, we found that the electric field *inside* a uniformly charged sphere is given by

$$E_2 = \frac{\rho r}{3\epsilon_0}\hat{r} \quad (r < R) \qquad (2)$$

The potential at point d (see Figure 3-6) can be calculated by noting that

$$V_d - V_\infty = (V_d - V_c) + (V_c - V_\infty)$$

Since $V_\infty = 0$, and $V_c = \frac{\rho R^2}{3\epsilon_0}$ [which follows from (1) for r = R], we have

$$V_d = -\int_R^r \mathbf{E}_2 \cdot d\mathbf{r} + \frac{\rho R^2}{3\epsilon_0}$$

where the integral term is $V_d - V_c$ as determined from Equation (3.5). Substituting (2) into the integral gives

$$-\int_R^r \mathbf{E}_2 \cdot d\mathbf{r} = -\frac{\rho}{3\epsilon_0}\int_R^r r\,dr = -\frac{\rho}{6\epsilon_0}[r^2 - R^2]$$

By combining terms in R^2, the expression for V_d reduces to

$$V_d = \frac{\rho}{6\epsilon_0}[3R^2 - r^2] \quad (r < R) \tag{3}$$

A plot of V *vs* r is given in Figure 3-7. Note that although the electric field is *zero* at r = 0 [see Equation (3.2)], $V(0) \neq 0$. Also, for $r \geqslant R$, you must use (1). The student should verify that (1) and (3) are *equal* at r = R.

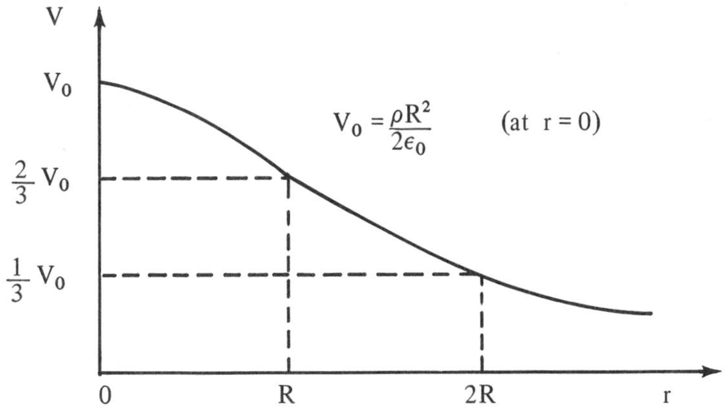

Figure 3-7 The electric potential of a uniformly charged solid sphere of radius R and charge density ρ.

3.5 POTENTIAL OF AN ISOLATED CHARGED CONDUCTOR

In the last example of a uniformly charged *insulating sphere,* we found that the potential inside and outside the sphere had different variations with r. Outside the sphere, the potential is equivalent to a point charge and varies as 1/r. Inside, the potential increases with decreasing r, since it takes work to move a test charge from the surface of the sphere to the center. That is, the charge within the sphere produces a nonzero electric field for r < R, and hence a force acts on a test charge placed within the sphere.

Now consider an isolated charged conductor in equilibrium. To make a comparison with Example 3.4, let us assume that the conductor is a solid metal sphere of radius R, with total charge Q. In such a system, recall that *all* of the charge resides at the surface of the conductor. This follows from the fact that there can be no currents in equilibrium, and the lowest energy configuration corresponds to the case where all the charges are at the surface. In Chapter 2 we showed, using Gauss's law, that $E \sim 1/r^2$ *outside* a charged conducting sphere, and $E = 0$ *inside*. The latter statement is equivalent to the statement that V = constant inside a charged conductor, since there can be no work done in moving a test charge from one point to another in a region where $E = 0$. Following Example 3.4, we see that the potential at the surface of the charged *metal* surface is kQ/R (with $V_\infty = 0$). Since there is no work done in moving a test charge from the surface to any interior point, the potential inside *must equal* the potential at the surface. That is,

Charged Metal Sphere

$$V = k\frac{Q}{R} \quad (r \leqslant R) \tag{3.16}$$

$$V = k\frac{Q}{r} \quad (r > R) \tag{3.17}$$

Since V is constant on the surface, we can conclude that the surface of a charged conductor is an equipotential surface. Figure 3-8 gives plots of V *vs* r and E *vs* r for a charged metal sphere. The results are also applicable to a charged *spherical shell.*

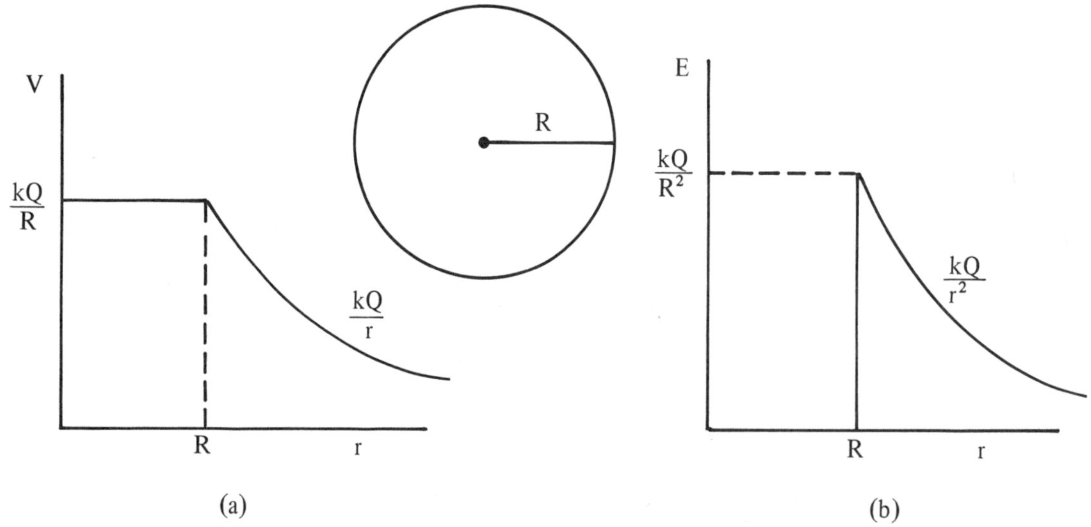

Figure 3-8 (*a*) The potential V *vs* r for a charged metal sphere (or spherical shell) of radius R. (*b*) The electric field *vs* r for a charged metal sphere of radius R.

In a more general situation, it can be shown that *any net static charge on an arbitrarily shaped conductor body will reside on its surface.* This follows from Gauss's law applied to an arbitrary closed surface within the conductor, as in Figure 3-9. For any such surface, since $E = 0$ everywhere, then $\phi_E = 0$, so $q = 0$. That is, no net charge is enclosed by the surface, and since the surface can be made arbitrarily small, then the net charge is zero throughout the interior of the conductor.

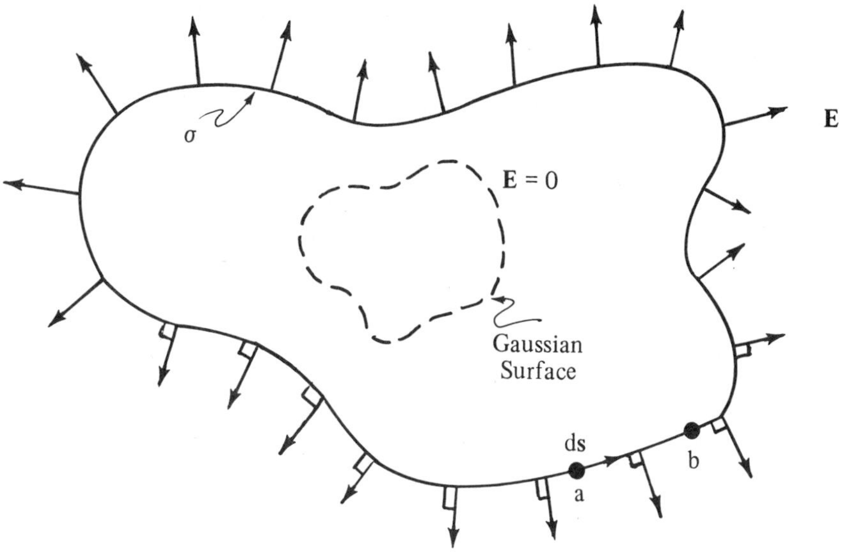

Figure 3-9 An arbitrarily shaped *conductor*. Under static conditions, the charge resides at the surface, **E** = 0 inside, and **E** is perpendicular to the surface immediately outside the conductor. The potential inside is constant and equals the potential at the surface. Note that the surface charge density is not constant; therefore, **E** is not constant in magnitude.

Another important property of a charged conductor is the fact that **E** just outside the conductor is *perpendicular* to the surface. This must be true, since any parallel component would cause charges to move on the surface; hence, a current and a nonequilibrium situation. Consider two points a and b on the surface of the conductor, as in Figure 3-9. Along a surface path connecting a and b, **E** is always perpendicular to ds, so **E**•ds = 0. Therefore, the potential difference between a and b is necessarily zero, or

$$V_b - V_a = -\int_a^b \mathbf{E} \cdot d\mathbf{s} = 0$$

This is valid for *any* two points on the surface, so we have proved that V = constant *everywhere* on the surface of a charged conductor. In other words, the surface of *any* conducting body in static equilibrium is an equipotential surface. In addition, all interior points of the conductor have the same potential as the surface.

3.6 OBTAINING E FROM THE ELECTRIC POTENTIAL

Another useful aspect of the potential concept is the fact that the electric field can be derived from a knowledge of the electric potential. In a general situation, the electric field can have three components, E_x, E_y and E_z. These are related to V by "inverting" Equation (3.5), giving the expressions

$$E_x = -\frac{\partial V}{\partial x} \tag{3.18}$$

$$E_y = -\frac{\partial V}{\partial y} \tag{3.19}$$

and

$$E_z = -\frac{\partial V}{\partial z} \tag{3.20}$$

In these equations, $\partial/\partial x$, $\partial/\partial y$, and $\partial/\partial z$ are *partial derivatives.* In these operations, the derivative is taken with respect to the variable indicated, while holding the other variables constant.*

Example 3.5

The electric potential in a certain region of space has a spatial dependence given by $V = Ax^2 + By^3$, where A and B are constants. Find the components of the electric field.

Applying Equations (3.18, 19, 20) to this function, we have

$$E_x = -\frac{\partial V}{\partial x} = -\frac{\partial}{\partial x}(Ax^2 + By^3) = -2Ax$$

$$E_y = -\frac{\partial V}{\partial y} = -\frac{\partial}{\partial y}(Ax^2 + By^3) = -3By^2$$

$$E_z = -\frac{\partial V}{\partial z} = 0$$

Example 3.6

Let us start with the potential function for a point charge q, Equation (3.4), to find the electric field intensity as a function of r.

$$V = k\frac{q}{r}$$

Since the problem has spherical symmetry, **E** only has a component in the *radial* direction. In this special case, E can be written as

$$E = -\frac{dV}{dr} = -\frac{d}{dr}\left[k\frac{q}{r}\right]$$

or

$$E = k\frac{q}{r^2}$$

This agrees with the value of E obtained by using Gauss's law (or Coulombs' law).

---

*A shorter notation which is often used is to write **E** as

$$\mathbf{E} = -\nabla V = -\left(\mathbf{i}\frac{\partial}{\partial x} + \mathbf{j}\frac{\partial}{\partial y} + \mathbf{k}\frac{\partial}{\partial z}\right) V$$

where ∇ is the gradient operator.

3.7 CAPACITANCE AND CAPACITORS

When two conductors are maintained at some potential difference V, say by means of a battery, then one conductor will be charged positive, while the other will have equal negative charge. Such a combination of charged conductors is a device called a *capacitor.* That is, capacitors are electrical devices which have the ability to store charge when a potential difference is maintained between the two conductors. The capacitance, C, of a capacitor is defined as the ratio of the charge on either plate divided by the potential difference between them:

$$C = \frac{q}{V} \qquad \text{General Definition of Capacitance} \tag{3.21}$$

where V is the potential difference between the + and − terminals, and is *positive.* Consequently, an increase in V provides an increase in the charge q for a given capacitor. There is a limit to the maximum charge that can be realized, since the medium between the two conductors will ultimately "break down," and conduction takes place through the medium. Note that C is *always positive.*

One can also speak of capacitance of an *isolated* body. For example, an isolated spherical shell of radius R and charge q has a capacitance given by Equation (3.21), where the potential difference V is simply kq/R (with $V_\infty = 0$). Therefore, its capacitance is simply R/k or $4\pi\epsilon_0 R$.

Capacitance, by definition, has units of coulombs per volt in SI units. The ratio of one coulomb per volt is *defined* to be one farad (F), or

$$[\text{Capacitance}] = \frac{\text{Coulomb}}{\text{Volt}} = \frac{C}{V} = 1\ F$$

The farad is a very large unit of capacitance. In practice, capacitances for typical devices range from microfarads (1 μF = 10^{-6} F) to picofarads (1 pF = 10^{-12} F).

The capacitor is commonly used (a) in electrical circuits for varying the frequency of resonance devices, (b) as a filter in power supplies, (c) as a device to eliminate sparking in ignition systems and (d) as an energy-storing device in electronic flashing units used by photographers.

The symbol that is commonly used to represent a capacitor is 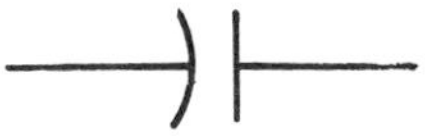 or sometimes , whereas a *variable* capacitor is represented by  or .

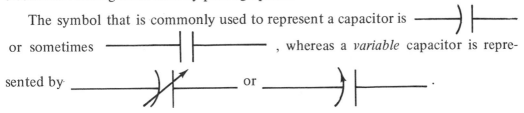

Suppose two parallel conducting plates, of equal area A and separation d, are connected to opposite terminals of a battery, as in Figure 3–10. In equilibrium, the plate connected to the positive terminal will have a net positive charge q, while the plate connected to the negative terminal will have a charge −q. Since the battery supplies the potential difference, V, then the electric field between the two parallel plates is uniform and is given by Equation (3.7). In addition, by Gauss's law, we found in Chapter 2 that $E = \sigma/\epsilon_0$ in the vicinity of a charged metal surface, where σ is the charge per unit area of either plate. Therefore, V can be written as

$$V = Ed = \frac{\sigma d}{\epsilon_0} \tag{3.22}$$

Since $q = \sigma A$, Equation (3.22) can be substituted into Equation (3.21) to give

$$C = \frac{\sigma A}{\sigma d/\epsilon_0} = \epsilon_0 \frac{A}{d} \quad \text{Parallel Plate Capacitor} \tag{3.23}$$

Note that Equation (3.23) applies *only* to a parallel plate capacitor with vacuum separating the two plates. In practice, capacitors are rarely constructed in this geometric form. However, Equation (3.21) can be used as a starting point for evaluating C for other geometries, such as cylindrical or spherical capacitors. When the space between the conducting plates is a vacuum, C is only a function of the geometry of the device. However, when the medium is other than a vacuum, we will see that C depends on the electrical properties of this medium.

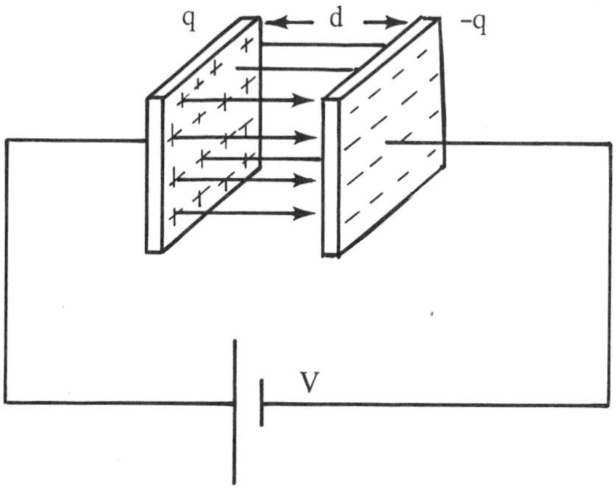

Figure 3-10 A parallel plate capacitor.

Example 3.7

A parallel plate capacitor has a plate separation of 2 mm, and each plate has a cross-sectional area of 1 cm^2. (a) Find the capacitance of this device.

We apply Equation (3.23), and recall that $\epsilon_0 = 8.85 \times 10^{-12} C^2/Nm^2$.

$$C = \epsilon_0 \frac{A}{d} = 8.85 \times 10^{-12} \frac{C^2}{Nm^2} \frac{(10^{-4} m^2)}{2 \times 10^{-3}\ m} = 4.43 \times 10^{-11}\ F$$

or

$$C = 44.3\ pF$$

The student should show that the units C^2/Nm reduce to farads.

(b) What is the charge on each plate of the capacitor if a potential different of 5 volts is maintained across the plates?

Here, we apply Equation (3.21) and the results of part (a).

$$q = CV = (4.43 \times 10^{-11})F \times 5\ V$$

$$q = 2.22 \times 10^{-10}\ C$$

Example 3.8

A capacitor is constructed from two conducting spherical shells having a common center, as in Figure 3-11. The inner and outer spheres have radii of a and b, respectively, and the inner shell is assumed to be positive. Let us calculate the capacitance of this device.

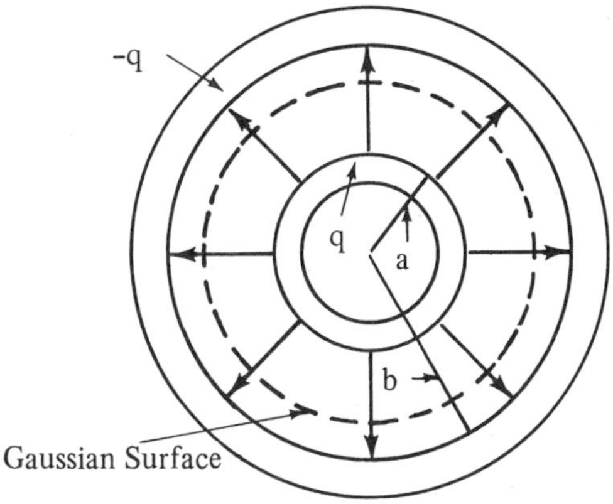

Figure 3-11 Cross-section of a spherical capacitor.

First, let us find the electric field in the region $a < r < b$. We construct a spherical Gaussian surface of radius r, and note that **E** is radially outwards. Therefore, since the charge enclosed by this surface is q, Gauss's law gives

$$\oint \mathbf{E}\cdot d\mathbf{A} = E\,(4\pi r^2) = \frac{q}{\epsilon_0}$$

or,

$$E = \frac{q}{4\pi\epsilon_0 r^2} = k\frac{q}{r^2} \qquad (a < r < b)$$

Now we must calculate the potential difference between the two surfaces. Using Equation (17.5) and the result for E gives

$$V_b - V_a = -\int_a^b k\frac{q}{r^2}\,dr = kq\left[\frac{1}{b} - \frac{1}{a}\right]$$

Since $b > a$, $V_b - V_a$ is negative. This is consistent with the assumption that the inner sphere is positive, or at a *higher* potential than the outer sphere.

Using the definition of C, that is, Equation (3.21), and noting that C is necessarily positive, gives

$$C = \frac{q}{V_a - V_b} = \frac{ab}{k\,[b - a]}$$

3.8 COMBINATION OF CAPACITORS

It is useful in practice to reduce two or more capacitors connected in a particular fashion to a single "equivalent" capacitor. A *parallel* combination of two capacitors is illustrated in Figure 3–12. A potential difference V is the same across *each* capacitor when connected in this fashion. Therefore, we can write the charge on each capacitor as

$$q_1 = C_1 V \qquad q_2 = C_2 V$$

The total charge on both capacitors is given by

$$q = q_1 + q_2 = (C_1 + C_2)\,V$$

Therefore, the "equivalent" capacitance C is the total charge supplied by the source over the potential difference

$$C = \frac{q}{V} = C_1 + C_2 \qquad \text{Parallel Combination} \tag{3.24}$$

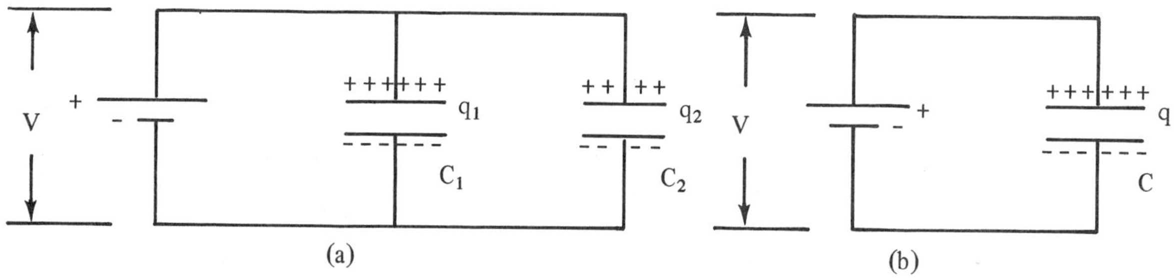

Figure 3–12 (*a*) Parallel combination of two capacitors and (*b*) the equivalent circuit. $q = q_1 + q_2$ and $C = C_1 + C_2$.

Likewise, for an arbitrary number of capacitors connected in *parallel* we have

$$C = C_1 + C_2 + C_3 + \ldots \qquad \text{Parallel Combination} \tag{3.25}$$

A *series* combination of two capacitors is illustrated in Figure 3–13. For this combination, note that both capacitors have the *same* charge q. This must be true, since the charge on the plates enclosed by the dotted lines is initially zero (before connecting the source of V); after the battery is connected, no charge "jumps across" the capacitors. However, charges rearrange themselves to give a +q on one of the plates, and −q on the other plates enclosed by the dotted line, giving a net charge of zero.

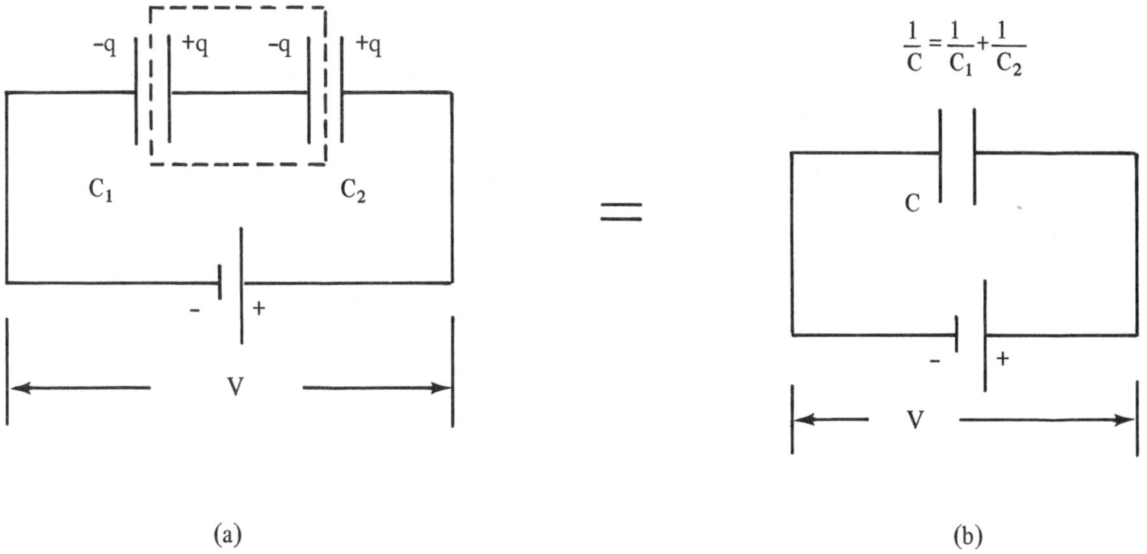

Figure 3-13 (*a*) A series combination of two capacitors and (*b*) the equivalent circuit.

The potential difference across each capacitor is given by

$$V_1 = q/C_1$$

$$V_2 = q/C_2$$

But the potential difference across the combination is simply V, the battery potential, so

$$V = V_1 + V_2 = q\left(\frac{1}{C_1} + \frac{1}{C_2}\right)$$

The equivalent capacitance C is q/V; therefore, from the previous expression, we get

$$\frac{1}{C} = \frac{1}{C_1} + \frac{1}{C_2} \quad \text{Series Combination} \tag{3.26}$$

For an arbitrary number of capacitors connected in *series* we have

$$\frac{1}{C} = \frac{1}{C_1} + \frac{1}{C_2} + \frac{1}{C_3} + \ldots \quad \text{Series Combination} \tag{3.27}$$

We conclude that the equivalent capacitance of a *series* combination is always *less* than any single capacitance in the combination. On the other hand, from Equation (3.25), we see that the equivalent capacitance of a parallel combination is *larger* than any single capacitance in the chain.

Example 3.9

(a) Two capacitors of 4 μF and 12 μF are connected in parallel, as in Figure 3-12. A potential difference of 6 V is maintained across each capacitor.

The equivalent capacitance of the parallel combination is given by Equation (3.24):

$$C = C_1 + C_2 = 4\ \mu F + 12\ \mu F = 16\ \mu F$$

Since the potential across each capacitor is 6 V, the charge on each capacitor can be calculated.

$$q_1 = C_1 V = 4\ \mu F \times 6\ V = 24\ \mu C$$

$$q_2 = C_2 V = 12\ \mu F \times 6\ V = 72\ \mu C$$

The total charge supplied by the battery is $q_1 + q_2$, or 96 μC. If we divide this by V, we get the equivalent capacitance C.

$$C = \frac{q_1 + q_2}{V} = \frac{96\ \mu C}{6\ V} = 16\ \mu F$$

(b) Suppose the same two capacitors are connected in *series,* as in Figure 3-13, and a potential difference of 6 V is supplied.

The equivalent capacitance of the series combination is given by Equation (3.26):

$$\frac{1}{C} = \frac{1}{C_1} + \frac{1}{C_2} = \frac{1}{4\ \mu F} + \frac{1}{12\ \mu F}$$

$$C = 3\ \mu F$$

The charge on each capacitor is the *same,* and is given by

$$q = CV = 3\ \mu F \times 6\ V = 18\ \mu C$$

The potential difference across each capacitor can be calculated:

$$V_1 = \frac{q}{C_1} = \frac{18\ \mu C}{4\ \mu F} = 4.5\ V$$

$$V_2 = \frac{q}{C_2} = \frac{18\ \mu C}{12\ \mu F} = 1.5\ V$$

Note that $V_1 + V_2 = 6$ V, which checks with the fact that a potential difference of 6 V is maintained across the series combination. Also, a *larger* potential difference occurs across the *smaller* capacitor.

3.9 DIELECTRIC-FILLED CAPACITOR

A common technique that is used to increase the capacitance of a capacitor is to insert an insulating or *dielectric* material between the conducting plates, as in Figure 3-14. This increase in capacitance with the introduction of the dielectric can be understood by considering the behavior of the molecules in the dielectric in the presence of an electric field. These molecules become "distorted" when an electric field is applied. That is, positive portions of the molecules are slightly shifted towards the negative plate, whereas negative portions are shifted towards the positive plate. The net result is the formation of surface charges on the dielectric, as in Figure 3–14(*b*). In such a situation, the dielectric has an electric dipole moment and is said to be *polarized.* The surface charges created in this fashion set up an *induced* electric field, E_i, which opposes the external field, E. Therefore, the *net electric field in the dielectric* is given by

$$E_d = E + E_i$$

Obviously, from Figure 3–14(*b*), since E_i is *opposite* in direction to E, the field E_d is always less than the field E. Since the charges in the dielectric are bound to their respective atoms or molecules, the induced charge on the dielectric surface is always less than the charge density on the conducting plates of the capacitor. Therefore, E_d is never zero. However, if the dielectric is replaced by a *conductor,* the induced surface charges on the conductor will produce an electric field which will exactly cancel the field produced by the capacitor plates. This results in a net field of *zero* in the conductor.

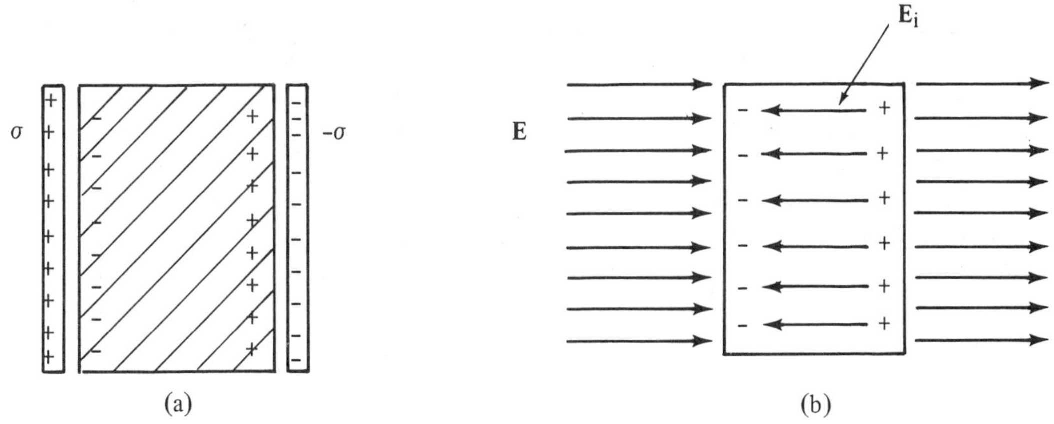

Figure 3-14 (*a*) A parallel plate capacitor filled with a dielectric material. (*b*) The electric field in the dielectric is the sum of the external field, E, and the field due to the induced charges, E_i.

We define the *dielectric constant* κ as the factor by which the electric field is reduced in the dielectric. In terms of the external field E, and the dielectric field E_d, κ is given by

$$\kappa = \frac{E}{E_d} \tag{3.28}$$

Since the electric field is reduced in the dielectric by the factor $1/\kappa$, the potential difference V across the parallel plate capacitor is also *reduced* by the same factor. That is,

$$V = E_d d = \frac{E}{\kappa} d \tag{3.29}$$

The capacitance can be written as

$$C = \frac{q}{V} = \frac{\sigma A}{V} \tag{3.30}$$

where σ is the charge density on the conducting plates of the capacitor. Subsituting Equation (3.29) into Equation (3.30), and noting that $E = \sigma/\epsilon_0$, gives

$$C = \frac{\sigma A}{Ed/\kappa} = \frac{\sigma A}{\sigma d/\kappa\epsilon_0}$$

or,

$$C = \kappa\epsilon_0 \frac{A}{d} \qquad \text{Dielectric-Filled Parallel Plate Capacitor} \tag{3.31}$$

The capacitance of the device with no medium present was found to be $\epsilon_0 A/d$, that is, Equation (3.23). Therefore, we can write the capacitance of a dielectric-filled capacitor as

$$C = \kappa\, C_0 \tag{3.32}$$

Equation (3.32) applies to *any* dielectric-filled parallel plate capacitor, C_0 being the capacitance of the device with no dielectric present (that is, a vacuum). In other words, the dielectric diminishes the electric field by the factor κ, which, in turn, causes a decrease in potential difference and an increase in capacitance. Consequently, more charge can be stored for a given potential difference across the plates of the capacitor.

The dielectric constants of some selected materials are listed in Table 3–1. The student should be careful not to confuse the dielectric constant with the *dielectric strength* of a material. The latter represents the maximum potential difference that can be applied to the material for a given thickness, before the material breaks down.

TABLE 3-1 DIELECTRIC CONSTANTS OF SOME MEDIA

Dielectric	κ
Vacuum	1.00000
Air (100°C, 1 atm)	1.00054
Water (20°C)	81
Paper	3.5
Bakelite	7
Porcelain	6–8
Pyrex Glass	4.5
Fused Quartz	3.8
Polyethylene	2.3
Rubber	2.2
Teflon	2.1
Nylon	3.5
Paraffin	2.0–2.5
Rutile (TiO_2)	100

3.10 ENERGY STORED IN A CHARGED CAPACITOR

Since a capacitor stores charge, it also is a device which stores energy. This energy can be released or given up by discharging the capacitor through some other circuit element. We can calculate the energy stored by considering the work done in transferring a charge dq′ from one plate to another plate at a higher potential. If the potential difference between the plates is V, the work done, dW, corresponds to an *increase* in potential energy, dU, given by

$$dU = V\,dq'$$

The work is done by the battery, which effectively takes electrons from the positive plate and moves them to the negative plate (a higher potential for electrons). Since V = q′/C, the total increase in energy of the capacitor for an increase in charge ranging from zero to q is given by

$$U = \int_0^q \frac{q'}{C}\,dq' = \frac{q^2}{2C} \tag{3.33}$$

Since q = CV, Equation (3.33) can also be written as

$$U = \frac{1}{2}CV^2 \qquad \text{Energy Stored in a Capacitor} \tag{3.34}$$

Equations (3.33) and (3.34) represent expressions for the energy stored in a capacitor for a potential difference V.

For a parallel plate capacitor filled with a dielectric, we can substitute Equation (3.31) into Equation (3.34) and use the fact that V = Ed. This gives

$$U = \frac{\kappa\epsilon_0 E^2 Ad}{2} \tag{3.35}$$

Since Ad is the volume of the capacitor, we can also express the energy density, or energy per unit volume, as

$$u = \frac{U}{Ad} = \frac{\kappa\epsilon_0 E^2}{2} \tag{3.36}$$

Although Equation (3.36) was obtained for a parallel plate capacitor, it also represents a *general* result. In other words, the energy density in a medium whose dielectric constant is κ is proportional to the square of the electric field intensity at a given point.

3.11 PROGRAMMED EXERCISES

1 A uniform electric field **E** is directed along the +x axis, as shown below. A particle of charge q moves from a to b along some path.

1.A

Write a general expression for the potential difference between points b and a.

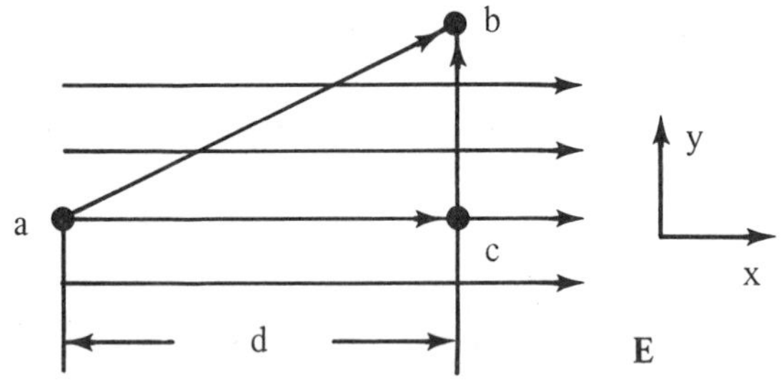

$$V_b - V_a = -\int_a^b \mathbf{E}\cdot d\mathbf{s} \qquad (1)$$

where ds is a differential displacement along the path of integration.

1.B

Since **E** is a uniform electric field, how can (1) be simplified?

E can be taken outside of the integral.

$$V_b - V_a = -\mathbf{E}\cdot\int_a^b d\mathbf{s} \qquad (2)$$

1.C

Write vector expressions for ds and **E**.

$$d\mathbf{s} = \mathbf{i}\,dx + \mathbf{j}\,dy \qquad (3)$$

$$\mathbf{E} = E\mathbf{i} \qquad (4)$$

1.D

Now evaluate $V_b - V_a$, assuming that the displacement in the x direction is d.

Substituting (3) and (4) into (2), and noting that $\mathbf{i}\cdot\mathbf{j} = 0$, gives

$$V_b - V_a = -E\int_0^d dx = -Ed \qquad (5)$$

1.E

Does the result $V_b - V_a$ depend on the path of integration? Explain.

No. We are dealing with a conservative field; therefore, the line integral is independent of the path.

1.F

We arrived at (5) without specifying a path, although the integral reduces to one along x. Find $V_c - V_a$, where c is a point a distance d from a, along the x axis.

Along the path ac, $ds = \mathbf{i}\,dx$, so (2) gives

$$V_c - V_a = -E\int_0^d dx = -Ed \qquad (6)$$

1.G

Now calculate $V_b - V_c$. Note that along this path, $ds = \mathbf{j}\,dy$.

In this case, we can write (2) as

$$V_b - V_c = -E\mathbf{i}\cdot\int\mathbf{j}\,dy = 0 \qquad (7)$$

Therefore,

$$V_b - V_a = V_c - V_a = -Ed \qquad (8)$$

1.H

We have shown that $V_b - V_a$ depends *only* on the displacement measured *parallel* to **E**. Sketch a curve of $V_b - V_a$ *vs* x.

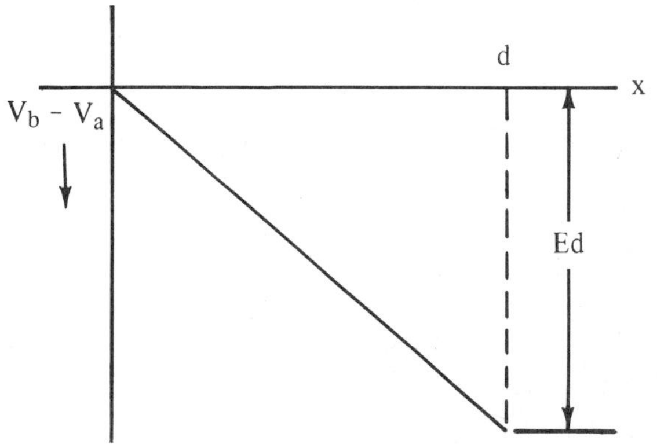

1.I

Is point b at a higher or lower potential than a?

From (5), we see that $V_b - V_a < 0$; ∴, $V_b < V_a$.

1.J

What is the *change in potential energy* of the charge q as it moves from a to b?

$$\Delta U = U_b - U_a = -q\int_a^b \mathbf{E}\cdot ds$$

$$\Delta U = -qEd \qquad (9)$$

1.K

From (9), what can you conclude about the change in potential energy of a *positive* charge q? What if q is negative?

If q is positive, $\Delta U < 0$, which corresponds to a *decrease* in potential energy. If q is negative, $\Delta U > 0$, meaning that a negative charge *increases* in potential energy as it moves from a to b.

1.L

Suppose the **E** field is produced by two parallel plates separated by 1 cm, and assume $\mathbf{E} = 3 \times 10^5 \frac{V}{m}\,\mathbf{i}$. An *electron* is ejected through a hole on the positive plate, with an initial velocity, $v_0\mathbf{i}$. What is the potential difference between b and a?

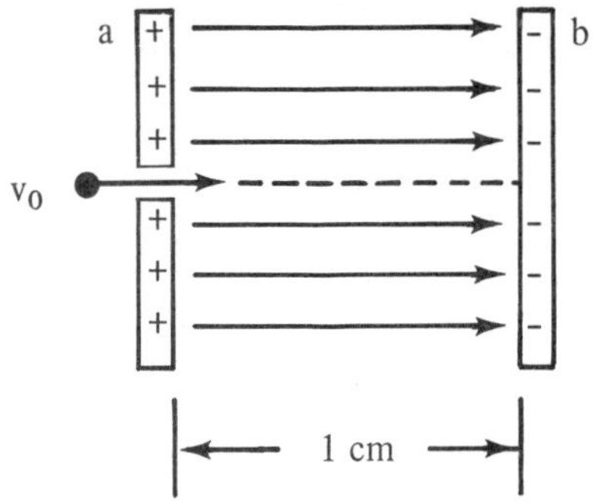

Using (5), we have

$$V_b - V_a = -Ed$$

$$V_b - V_a = -3 \times 10^5 \frac{V}{m}(10^{-2}\ m)$$

$$V_b - V_a = -3 \times 10^3\ V \qquad (10)$$

1.M

What is the change in potential energy of the electron as it moves from a to b? Recall that $q = e = -1.6 \times 10^{-19}$ C for an electron.

Using (9) and (10) gives

$$\Delta U = eEd = 1.6 \times 10^{-19}\,C\,(3 \times 10^3\,V)$$

$$\Delta U = 4.8 \times 10^{-16}\ J \qquad (11)$$

1.N

Write an expression for the work-energy theorem for the electron as it moves from a to b. Let its velocity be v when it reaches plate b.

$$W = \Delta K + \Delta U = 0$$

$$\frac{1}{2}mv^2 - \frac{1}{2}mv_0{}^2 + 4.8 \times 10^{-16}\,J = 0 \qquad (12)$$

1.O

From (12), we see that v can be determined if v_0 is known. Is v less or greater than v_0? Argue the result in terms of the force on the electron.

Obviously, from (12) we see that $v < v_0$. The *electric* force on the electron is to the *left;* hence, it slows down.

2 A point charge q is fixed at some point in space. Consider two points a and b as in the figure below.

2.A

What is *potential difference* between points b and a?

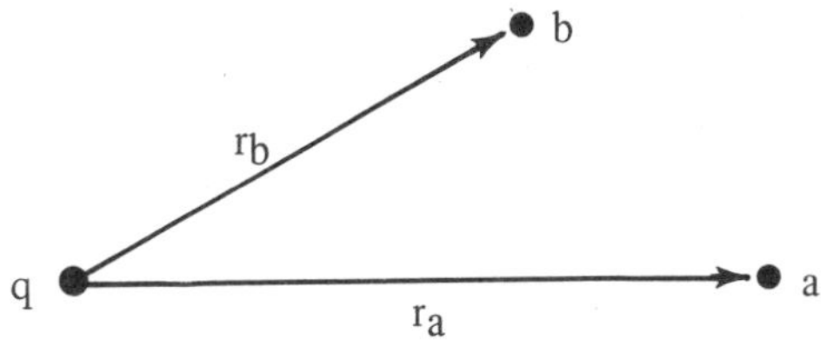

$V_b - V_a$ is a function only of the initial and final radial coordinates.

$$V_b - V_a = kq\left[\frac{1}{r_b} - \frac{1}{r_a}\right] \qquad (1)$$

where $k = \dfrac{1}{4\pi\epsilon_0}$.

2.B

Sometimes it is convenient to choose V = 0 at $r = \infty$. In this case, write an expression for V at some arbitrary radial coordinate r.

In this case, (1) becomes

$$V(r) - V(\infty) = k\frac{q}{r}$$

or,

$$V(r) = k\frac{q}{r} \qquad (2)$$

2.C

In reality, what does (2) represent?

V(r) represents the potential difference between the point r and the point at ∞.

2.D

What are the units of electric potential in SI units?

Volts (V);

1 volt = 1 Joule per Coulomb;

or,

1 V = 1 J/C.

2.E

Suppose $q = 5\mu C$, and $r_a = 1$ m. Find V_a taking $V(\infty) = 0$. Note that

$$k = 9 \times 10^9 \text{ N m}^2/\text{C}^2$$

$$V_a = k\frac{q}{r_a}$$

$$V_a = 9 \times 10^9 \text{ N}\frac{\text{m}^2}{\text{C}^2} \times \frac{5 \times 10^{-6}\text{ C}}{1 \text{ m}}$$

$$V_a = 4.5 \times 10^4 \text{ N}\frac{\text{m}}{\text{C}}$$

$$V_a = 4.5 \times 10^4 \text{ V} \qquad (3)$$

2.F

Show that the unit $N\frac{m}{C}$ reduces to 1 V.

By definition, 1 J = 1 Nm, and 1 V = 1 J/C;

$$\therefore \quad N\frac{m}{C} = \frac{J}{C} = V$$

2.G

Suppose $r_b = 0.25$ m. Find V_b, taking $V(\infty) = 0$, and use the numerical data in 2.E.

$$V_b = k\frac{q}{r_b}$$

$$V_b = 9 \times 10^9 \; N\frac{m^2}{C^2} \times \frac{5 \times 10^{-6} C}{0.25 \text{ m}}$$

$$V_b = 18 \times 10^4 \text{ V} \qquad (4)$$

2.H

What is the potential difference $V_b - V_a$?

From (3) and (4), we get

$$V_b - V_a = 13.5 \times 10^4 \text{ V} \qquad (5)$$

2.I

Is the result $V_b - V_a$ unique? Explain.

Yes. Potential differences are always unique. Absolute potentials are not.

2.J

A second charge, q′, is placed at a distance r from q. What is the *potential energy* of the system?

Making use of (2), we have

$$U = q' \, V(r) = k\frac{qq'}{r} \qquad (6)$$

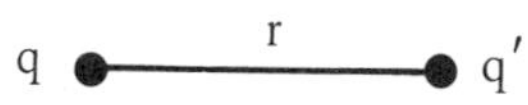

2.K

Find a numerical value for U, if $q' = -2 \, \mu C$, assuming q′ is located at $r_a = 1$ m. Make use of (3) in your calculation.

$$U_a = q'V_a$$

$$U_a = -2 \times 10^{-6} C \times 4.5 \times 10^4 \frac{J}{V}$$

$$U_a = -0.09 \text{ J} \qquad (7)$$

2.L

What is the physical meaning of the negative sign in (7)?

The negative potential energy implies that it would take positive work by an external agent to separate the pair to $r = \infty$.

2.M

What is the potential energy of the $-2\ \mu C$ charge when it is located at $r_b = 0.25$ m? Make use of (4) in your calculation.

$U_b = q' V_b$

$U_b = -2 \times 10^{-6} C \times 18 \times 10^4 \frac{J}{C}$

$U_b = -0.36$ J (8)

2.N

What is the change in the potential of the system as q' goes from a to b?

Making use of (7) and (8) gives

$$\Delta U = U_b - U_a$$

$$\Delta U = -0.27 \text{ J} \quad (9)$$

2.O

What is the *minimum* work required to bring the $-2\ \mu C$ charge from $r_b = 0.25$ m to ∞? Assume that the 5 μC is fixed.

The minimum work required would be simply U_b, or 0.36 J. That is, since $U_\infty = 0$,

$$W = \Delta U + \Delta K = U_b + \Delta K$$

But $\Delta K = 0$ gives U_{min},

$$\therefore \quad W = U_b$$

3 Three charges $q_1 = 3\ \mu C$, $q_2 = -6\ \mu C$, and $q_3 = 9\ \mu C$ are located at three points on a circle whose radius is $R = 0.3$ m. The points are as shown in the figure below.

3.A

Write a *general* expression for the potential at point a (the center of the circle), taking $V(\infty) = 0$, Recall that V is a *scalar* quantity.

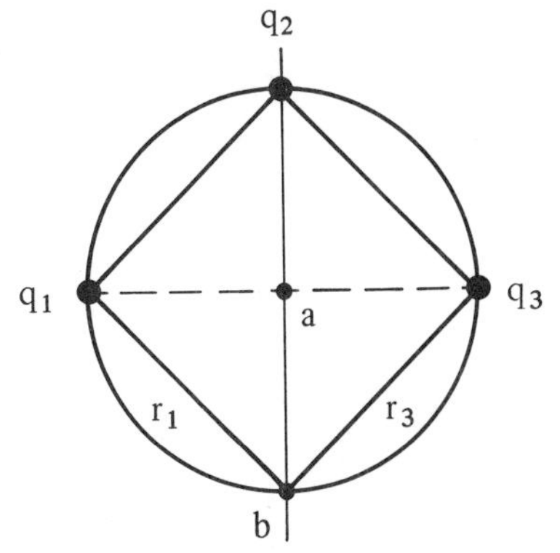

In general, we have

$$V = k \sum_i \frac{q_i}{r_i}$$

Since the distances to all charges are the same for point a, the scalar sum gives

$$V_a = \frac{k}{R} [q_1 + q_2 + q_3] \quad (1)$$

where $k = \frac{1}{4\pi\epsilon_0}$.

3.B

Write a *general* expression for the potential at point b, the bottom of the circle, with $V(\infty) = 0$.

$$V_b = k\left(\frac{q_1}{r_1} + \frac{q_2}{2R} + \frac{q_3}{r_3}\right)$$

But $r_1 = r_3 = \sqrt{2}\,R$, so this can be written as

$$V_b = \frac{k}{2R}\left(\sqrt{2}\,q_1 + q_2 + \sqrt{2}\,q_3\right) \qquad (2)$$

3.C

Obtain a numerical value for V_a with the information given.

$$k = 9 \times 10^9\ \mathrm{N}\frac{\mathrm{m}^2}{\mathrm{C}^2}$$

$$V_a = \frac{9 \times 10^9}{0.3}(3 - 6 + 9) \times 10^{-6}$$

or,

$$V_a = 18 \times 10^4\ \mathrm{V} \qquad (3)$$

3.D

Obtain a numerical value for V_b with the information given.

$$V_b = \frac{9 \times 10^9}{2\,(0.3)}\left(3\sqrt{2} - 6 + 9\sqrt{2}\right) \times 10^{-6}$$

$$V_b \cong 16 \times 10^4\ \mathrm{V} \qquad (4)$$

3.E

Write a general expression for the *potential energy* of the system of three charges. Recall that you must sum over *all* pairs, and the potential energy of *one* pair (say q_1 and q_2) is $k\,q_1 q_2/r$.

$$U = k\left(\frac{q_1 q_2}{r_1} + \frac{q_1 q_3}{2\,R} + \frac{q_2 q_3}{r_3}\right)$$

But $r_1 = r_3 = \sqrt{2}\,R$, so this reduces to

$$U = \frac{k}{2\,R}\left(\sqrt{2}\,q_1 q_2 + q_1 q_3 + \sqrt{2}\,q_2 q_3\right) \qquad (5)$$

3.F

Obtain a numerical value for U. (You should carry out the units and show they reduce to joules.)

$$U = \frac{9 \times 10^9}{2\,(0.3)}\left(-18\sqrt{2} + 27 - 54\sqrt{2}\right)10^{-12}$$

$$U \cong -1.1\ \mathrm{J} \qquad (6)$$

†4 *Potential of a Continuous Charge Distribution.* A long, thin filament of length l, carries a *uniform* charge per unit length λ. We wish to calculate the potential at the point p along the axis of the filament.

4.A

What is the charge, dq, on the infinitesimal element whose length is dx?

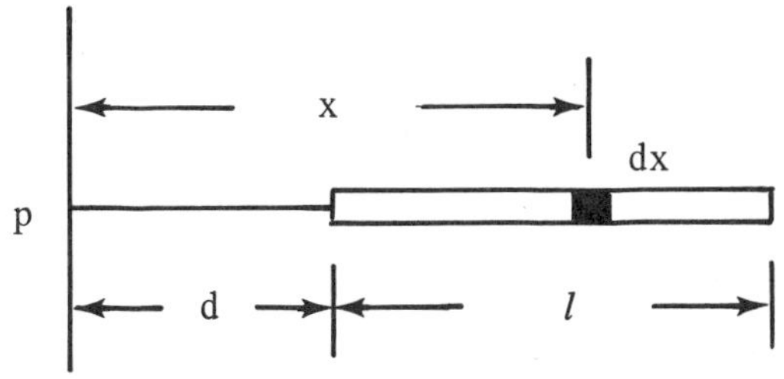

$$dq = \lambda\, dx \qquad (1)$$

where λ is the charge per unit length.

4.B

What is the potential at p due to the charge element dq?

Using (1), we have

$$dV = k\frac{dq}{r} = k\frac{\lambda dx}{x} \qquad (2)$$

4.C

Find the total potential at p due to the filament by integrating over x.

Since x ranges from d to d + l, Equation (2) can be integrated to give

$$V_p = k\lambda \int_a^{d+l} \frac{dx}{x} = k\lambda\, ln x \Big]_d^{l+d}$$

$$V_p = k\lambda\, ln\left(\frac{d+l}{d}\right) \qquad (3)$$

Note: $ln x\Big]_a^b = ln b - ln a = ln \frac{b}{a}$

4.D

Let the total charge be Q, so $\lambda = Q/l$. What does (3) reduce to when $d \gg l$? Note that in this case, l/d is small, so we can use the approximation

$$ln(1 + x) \cong x \quad (\text{for } x \ll 1)$$

where $x = l/d$.

We can write

$$ln\left(\frac{d+l}{d}\right) = ln\left(1 + \frac{l}{d}\right)$$

so $ln\left(1+\frac{l}{d}\right) \cong \frac{l}{d}$ and (3) reduces to

$$V_p \cong k\frac{Q}{l}\frac{l}{d} = k\frac{Q}{d} \quad (d \gg l) \qquad (4)$$

4.E

Does your result for $d \gg l$ make sense? Explain.

Yes. At distances d, large compared to l, the filament "looks like" a point charge. (4) is the potential of a point charge Q at a distance d from the charge.

†5 *Obtaining V for a Continuous Charge Distribution when E is Known.* Consider a *very long,* straight filament which carries a positive charge per unit length λ. A section of the filament is shown below. We wish to calculate the electric potential at P, a distance r from the wire.

5.A

In Chapter 2, Example 2.5, we found that the electric field could be calculated using Gauss's law. Write the expression for E, and describe its direction.

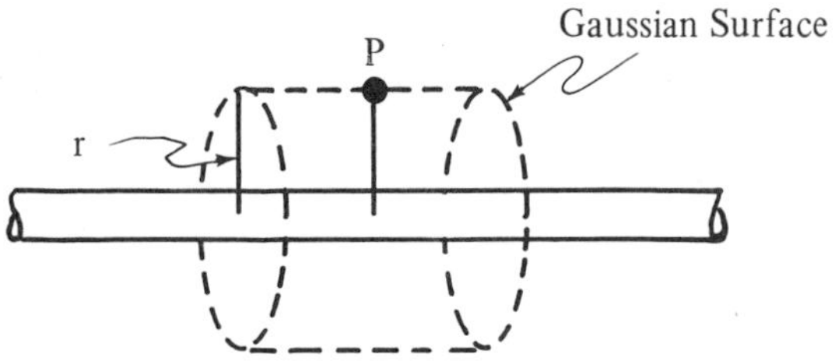

Using Gauss's law, we found that

$$E = 2k\frac{\lambda}{r} \qquad (1)$$

where $k = 1/4\pi\epsilon_0$. The direction of E is radially *outwards.*

5.B

Write a general expression for the *potential difference* $V_b - V_a$ between any two exterior points at radial distances r_a and r_b from the filament.

For any two points a and b,

$$V_b - V_a = -\int_{r_a}^{r_b} \mathbf{E} \cdot d\mathbf{s} \qquad (2)$$

5.C

Integrate (2) along a radial direction. Note that for this path, ds = dr.

Since $\mathbf{E} = 2k\frac{\lambda}{r}\hat{\mathbf{r}}$, and $d\mathbf{s} = d\mathbf{r}$, (2) reduces to

$$V_b - V_a = -2k\lambda \int_{r_a}^{r_b} \frac{dr}{r}$$

or

$$V_b - V_a = 2k\lambda \ln\left(\frac{r_a}{r_b}\right) \qquad (3)$$

5.D

Note that the point at ∞ is not a suitable reference, since V_b would be ∞ at $r_a = \infty$. However, we can set $V_a = 0$ at some arbitrary reference point, say R. Now write an expression for V(r) using this latter condition.

Let $V_a = 0$ at $r_a = R$. Applying this to (3) gives

$$V(r) = 2k\lambda \ln \left(\frac{R}{r}\right) \quad (4)$$

or

$$V(r) = -2k\lambda \ln r + D \quad (5)$$

where $D = 2k\lambda \ln R$

†6 The electric field in a certain region of space varies according to the expression $\mathbf{E} = (10x\mathbf{i} + 4y\mathbf{j})$ V/m.

6.A

Write a general expression for the potential difference between two points a and b in the xy plane.

$$V_b - V_a = -\int_a^b \mathbf{E}\cdot d\mathbf{s}$$

$$V_b - V_a = -\int_a^b (10x\mathbf{i} + 4y\mathbf{j})\cdot d\mathbf{s} \quad (1)$$

6.B

We would like to integrate (1) along a path located in the xy plane. Write a vector expression for ds, and reduce the *path integral* to a more simple form. In this case, the path integral can be separated into one line integral along x and one along y.

Since $d\mathbf{s} = \mathbf{i}dx + \mathbf{j}dy$, (1) can be written as

$$V_b - V_a = -\int_{x_a}^{x_b} 10x\,dx - \int_{y_a}^{y_b} 4y\,dy \quad (2)$$

Note that the first integral depends only on the x coordinates, while the second depends only on the y coordinates.

6.C

Obtain a numerical value for $V_b - V_a$ if $(x_a, y_a) = (0, 0)$ m and $(x_b, y_b) = (2, 3)$ m.

$$V_b - V_a = -\int_0^2 10x\,dx - \int_0^3 4y\,dy$$

$$V_b - V_a = -\left.\frac{10x^2}{2}\right]_0^2 - \left.\frac{4y^2}{2}\right]_0^3$$

$$V_b - V_a = -38 \text{ V} \quad (3)$$

6.D

What is the work done in carrying a $-5\ \mu$C charge from (0, 0) to (2, 3) m?

$$W = q(V_b - V_a)$$

$$W = -5 \times 10^{-6} \text{ C} \times (-38 \text{ V})$$

$$W = 1.9 \times 10^{-4} \text{ J} \quad (4)$$

†7 *Obtaining* **E** *from V.* Consider the electric dipole, that is, two equal and opposite charges separated by a distance 2d, as shown below. First, we will find the potential of the dipole at p, and from this we will derive the electric field.

7.A

What is the potential at p due to q? Let the origin of coordinates be midway between the two charges.

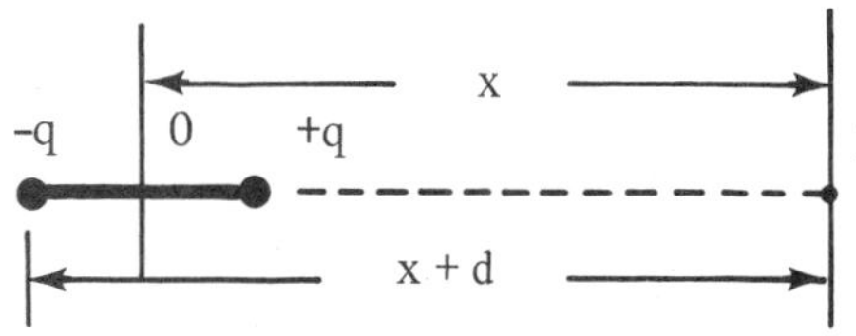

The distance from q to p is x − d, so

$$V_1 = k\left[\frac{q}{x-d}\right] \qquad (1)$$

7.B

What is the potential at p due to −q? [Take V (∞) = 0 as usual.]

$$V_2 = k\left[\frac{-q}{x+d}\right] \qquad (2)$$

7.C

What is the *total* potential at p?

$$V_p = V_1 + V_2$$

$$V_p = k\left[\frac{q}{x-d} - \frac{q}{x+d}\right]$$

$$V_p = \frac{2kqd}{x^2 - d^2} \qquad (3)$$

7.D

What is the x component of **E** at p? Use the fact that

$$E_x = -\frac{\partial V}{\partial x}$$

where V is given by (3).

$$E_x = -\frac{\partial V}{\partial x} = -2\,kqd\frac{\partial}{\partial x}(x^2 - d^2)^{-1}$$

$$E_x = 4kqd\frac{x}{(x^2 - d^2)^2} \qquad (4)$$

7.E

What is the y component of **E** at p? Use the relation

$$E_y = -\frac{\partial V}{\partial y}$$

Since V does not depend on y,

$$E_y = -\frac{\partial V}{\partial y} = 0$$

Of course, this can be seen by symmetry.

7.F

Write expressions for V_p and E_p for large distances from the dipole, that is, for $x \gg d$. Write the expressions in terms of the *dipole moment* $p = 2qd$.

For the $x \gg d$, we can neglect d in the denominators of Equations (3) and (4). This gives

$$V_p \cong k\frac{p}{x^2} \qquad (5)$$

$$E_p \cong 2k\frac{p}{x^3} \qquad (6)$$

8 *Capacitance.* Consider two parallel conducting plates, each of area A, separated by a distance d. A battery is connected to the plates, and a charge +q is produced on one plate, −q on the other. This difference in charge gives rise to a potential difference V between the plates.

8.A

Define the capacitance of this device.

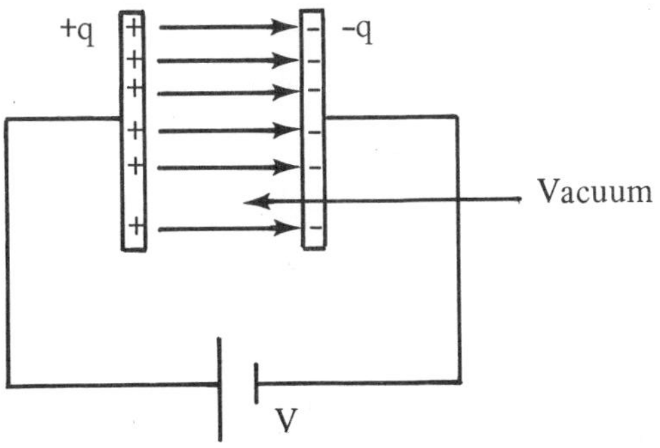

$$C = \frac{q}{V} \qquad (1)$$

8.B

What are the units of capacitance?

$$1 \text{ Farad (F)} = \frac{1 \text{ Coulomb}}{1 \text{ Volt}}$$

or,

$$1 \text{ F} = 1\frac{\text{C}}{\text{V}}$$

since 1 V = 1 J/C, $F = C^2/J$

8.C

Write an expression for the potential difference V in terms of the charge per unit area, σ. Recall that $\sigma = q/A$, where A is the area of each plate.

Since the field is uniform, we see from Exercise 1 that

$$V = Ed$$

But from Gauss's law, $E = \sigma/\epsilon_0$, so

$$V = \frac{\sigma d}{\epsilon_0} \qquad (2)$$

8.D

Substitute $q = \sigma A$ and (2) into (1), to get a final form for C.

$$C = \frac{\sigma A}{\sigma d/\epsilon_0} = \epsilon_0 \frac{A}{d} \tag{3}$$

8.E

If the capacitor is filled with an insulating material whose dielectric constant is κ, how does this change the result for C?

In Equation (3.1), q is the same for a given source of potential, but V is diminished by the factor $1/\kappa$. This causes an *increase* in C by the factor κ.

$$C = \kappa\epsilon_0 \frac{A}{d} \tag{4}$$

8.F

A technician wishes to construct a parallel plate capacitor using *Rutile* as the dielectric ($\kappa = 100$). If he chooses the cross-sectional area to be 0.5 cm², what should he choose for the Rutile thickness to get a capacitance of 40 pF?

We use Equation (4) and recall that $\epsilon_0 = 8.85 \times 10^{-12}\ C^2/Nm^2$. Also, $A = 0.5\ cm^2 = 0.5 \times 10^{-4}\ m^2$.

$$d = \frac{\kappa\epsilon_0 A}{C} = \frac{100\ (8.85 \times 10^{-12})(0.5 \times 10^{-4})}{40 \times 10^{-12}}$$

$$d = 1.1 \times 10^{-3}\ m \cong 1\ mm$$

8.G

The *dielectric strength* of Rutile is 6 kV/mm, which corresponds to the *maximum* potential gradient that can exist in the dielectric before breakdown occurs. For the capacitor described in 8.F, what *maximum* charge can the capacitor hold?

Since the capacitor in 8.F has a thickness of 1 mm, the *maximum* potential difference that it can handle is 6 kV. Therefore, from (1), we have

$$q_{max} = CV_{max} = 40 \times 10^{-12} \times 6 \times 10^3$$

$$q_{max} = 0.24\ \mu C$$

8.H

What is the maximum energy that can be stored in this capacitor?

In general, $U = \frac{1}{2}CV^2$, so

$$U_{max} = \frac{1}{2}CV_{max}^2$$

$$U_{max} = \frac{1}{2}(40 \times 10^{-12})\ (6 \times 10^3)^2\ J$$

$$U_{max} = 7.2 \times 10^{-4}\ J$$

8.I

What is the maximum *energy density* or energy per unit volume stored in the capacitor?

$$U = \frac{U_{max}}{Ad} = \frac{7.2 \times 10^{-4}\ \text{J}}{(0.5 \times 10^{-4} \times 1.1 \times 10^{-3}\ \text{m}^3)}$$

$$U = 1.3 \times 10^4\ \text{J/m}^3.$$

9 A capacitor is constructed using two coaxial conducting *cylindrical* shells, whose inner and outer radii are a and b, respectively. Both cylinders have a length l, and l is large compared to a or b. We wish to find the capacitance of this device. A cross-section is shown below.

9.A

To find C, we first must know $V_b - V_a$. Let us assume the inner conductor has a charge q, and the outer a charge $-q$. Write a general expression for $V_b - V_a$.

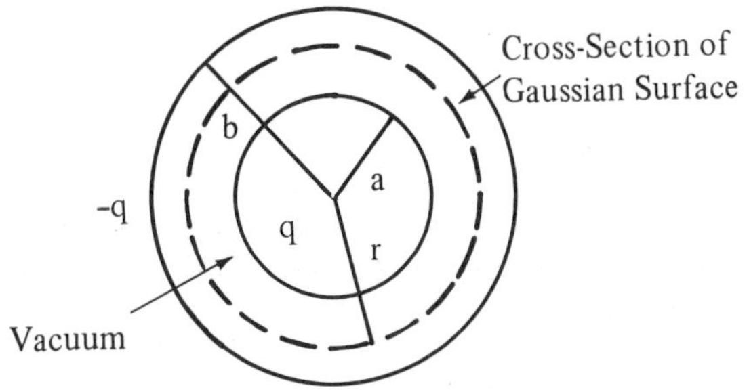

$$V_b - V_a = -\int_a^b \mathbf{E} \cdot d\mathbf{s} \qquad (1)$$

where **E** is the electric field in the region $a < r < b$.

9.B

In Chapter 2, Example 2.5, we found that **E** for a cylinder of charge per unit length λ is given by $E = 2k\lambda/r$. Use this in (1) to find a working equation for $V_b - V_a$. Note that $\lambda = q/l$.

Since **E** is radial *outward*, $\mathbf{E} = 2k\lambda/r\ \hat{r}$, and the path can be taken along r. So (1) becomes

$$V_b - V_a = -2k\lambda \int_a^b \frac{dr}{r}$$

$$V_b - V_a = 2k\frac{q}{l} \ln\left(\frac{a}{b}\right) \qquad (2)$$

9.C

Now write the expression for the capacitance, using (2) and the definition of C. Note that $V_b - V_a$ is *negative*, since $b > a$.

$$C = \frac{q}{V_a - V_b} = \frac{l}{2k \ln\left(\frac{b}{a}\right)} \qquad (3)$$

9.D

How would the result differ, if at all, for C, if a material of dielectric constant κ fills the space between the cylinders?

E is reduced by the factor $1/\kappa$, and $V_a - V_b$ is reduced by the same amount; therefore,

$$C = \frac{\kappa l}{2k\ ln\ (b/a)} \quad (4)$$

9.C

Obtain a numerical value for a long cylindrical capacitor of length l = 20 cm, a = 1 cm, b = 1.2 cm, filled with paper (κ = 3.5).

$$C = \frac{3.5 \times 0.2}{2\ (9 \times 10^9)\ ln \left(\frac{1.2}{1}\right)}$$

$$C = \frac{0.7}{18 \times 10^9 \times 0.18}\ F$$

$$C \cong 220\ pF$$

10 Six capacitors are connected as shown below. All capacitances are in μF. We wish to find the equivalent capacitance of the network and the charge on each capacitor.

10.A

What is the equivalent capacitance C_1 of the upper portion of the circuit enclosed by the dotted lines?

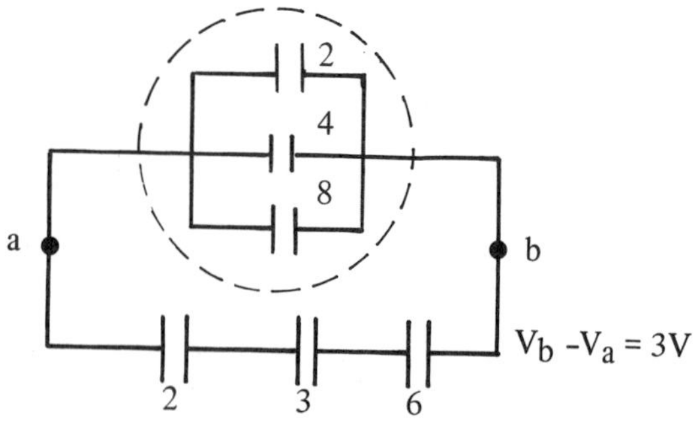

These represent three capacitors connected in *parallel*; therefore, we simply add the capacitances

$$C_1 = 2 + 4 + 8 = 14\ \mu F \quad (1)$$

10.B

What is the equivalent capacitance C_2 of the lower portion of the circuit, that is, for the 2, 3, and 6 μF capacitors?

These are connected in *series*,

$$\therefore \quad \frac{1}{C_2} = \frac{1}{2} + \frac{1}{3} + \frac{1}{6}$$

or,

$$C_2 = 1\ \mu F \quad (2)$$

10.C

The circuit has now reduced to

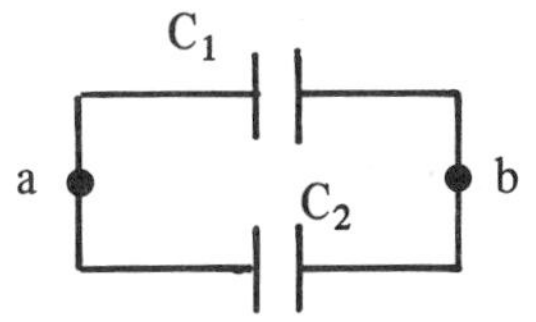

What is the equivalent capacitance of this combination?

This is a parallel combination, so

$$C_{eq} = C_1 + C_2$$

$$C_{eq} = 14 + 1 = 15\ \mu F \qquad (3)$$

10.D

Now calculate the charge on the three capacitors within the dotted line, with $V_b - V_a = 3$ V.

For these capacitors, the potential difference across each is 3 V, so q = CV can be applied to each:

$$2\ \mu F: q_1 = 2 \times 10^{-6} \times 3 = 6\ \mu C \qquad (4)$$

$$4\ \mu F: q_2 = 4 \times 10^{-6} \times 3 = 12\ \mu C \qquad (5)$$

$$8\ \mu F: q_2 = 8 \times 10^{-6} \times 3 = 24\ \mu C \qquad (6)$$

10.E

What is the charge on each of the capacitors in the lower portion of the circuit?

Since these are in series, the charge on each is the *same.* Since the equivalent capacitance of the lower portion is $C_2 = 1\ \mu F$, we have

$$q = C_2 V = 1 \times 10^{-6}\ (3) = 3\ \mu C. \qquad (7)$$

10.F

Find the potential difference across each of the capacitors in the lower portion of the circuit. Make use of (7) in your calculations.

$$2\ \mu F: V_2 = \frac{q}{C_2} = \frac{3\ \mu C}{2\ \mu F} = 1.5\ V \qquad (8)$$

$$3\ \mu F: V_3 = \frac{q}{C_3} = \frac{3\ \mu C}{3\ \mu F} = 1\ V \qquad (9)$$

$$6\ \mu F: V_6 = \frac{q}{C_6} = \frac{3\ \mu C}{6\ \mu F} = 0.5\ V \qquad (10)$$

10.G

Does the sum of the potential differences obtained in 10.F agree with your expectations?

Yes. $V_2 + V_3 + V_6 = 3$ V. This is precisely $V_b - V_a$. So we conclude that

$$V_b - V_a = V_2 + V_3 + V_6 = 3\ V$$

10.H

What is the energy stored in each of the capacitors within the dotted line?

$$U_1 = \frac{1}{2} C_1 V^2 = \frac{1}{2}(2 \times 10^{-6})\, 3^2 = 9\ \mu J$$

$$U_2 = \frac{1}{2} C_2 V^2 = \frac{1}{2}(4 \times 10^{-6})\, 3^2 = 18\ \mu J$$

$$U_3 = \frac{1}{2} C_3 V^2 = \frac{1}{2}(8 \times 10^{-6})\, 3^2 = 36\ \mu J$$

10.I

What is the total energy stored in the three capacitors in the *lower* portion of the circuit?

The equivalent capacitance of this portion is 1 μF, that is, (2);

$$\therefore \quad U = \frac{1}{2} C V^2$$

$$U = \frac{1}{2}(10^{-6})\, 3^2 = 4.5\ \mu J$$

10.J

What is the total energy stored in the system? Use the results to 10.H and 10.I in your calculation.

$$U_{total} = (9 + 18 + 36 + 4.5)\ \mu J$$

$$U_{total} = 67.5\ \mu J \qquad (11)$$

10.K

The equivalent capacitance of the circuit was found in (3) to be 15 μF. Calculate U_{total} using this value of C, and show that is agrees with (11).

$$U = \frac{1}{2} C_{eq}\, V^2$$

$$U = \frac{1}{2}(15 \times 10^{-6})\, 3^2 = 67.5\ \mu J$$

11 In the circuit shown below, $C_1 = 2\ \mu F$, $C_2 = 6\ \mu F$, and $C_3 = 4\ \mu F$. The battery supplies a potential difference of 16 V.

11.A

Initially, switch S_1 is closed and S_2 is opened. What is the potential difference across C_1 and C_2?

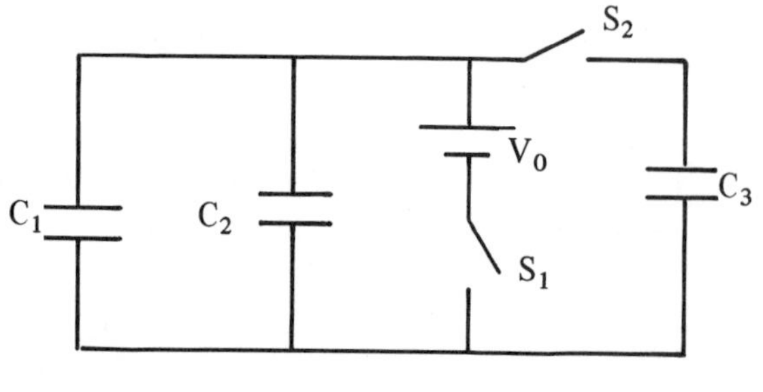

Since C_1 and C_2 are a parallel combination, the potential difference is the *same* and equal to 16 V.

11.B

What is the equivalent capacitance of C_1 and C_2?

$C_{eq} = C_1 + C_2 = 8\ \mu F$

11.C

What is the charge on C_1, C_2, and C_3 when S_1 is closed and S_2 is opened?

$q_1 = C_1 V_0 = 2\ \mu F \times 16\ V = 32\ \mu C$

$q_2 = C_2 V_0 = 6\ \mu F \times 16\ V = 96\ \mu C$

$q_3 = 0$

11.D

Now suppose S_1 is *opened* and S_2 is closed, in that sequence. The equivalent circuit now looks like

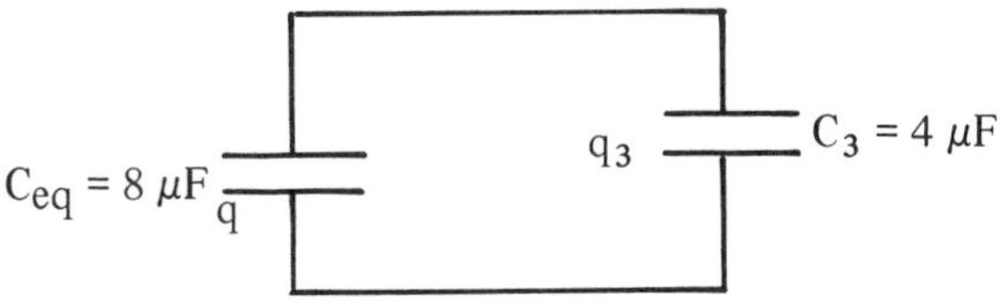

What is the sum $q + q_3$ equal to?

The initial total charge was

$$q_1 + q_2 = 128\ \mu C.$$

This charge is now shared by C_{eq} and C_3, that is, the charge is transferred when S_1 is opened and S_2 is closed.

$$\therefore \quad q + q_3 = 128\ \mu C \qquad (1)$$

11.E

What is the potential difference across C_{eq} and C_3? Note that they are now a parallel combination.

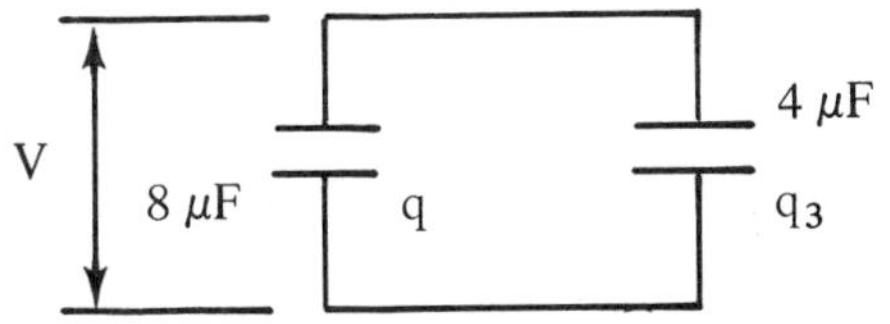

It must be the same, since they are a parallel combination, but it is *not* equal to 16 V. Applying the fact that $q = CV$ to the terms in (1) gives

$$C_{eq} V + C_3 V = 128\ \mu C$$

$$V = \frac{128\ \mu C}{12\ \mu F} = 10.7\ V \qquad (2)$$

11.F

Determine the values of q_1, q_2 and q_3 in the *final* configuration. Note that $q = q_1 + q_2$.

$q_1 = C_1 V = 2\ \mu F\ (10.7\ V) = 21.4\ \mu C$

$q_2 = C_2 V = 6\ \mu F\ (10.7\ V) = 64.2\ \mu C$

$q_3 = C_3 V = 4\ \mu F\ (10.7\ V) = 42.8\ \mu C$

11.G

Calculate the energy stored in the system initially, that is, when S_1 is closed and S_2 is opened.

In this case, C_3 is uncharged and stores no energy.

$$U_0 = \frac{1}{2}C_1 V_0^2 + \frac{1}{2}C_2 V_0^2$$

$$U_0 = \frac{1}{2} 2\ \mu F\ (16\ V)^2 + \frac{1}{2} 6\ \mu F\ (16\ V)^2$$

$$U_0 = 1.02 \times 10^{-3}\ J \qquad (3)$$

11.H

Now calculate the energy stored in the system when S_1 is opened and S_2 is closed. Make use of (2).

$$U = \frac{1}{2} C_{eq} V^2 + \frac{1}{2} C_3 V^2$$

$$U = \frac{1}{2} (8\ \mu F + 4\ \mu F)\ (10.7\ V)^2$$

$$U = 0.69 \times 10^{-3}\ J \qquad (4)$$

11.I

Note the $U < U_0$, that is, some energy is lost! What do you think happened to this lost energy?

In the process of opening S_1 and closing S_2, charges are transferred from C_{eq} to C_3. Therefore, the moving charges correspond to a current in the wires. Since the wires have a finite resistance, they *heat up,* and part of the missing energy is accounted for. Some energy is also radiated in the form of electromagnetic waves.

12 *The Classical Radius of the Electron.* In this exercise, we will use the concept of capacitance of an isolated charged sphere to estimate the radius of the electron.

12.A

Consider the electron as a conducting sphere of radius R, with its charge distributed uniformly on its surface. What is the potential at the surface of the electron?

$V = k\frac{q}{R}$, where $V_\infty = 0$. But, $q = -e$, so

$$V = -k\frac{e}{R} \qquad (1)$$

12.B

What is the effective capacitance of the electron?

$C = \frac{q}{V} = -\frac{e}{V}$ or,

$$C = \frac{R}{k} = 4\pi\epsilon_0 R \qquad (2)$$

12.C

Now calculate the electrical energy of the electron. Recall that

$$U = \frac{1}{2} CV^2$$

Using (1) and (2), we have

$$U = \frac{1}{2} CV^2 = \frac{1}{2} \frac{R}{k} \frac{k^2 e^2}{R^2}$$

$$U = k \frac{e^2}{2R} \qquad (3)$$

12.D

We can now equate (3) to the *rest mass energy* of the electron, $m_0 c^2$ (a relativistic formula). Do this, and obtain an expression for R.

$$U = m_0 c^2$$

$$k \frac{e^2}{2R} = m_0 c^2$$

$$R = \frac{ke^2}{2m_0 c^2} \qquad (4)$$

12.E

In practice, it is common to define the classical electron radius as *2R*. Use the fact that $m_0 = 9.11 \times 10^{-31}$ kg, and $c = 3 \times 10^8$ m/s to get a numerical value for this radius.

$$r_e = 2R = ke^2/m_0 c^2$$

$$r_e = \frac{9 \times 10^9 \ (1.6 \times 10^{-19})^2}{(9.11 \times 10^{-31})\ (3 \times 10^8)^2} \quad \text{(SI)}$$

$$r_e \cong 2.82 \times 10^{-15} \text{ m} \qquad (5)$$

12.F

Do you think that the result represents a strict geometric dimension?

No. The result can only be considered to be an estimate of the dimensions of the region over which the electron is concentrated.

3.12 SUMMARY

The *potential difference* between any two points a and b in a static electric field **E** is given by

$$V_b - V_a = -\int_a^b \mathbf{E}\cdot d\mathbf{s} \tag{3.5}$$

where V is potential energy per unit charge (that is, $V = U/q$). Hence, the *potential energy difference* between the points a and b is given by

$$\Delta U = U_b - U_a = -q\,(V_b - V_a)$$

The *electric potential* of a point charge q at a distance r from the charge is a scalar quantity given by

$$V(r) = k\frac{q}{r} \qquad [\text{where } V(\infty) = 0] \tag{3.11}$$

The potential of a continuous distribution of charge is obtained by integration over all segments of the distribution.

The *potential energy* of a pair of charges q_1 and q_2 separated by a distance r is

$$U = k\frac{q_1 q_2}{r} \tag{3.13}$$

The static charge on an arbitrarily shaped conducting body (metal) will always reside on its surface. The **E** field within a conductor in equilibrium is *zero*. The **E** field just *outside* the conductor is *perpendicular* to the surface. Therefore, a charged conductor in equilibrium is an *equipotential surface*, that is, V is the same at all points on the surface.

If the electric potential of a charge distribution is known, the electric field components can be obtained from the relations

$$E_x = -\frac{\partial V}{\partial x}, \text{ and so on, for the y and z components of } \mathbf{E}.$$

The capacitance of a pair of charged conductors with a potential difference V between them is defined as

$$C = \frac{q}{V} \tag{3.21}$$

where C is a positive quantity and q is the charge on either conductor. The capacitance of a parallel plate capacitor of area A, separation d, is given by

$$C = \epsilon_0 \frac{A}{d} \tag{3.23}$$

When capacitors C_1, C_2, C_3... are added in *parallel,* the equivalent capacitance of the combination is given by

$$C = C_1 + C_2 + C_3 + \ldots \tag{3.25}$$

When capacitors are added in *series,* the equivalent capacitance is obtained through the relation

$$\frac{1}{C} = \frac{1}{C_1} + \frac{1}{C_2} + \frac{1}{C_3} + \ldots \tag{3.27}$$

When a capacitor is filled with a dielectric material, whose dielectric constant is κ, the capacitance is given by

$$C = \kappa C_0 \tag{3.32}$$

where C_0 is the capacitance of the device with no dielectric.

The *total energy* stored in a capacitor is given by

$$U = \frac{q^2}{2C} = \frac{1}{2} CV^2$$

3.13 PROBLEMS

(Assume that $V_\infty = 0$ in all problems involving point charges.)

1. A point charge of 8 μC is located at the origin, while a second point charge of -5 μC is located at (0, 0.3) m. Find the total electric potential at the point (0.4, 0) m.

2. A point charge of -3 μC is located at (-0.3, 0) m, while a second point charge of 6 μC is located at (0.1, 0) m. (a) What is the electric potential at the origin? (b) Find a point along the x axis where V = 0.

3. Four charges are fixed at the corners of a square whose side is a, as in Figure 3-15. (a) Calculate the total electric potential at the center of the square. (b) What is the minimum work required to bring a test charge, q′, from ∞ to the center of the square?

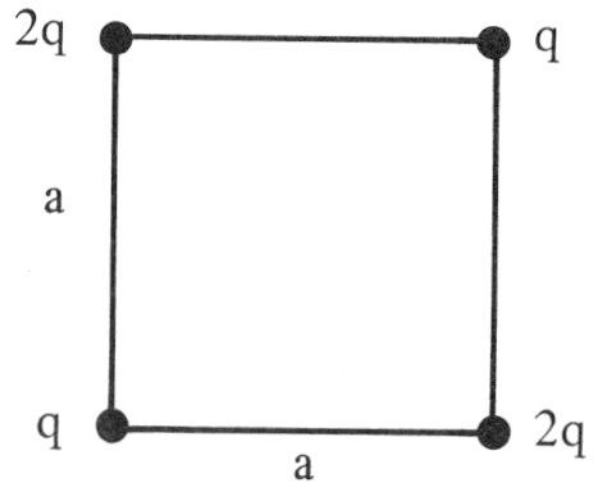

Figure 3-15

4. Find the total electric potential at the center of the equilateral triangle shown in Figure 3-16. Each side of the triangle has a length b.

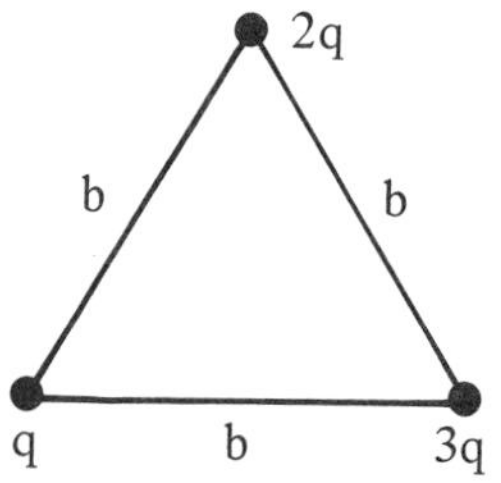

Figure 3-16

5. Determine the electric potential energy of the system of three charges shown in Figure 3-16.

6. Four charges are positioned at the corners of a rectangle, as in Figure 3-17. (a) Calculate the total electric potential at the center of the rectangle. (b) Determine the electric potential energy of the system of four charges.

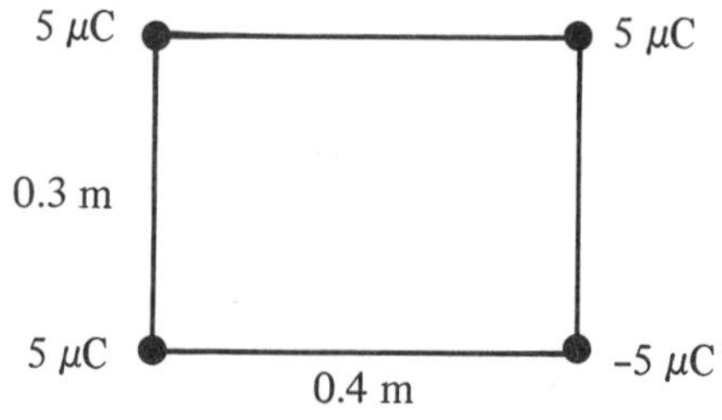

Figure 3-17

7. An electron and a proton are at rest and separated by 2×10^{-10} m. What is the minimum energy (in eV) required to separate these particles by an infinite distance? (Use the numerical data in Table 2-1.)

8. A charge of 8 μC is located at the origin, while a second charge of -3 μC is located at the point (2, 0) m. Calculate the *potential difference* $V_B - V_A$, where A is the point (2, 3) m and B is the point (0, 3) m.

9. A *conducting* sphere has a total charge Q and a radius R. (a) Find the electric potential at (1) $r < R$, (2) $r = R$, and (3) $r > R$. [Take $V(\infty) = 0$]. (b) What is the capacitance of the sphere? *Hint:* Recall that $\mathbf{E} = 0$ inside a conductor; therefore, V is *constant* inside.

10. A large metal sheet has a charge per unit area of $+5 \times 10^{-8}$ C/m^2. (a) Find the potential difference between a point A at its surface and a point B located 0.2 m from its surface. (b) A particle of mass 0.02 g and charge 2 μC is released from rest at A. Find the speed of the particle when it reaches B. (Neglect gravity.)

11. Determine the maximum charge that can be put on a metal sphere 50 cm in radius. Assume the maximum electric field in air before breakdown occurs is 3×10^6 V/m. (At fields greater than this, the air will conduct.)

12. (a) Find the electric potential at the point P for the system of three charges in Figure 2-17. (b) What does V become for large values of r?

13. Determine the electric potential at the center of curvature of the charged filament shown in Figure 2-18. Assume a uniform charge per unit length λ.

14. A charged ring has a uniform charge per unit length λ and radius R, as in Figure 3-5. Use the result to Example 3.3 for V_P, and obtain an expression for the electric field at P. *Hint:* Treat the distance d as a variable, x.

15. (a) A flat disc has a uniform charge per unit area σ, as in Figure 2-19. Calculate the electric potential at the point P along the axis of the disc. (b) Use the result for V to determine the electric field intensity at P. Compare the result with Problem 12 in Chapter 2.

16. (a) An insulating filament of length l has a *nonuniform* charge per unit length which varies as $\lambda = \lambda_0(x - d)$. The filament is oriented along the x axis, as in Figure 2-20. Find the electric potential at the origin. (b) Use this result to derive the electric field intensity at the origin. Compare the result with Problem 13 in Chapter 2.

17. Two concentric metal cylinders have radii of a and b, as in Figure 2-25. A battery of voltage V is connected to the cylinders, such that the inner cylinder is negative and the outer is positive. (a) Determine the charge per unit length on the cylinder. (b) Find the capacitance of the system, assuming the length of the cylinders is l (where $l \gg$ a or b).

18. An insulating solid sphere of radius R has a *nonuniform* charge density, given by $\rho = \rho_0 r$ C/m^3, and total charge Q. Find the electric potential at (a) a point $r > R$ and (b) a point $r < R$. (c) Plot the potential as a function of r. [*Hint:* Use the results to Programmed Exercise 2.12, namely, Equations (1) and (4).]

19. An electric field exists in a certain region, having both an x and a y component. The vector expression is $\mathbf{E} = (400\ \mathbf{i} + 600\ \mathbf{j})$ V/m. (a) Find the potential difference between the point b at (3, 2) m and the point a at (1, 0) m. (b) What is the work done in moving a $-5\ \mu$C charge from b to a?

20. Suppose the electric potential in a certain region varies according to the expression

$$V = 8x + 12x^2y - 20y^2$$

where V is in volts. (a) Find the x, y and z components of the electric field. (b) Determine a numerical value for these components at the point (2, 3, −8) m. (c) How much work is done in moving a 3 μC charge from (0, 0, 0) to (2, 3, −8) m in the presence of this field?

21. A particle of mass m and positive charge q_1 is headed on a "collision course" with a stationary target particle whose charge is q_2, where $q_2 > 0$. The incoming charge q_1 has a velocity v_0 when it is *very* far from q_2. Find the distance of closest approach, that is, the distance between q_1 and q_2 when q_1 finally comes to rest. (*Hint:* Assume q_2 remains at rest, and use energy principles. Note that the potential energy of the system is approximately zero when the particles are very far apart.)

22. (a) A proton is accelerated from rest through a potential difference of 100 V. Find the final speed of the proton. (b) Repeat the calculation for an electron accelerated through a potential difference of 100 V.

23. Three capacitors having values of 3 μF, 6 μF, and 9 μF are connected in *series.* A potential difference of 300 V is maintained across the combination. (a) What is the equivalent capacitance of the circuit? (b) What is the charge on each capacitor? (c) What is the potential difference across each capacitor?

24. Suppose the three capacitors in Problem 23 are connected in *parallel,* and a potential difference of 300 V is maintained across each capacitor. (a) Find the equivalent capacitance of the circuit. (b) Calculate the charge on each capacitor. (c) If you had the "choice" of being "zapped" by the charged capacitors, would you choose the parallel or series combination described in Problem 23? Why?

25. A parallel plate capacitor has *circular* plates whose radii are 5 cm. The plate separation is 2 mm. (a) Calculate the capacitance of this device if the space between the plates is a vacuum. (b) Calculate the capacitance if the space between the plates is filled with nylon, whose dielectric constant is 3.5.

26. A capacitor is constructed from two spherical shells, as in Figure 3-11. The inner and outer spheres have radii of 5 mm and 10 mm, respectively. (a) Calculate the capacitance of this device if the medium between the shells is a vacuum. (b) Find the energy stored in the capacitor if the potential difference between the inner and outer spheres is 80 V.

27. In the circuit shown in Figure 3-18, find (a) the equivalent capacitance of the circuit, (b) the charge on each capacitor and (c) the voltage across each capacitor.

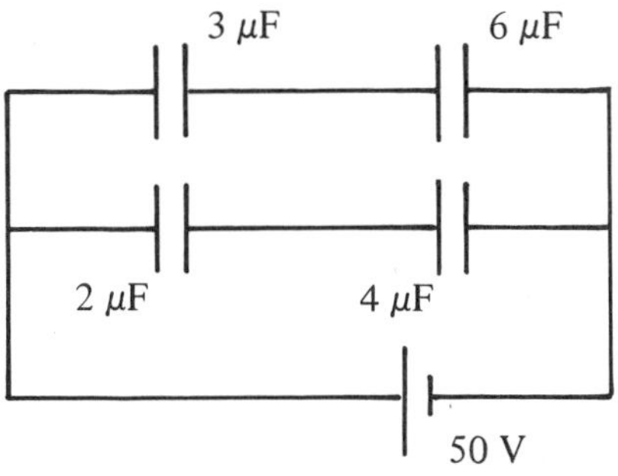

Figure 3-18

28. A 6 μF capacitor is charged with a battery whose potential difference is 22 V. The battery is removed from the circuit and the leads from the charged capacitor are connected to the terminals of an uncharged 4 μF capacitor. Find (a) the initial charge on the 6 μF capacitor, (b) the final potential difference across each capacitor after they are connected, (c) the initial energy of the system before the battery is removed and (d) the loss in energy after the capacitors are connected. Where does this energy go? *Hint:* The original charge on the 6 μF capacitor is shared by the two capacitors after they are connected.

29. A 3 μF capacitor is connected in series to a 12 μF capacitor. A potential difference of 30 V is maintained across the combination by means of a series-connected battery. (a) If the wires to the battery are disconnected and attached to each other, what is the final charge on each capacitor? (b) Suppose the wires to the battery are disconnected, and the charged capacitors are disconnected from each other. The positive side of *each* charged capacitor is then connected to the positive side of the other. Likewise, the negative terminals are connected. Now find the final charge on each capacitor.

30. Show that the energy of a conducting sphere of radius R and charge Q, surrounded by a vacuum, is given by

$$U = \frac{kQ^2}{2R}$$

Hint: Use Equation (3.33) and the definition of capacitance.

31. Show that the energy required to "build" a uniformly charged solid sphere of radius R and total charge Q is given by

$$U = \frac{3}{5}\frac{kQ^2}{R}$$

Hint: Use the fact that $dU = Vdq$, where V is the electric potential at the surface of a sphere of radius r, charge q. The charge q contained within the sphere of radius $r < R$ is given by $q = Qr^3/R^3$; therefore, $dq = 3Qr^2\,dr/R^3$.

32. A parallel plate capacitor has a charge q and area A. (a) Show that the force on one plate due to the charge on the other is given by

$$F = \frac{q^2}{2\,\epsilon_0 A}$$

Hint: Start with Equation (3.33), and express the capacitance in terms of A and x, an arbitrary plate separation. Then use the fact that the work done in moving the plates from x to x + dx is given by Fdx. (b) Obtain a numerical value for F for a 2 pF capacitor, whose plate area is 2 cm^2, with a potential difference of 100 V.

33. Two concentric *metal* spherical *shells* have radii of a and b, respectively, as in Figure 3-19. When the switch S is open, the inner sphere has a positive charge q, while the outer has a negative charge -Q. (a) Find the potential of each sphere when S is opened. (b) Find the charge on each sphere when S is closed. (c) What is the potential difference between the spheres after S is closed and equilibrium is obtained?

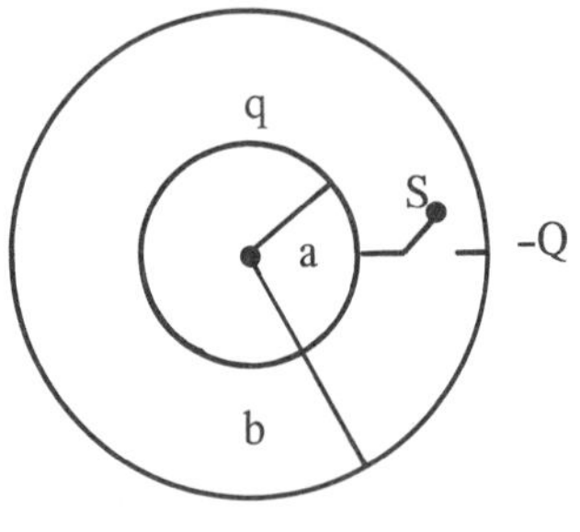

Figure 3-19

34. *Gauss's Law in a Dielectric.* Consider a dielectric which fills the plates of a capacitor, as in Figure 3-20. The free charge on the metal plates is q, and the induced charge on the dielectric is $-q_i$. (a) Using the Gaussian surface shown in the figure, and Gauss's law, show that the electric field in the dielectric is given by

$$E_d = \frac{1}{\epsilon_0 A} [q - q_i]$$

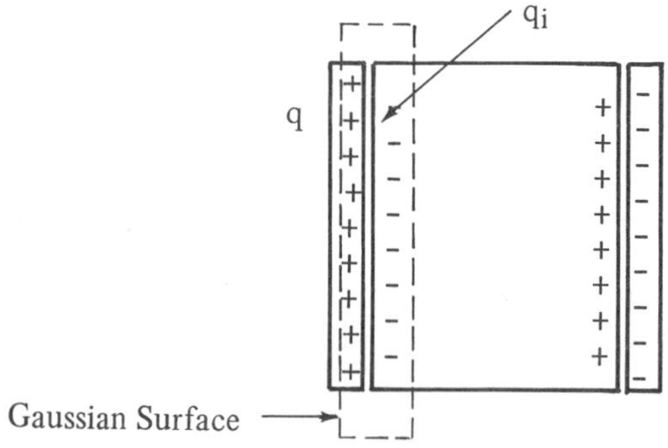

Figure 3-20

(b) Use the definition of the dielectric constant, namely, Equation (3.28), and verify that the induced charge is related to the free charge through the relation

$$q_i = q \left[1 - \frac{1}{\kappa}\right]$$

(c) Use the results for (a) and (b), and show that Gauss's law in a dielectric can be written as

$$\kappa \oint \mathbf{E}_d \cdot d\mathbf{A} = \frac{q}{\epsilon_0}$$

This result is valid for any medium whose dielectric constant is κ, q being the *free* charge *enclosed* by the Gaussian surface.

†35. A capacitor is constructed from two square plates of side l and separation d. A material whose dielectric constant is κ is inserted a distance x into the capacitor, as in Figure 3-21. (a) Find the equivalent capacitance of the device. (b) Calculate the energy stored in the capacitor if the potential difference is V. (c) Find the direction and magnitude of the force exerted on the dielectric, assuming a constant potential difference V. Neglect friction and edge effects. (d) Obtain a numerical value for F, assuming l = 5 cm, V = 2000 V, d = 2 mm and the dielectric is glass (κ = 4.5). *Hint:* The system can be considered as two capacitors connected in *parallel.* Also, recall that F = -dU/dx.

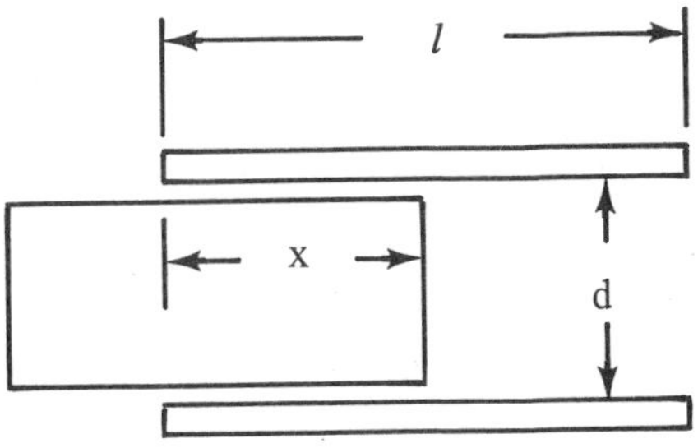

Figure 3-21

4

CURRENT, RESISTANCE AND DC CIRCUITS

4.1 ELECTRICAL CURRENT

The term electrical current or simply *current* is used to describe the flow of charges through a cross-section of a conductor. In an *isolated* conductor, the "free electrons" undergo random motion much like atoms or molecules in a closed container. However, if a potential difference is applied across the conductor by means of a battery, an electric field will be established which will produce a driving force on the "free electrons." Consequently, a continuous current will be maintained. In reality, the charges do not move in straight lines down the conductor, but undergo inelastic collisions with metal atoms. In effect, there is an energy transfer between the moving charges and the metal atoms. This results in an increase in the vibrational energy of the atoms and a corresponding increase in the temperature of the conductor. However, on the average, the charges are said to *drift* along the conductor with an average velocity called the *drift velocity*, $\mathbf{v}_d$. Electrons, because they are negatively charged, move in the direction $-\mathbf{E}$, whereas positive charge carriers (which can exist, say, in a semiconductor) move in the direction $\mathbf{E}$. A schematic representation of the random motion exhibited by electrons in a conductor under the action of an applied $\mathbf{E}$ field is shown in Figure 4–1. Although the electrons have rather large speeds between collisions $\left(\sim 10^5 \frac{\text{m}}{\text{sec}}\right)$, the drift velocities will be of *very small* magnitude $\left(\sim 10^{-4} \frac{\text{m}}{\text{sec}}\right)$.

Current is *defined* as the net charge Δq that passes through a given area per unit time Δt. This can be written as

$$I = \frac{\Delta q}{\Delta t} \qquad (4.1)$$

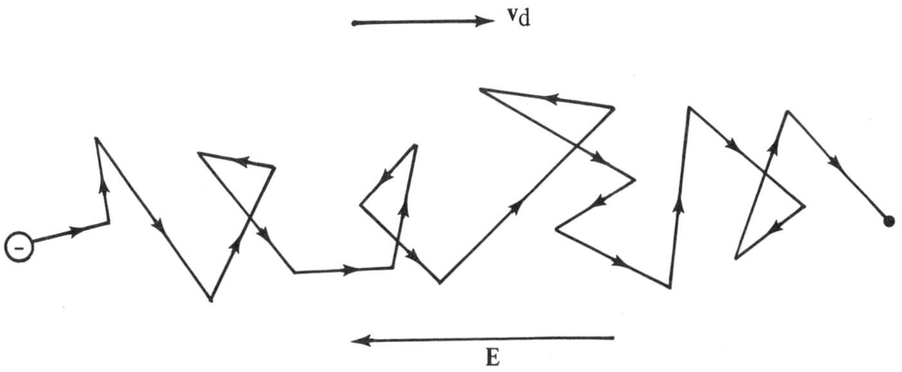

Figure 4–1 Drift velocity of electrons in a conductor under the action of an applied field, **E**.

When the rate of flow of charge varies in time, we must write I as

$$I = \lim_{\Delta t \to 0} \frac{\Delta q}{\Delta t} = \frac{dq}{dt} \tag{4.2}$$

Therefore, current is a *scalar quantity* and, in SI units, current has dimensions of coulomb per second. One coulomb per second is defined to be one ampere (1 A).

$$1\ \text{A} = 1\ \frac{\text{C}}{\text{sec}} \tag{4.3}$$

In practice, smaller units of current are often used, such as the milliampere ($1\ \text{mA} = 10^{-3}$ A) and the microampere ($1\ \mu\text{A} = 10^{-6}$ A).

It is useful to introduce the current density, **J**, whose magnitude represents the current divided by the cross-sectional area of the conductor. We will show that **J** is proportional to the drift velocity $\mathbf{v}_d$, which in turn is proportional to **E**. Consider a conductor such as copper, where the charge carriers are electrons. In an applied **E** field, the "free electrons" move in the opposite direction to **E**, as in Figure 4–2. The volume of a segment of the conductor in the shaded region of Figure 4–2 is $A\Delta x$. If n is the number of "free electrons" per unit volume, then $nA\Delta x$ is the number of electrons in the volume element $A\Delta x$. Therefore, the charge Δq in this region is given by

$$\Delta q = (nA\Delta x)e$$

We can express the distance Δx covered by the electrons in the time Δt as

$$\Delta x = v_d \Delta t$$

Therefore, the charge Δq reduces to

$$\Delta q = (nAv_d\Delta t)e$$

Applying the definition of current, namely Equation (4.1), gives*

$$I = \frac{dq}{dt} = (nAv_d)e \qquad (4.4)$$

Current density, on the other hand, is a *vector* quantity which is defined by the relation

$$\mathbf{J} = ne\mathbf{v}_d \qquad (4.5)$$

Substituting Equation (4.4) into Equation (4.5), we see that the magnitude of **J** for a uniform conductor is given by

$$J = \frac{I}{A} \qquad (4.6)$$

In addition, if the charge carriers are electrons or other *negative charge carriers,* **J** will be *opposite* to $\mathbf{v}_d$, whereas for *positive charge carriers* **J** is in the *same* direction as $\mathbf{v}_d$. Therefore, the direction of the conventional current in a metal is *opposite* to the direction of the electron motion. We can think of the direction of I effectively as the direction of motion of positive charge.

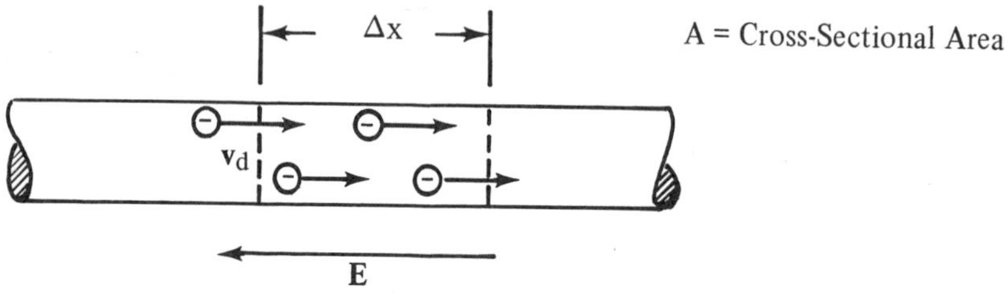

Figure 4–2 Motion of electrons in a conductor in an electric field.

Example 4.1

A conductor has a uniform cross-sectional area of $3 \times 10^{-6}\,m^2$, and carries a current of 21 A. (a) Determine the charge density in the wire.

$$J = \frac{I}{A} = \frac{21\ A}{3 \times 10^{-6}\,m^2} = 7 \times 10^6\ \frac{A}{m^2}$$

*In general, there may be positive and negative charge carriers, in which case the current I would contain a second term, $(n'Av'_d)q$. This term arises from the motion of positive charges to the left in Figure 4–2. This occurs, for example, in semiconductors, where both electrons and positive charge carriers (holes) contribute to I and **J**.

(b) If the charge carriers are electrons, determine the drift velocity, assuming that the number of free electrons per unit volume is 10^{29} electrons per m^3.

From Equation (4.5) and the results to (a), we have

$$v_d = \frac{J}{ne} = \frac{7 \times 10^6 \ \frac{A}{m^2}}{10^{29} m^{-3} \times 1.6 \times 10^{-19} C}$$

$$v_d \cong 4.4 \times 10^{-4} \ \frac{m}{sec}$$

Note that this is extremely slow compared to the average speed of the electrons between collisions, which is of the order of $10^5 \ \frac{m}{sec}$ at room temperature.

4.2 RESISTANCE AND OHM'S LAW

A current is maintained in a conductor when an electric field is present. Therefore, a potential difference must be maintained across the conductor to establish the **E** field and a current density **J**. The current density is proportional to the electric field, according to the relation

$$\mathbf{J} = \sigma \mathbf{E} \tag{4.7}$$

where the proportionality constant σ is the electrical conductivity of the material. Equation (4.7) is known as *Ohm's law,* first proposed by George Ohm in 1826. Note that **J** is in the same direction as **E**, consistent with the convention that **J** is in the direction of motion of positive charges.

To obtain a more useful form of Equation (4.7), consider a straight wire of length l and cross-sectional area A, as in Figure 4–3. The wire is coincident with the x-axis. An electric field exists in the conductor only when a potential difference $V_b - V_a$ is maintained, say by means of a battery. In one dimension, E can be written as $-\frac{dV}{dx}$, that is, E is the negative gradient of V. Since $J = \frac{I}{A}$ in magnitude, we can write Equation (4.7) in one dimension as

$$J = \frac{I}{A} = -\sigma \frac{dV}{dx}$$

or,

$$I\,dx = -\sigma A dV \tag{4.8}$$

Since I, σ and A are constant over the length of the wire, Equation (4.8) can be integrated to give

$$I \int_0^l dx = -\sigma A \int_{V_a}^{V_b} dV$$

or,

$$V = \left(\frac{l}{\sigma A}\right) I \tag{4.9}$$

where $V = V_a - V_b$. The constant $\frac{l}{\sigma A}$ is defined as the *resistance* R of the conductor, or

$$R = \frac{l}{\sigma A} \tag{4.10}$$

Therefore, the relation between the potential difference across the conductor and its resistance reduces to

$$V = IR \tag{4.11}$$

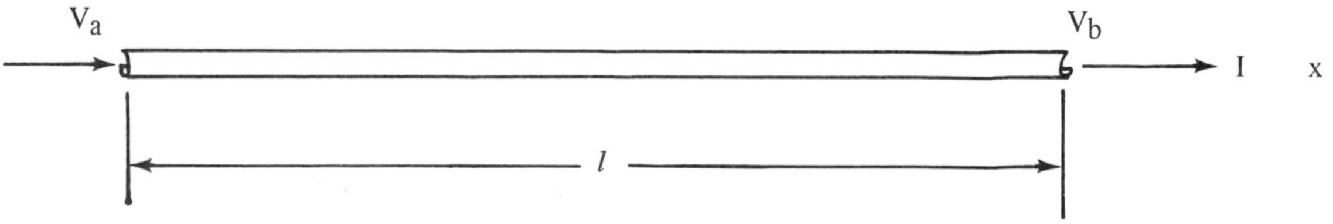

Figure 4-3 Current-carrying conductor.

The *inverse* of the conductivity is called the *resistivity, ρ*, of the material.

$$\rho = \frac{1}{\sigma} \tag{4.12}$$

Therefore, the resistance can also be expressed as

$$R = \frac{\rho l}{A} \tag{4.13}$$

From Equation (4.11) we see that R has units of volts per ampere in SI units. One volt per ampere is defined to be one ohm (Ω). That is, if a potential difference of one volt across a conductor provides a current of one ampere, the conductor has a resistance of one ohm.

$$1\ \Omega = 1\ \frac{V}{A}$$

Likewise, from Equation (4.10), we see that σ has units of $(\text{ohm m})^{-1}$. (This unit is sometimes referred to as $\frac{\text{mho}}{\text{m}}$. A mho is the reciprocal of ohm, and is the unit of a property called *conductance.*)

In circuits, a resistor is represented by the symbol , whereas a variable resistor is represented as  or .

The conductivity of a number of rather standard materials is listed in Table 4–1. Note the enormous range in conductivity from good conductors, such as silver ($\sigma \sim 10^7$), to good insulators, such as glass ($\sigma \sim 10^{-14}$). Intermediate values of σ correspond to materials called semiconductors, such as silicon and germanium. These are used in fabricating many solid-state devices, such as transistors and integrated circuits.

TABLE 4–1 ELECTRICAL CONDUCTIVITY OF A NUMBER OF MATERIALS

Material	Conductivity σ $(\Omega\ \mathrm{m})^{-1}$
Silver	6.8×10^7
Copper	5.8×10^7
Gold	4.1×10^7
Aluminum	3.8×10^7
Tungsten	1.8×10^7
Nickel	1.3×10^7
Iron	1.0×10^7
Steel	5.6×10^6
Lead	5×10^6
Carbon	2.9×10^4
Germanium	1.7
Silicon	1.6×10^{-3}
Glass	10^{-13} - 10^{-15}

Example 4.2

(a) Calculate the resistance of a copper wire whose length is 10 m and whose cross-sectional area is 5 mm^2.

We can use Equation (4.10) together with the value of σ for copper in Table 4-1. Since 1 mm^2 = 10^{-6} m^2, we get

$$R = \frac{l}{\sigma A} = \frac{10\ \mathrm{m}}{6.8 \times 10^7\ (\Omega\mathrm{m})^{-1}\ 5 \times 10^{-6}\,\mathrm{m}^2}$$

$$R \cong 3 \times 10^{-2}\ \Omega$$

(b) If a potential difference of 1 V is maintained over the length of the wire, calculate the current in the wire.

Using Equation (4.11), we have

$$I = \frac{V}{R} = \frac{1\ V}{3 \times 10^{-2}\ \Omega} = 33\ A$$

4.3 THE RESISTIVITY OF DIFFERENT CONDUCTORS

If a conductor behaves according to Ohm's law, then a plot of I *vs* V will be linear, as in Figure 4–4(*a*). The slope of this curve will yield a value for R. Such conductors are said to be *ohmic* in nature and therefore obey Ohm's law. However, the relation $R = \frac{V}{I}$ is true for *any* material. When the ratio is *not* constant, we say it is *nonohmic;* that is, there is a *nonlinear* relation between V and I for such materials. An example of such a nonlinear behavior is shown in Figure 4–4(*b*), which represents typical data for a semiconducting p–n junction.

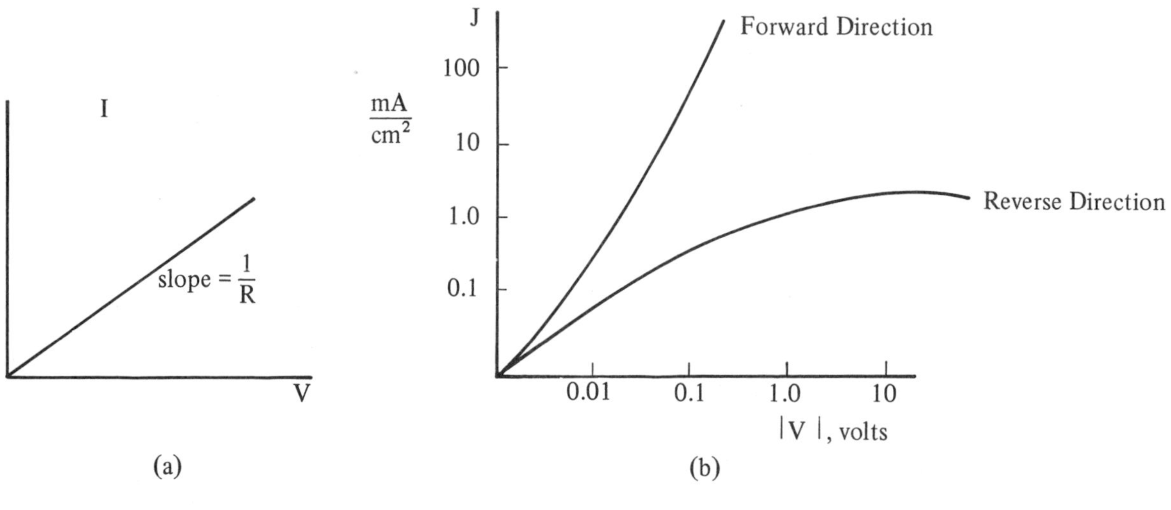

Figure 4–4 (*a*) Graphical representation of a material for which $\frac{V}{I}$ is constant, that is, one which obeys Ohm's law. (*b*) Typical current-voltage plot for a p-n junction, which is a semiconducting rectifier, a *nonohmic* device. Note that a large current occurs when V is in the "forward" direction, corresponding to the p region positive and the n region negative.

It is important to recognize that the resistivity of a given substance generally depends on several factors, such as *temperature,* mechanical strains and, in some cases, external magnetic fields. For *most* conductors, the resistivity, ρ, varies *linearly* with temperature according to the expression

$$\rho = \rho_0 [1 + \alpha (T - T_0)] \qquad (4.14)$$

where T_0 is some reference temperature, ρ_0 is the resistivity at that temperature and α is a quantity called the *temperature coefficient of resistivity.* Typical values of α

range from 10^{-5} to 10^{-3} per °C. This coefficient is *positive* for most metals, corresponding to an increase in ρ with increasing temperature. On the other hand, α is *negative* for nonmetals such as carbon and electrolytes. In such cases, ρ decreases with increasing temperature.

In semiconductors (such as silicon, germanium and gallium arsenide), the resistivity decreases rapidly with increasing temperature as a result of the increase in the number of charge carriers at the higher temperatures. Since the charge carriers in a semiconductor are associated with impurity atoms (called donors and acceptors), the resistivity is very sensitive to the type and concentration of such impurities. A *thermistor* is a thermometer which makes use of the large changes in resistivity of a semiconductor with temperature.

In reality, the resistivity of most metals does *not* go to zero at absolute zero, as in Figure 4–5(*a*). Generally, the high temperature resistivity is dominated by the collision of electrons with the metal atoms, whereas the low temperature resistivity is dominated by the collision of electrons with impurities and imperfections. Therefore, there is a *residual* resistivity at absolute zero, as in the case of sodium metal. However, there are a number of substances called *superconductors* whose resistivity goes to *zero* at absolute zero. The phenomenon of *superconductivity,* discovered by Kammerlingh Onnes in 1911, is a sudden decrease in resistance of a material at some finite temperature (called the critical temperature), typically near 4.2°K (the boiling point of liquid helium). The superconducting transition for mercury is shown in Figure 4–5(*b*). Later, it was shown that lead, tin and indium become superconductors at low temperatures. Today, there are hundreds of superconducting materials available, with critical temperatures as high as 21°K. They are used, for example, in generating high intensity magnetic fields in devices called superconducting magnets. One of the truly remarkable features of a superconductor is the fact that once a current is established, it will persist *without* any applied voltage.

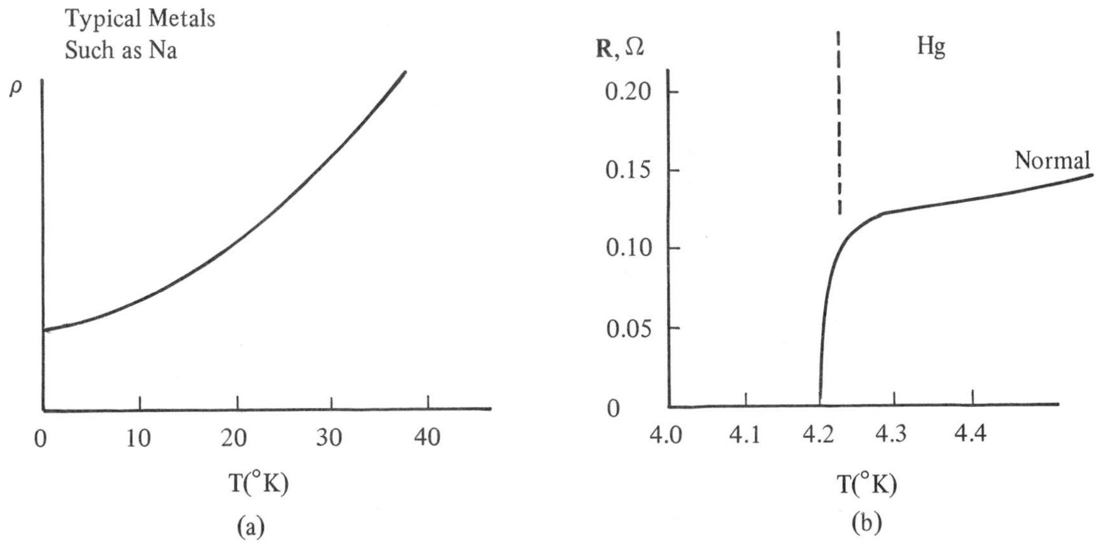

Figure 4–5 (*a*) Typical resistivity curve *vs* temperature for a metal such as sodium. The curve is nonlinear at low temperatures (<20° K) and a residual resistivity exists at absolute zero. (*b*) The transition from finite resistance to zero resistance for mercury at ~4.2° K. This substance is a superconductor below 4.2° K.

4.4 SIMPLE CIRCUITS CONTAINING RESISTORS

In the circuits that we will treat in this section, we will assume that the resistors obey Ohm's law; hence, current will be given by Equation (4.11). In order to establish a current in a circuit, a driving force is required to move charges and to set up a potential difference. Batteries and electric generators are devices which are commonly used for this purpose, and are sometimes referred to as seats of electromotive force (or seats of emf). In effect, the battery does work in taking positive charges from a low potential to a location of higher potential (or equivalently, it "lifts" electrons from a higher potential to a low potential). We will *assume* that the battery is a fixed source of electric potential, and, unless stated otherwise, we will neglect the internal resistance of the battery. Usually, the internal resistance of the battery is small, compared to external resistances, so the actual potential difference across the battery is approximately equal to the emf of the battery.

Series Circuit

Suppose a circuit consists of two resistors, R_1 and R_2, connected in *series* to a battery, as in Figure 4–6(*a*). The symbol for the battery is —| |— (− +), where the left side (−) is at a lower potential than the right side (+). Obviously, the current through R_1 must equal the current through R_2 in this case, so the potential difference across each resistance is given by

$$V_{ab} = IR_1$$

$$V_{bc} = IR_2$$

These potential differences are precisely the voltages one would measure with a voltmeter (an infinite resistance device), when connected between these points. Since the potential difference across the combination of resistors, V, is equal to the sum of the individual potential differences, we have

$$V = V_{ab} + V_{bc} = I\,(R_1 + R_2)$$

Therefore, we can replace the two resistances by one effective resistance R, representing the total resistance of the circuit.

$$R = R_1 + R_2 \qquad \text{Series Combination} \qquad (4.15)$$

Of course, the equivalent resistance of *more* than two resistors connected in series is equal to the sum of the individual resistances. That is,

$$R = R_1 + R_2 + R_3 + \ldots \qquad \text{Series Combination} \qquad (4.16)$$

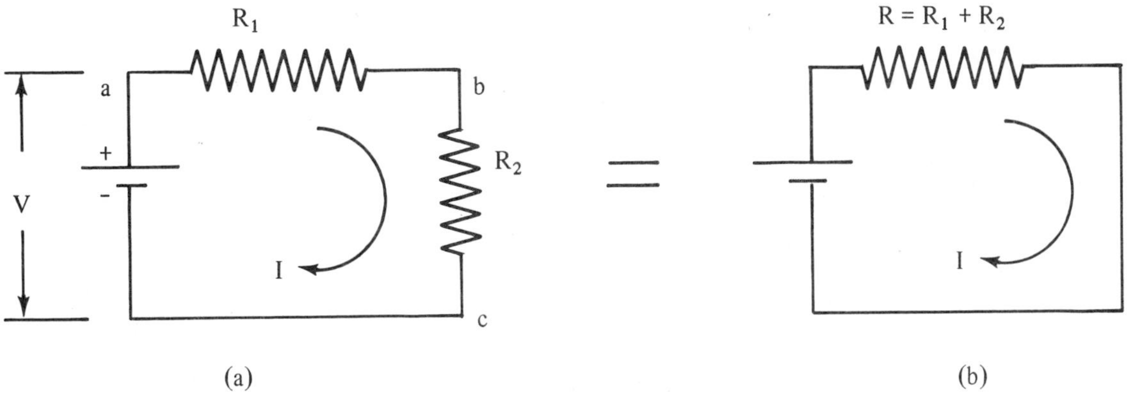

Figure 4–6 (*a*) Series connection of two resistors. (*b*) The equivalent circuit of two series-connected resistors.

Parallel Circuit

If two resistors are connected in *parallel* to a battery, as in Figure 4–7(*a*), we see that the potential difference across each must be the *same.* However, if $R_1 \neq R_2$, the current in each is not the same. The current that enters the junction at a "splits" into two parts, one going through R_1 and the rest through R_2. If $R_1 > R_2$, then $I_1 < I_2$. That is, the current will be small through a large resistor, and vice versa. Since charge is not created or destroyed (conservation of charge), we can write

$$I = I_1 + I_2$$

Applying Ohm's law to each resistor and noting that the potential difference across *each* is V gives

$$I_1 = \frac{V}{R_1}$$

$$I_2 = \frac{V}{R_2}$$

We see that the currents in each resistor are inversely proportional to their resistances.

The sum of the currents can be written as

$$I = I_1 + I_2 = V\left(\frac{1}{R_1} + \frac{1}{R_2}\right)$$

Therefore, the equivalent resistance of two resistors connected in parallel can be calculated from the expression

$$\frac{1}{R} = \frac{1}{R_1} + \frac{1}{R_2} \qquad \text{Parallel Combination} \qquad (4.17)$$

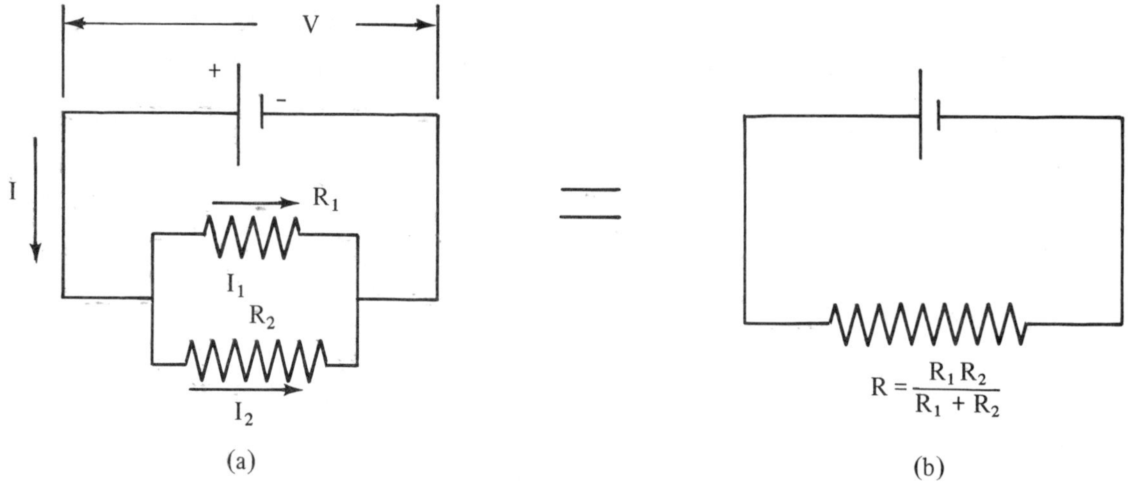

Figure 4–7 (*a*) Parallel connection of two resistors. (*b*) The equivalent circuit of two resistors connected in parallel.

or,

$$R = \frac{R_1 R_2}{R_1 + R_2}$$

If there are a number of resistors connected in parallel, the equivalent resistance is given by

$$\frac{1}{R} = \frac{1}{R_1} + \frac{1}{R_2} + \frac{1}{R_3} + \ldots \quad \text{Parallel Combination} \qquad (4.18)$$

This result shows that the equivalent resistance of two or more resistors connected in parallel is always less than the smallest resistance of the group. Circuits in homes and industry are ordinarily parallel circuits. Each "device," such as a lamp, radio, television or toaster, has a fixed potential difference across it. If more devices are added to a given circuit, the equivalent resistance *decreases* and the current *increases,* thereby raising the cost of the electric bill.

Example 4.3

(a) Find the equivalent resistance of the circuit shown in Figure 4-8. All resistance are in ohms.

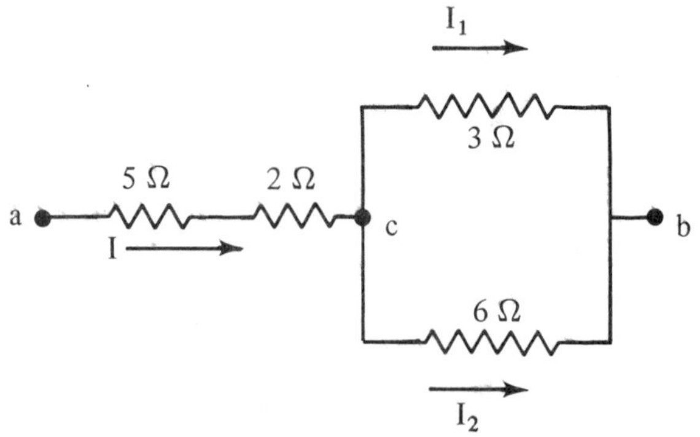

Figure 4–8

We can reduce the circuit in steps as follows. The 5Ω and 2Ω resistors are in series, so the equivalent resistance between a and c is 7Ω, or $R_{ac} = 7\Omega$. The 3Ω and 6Ω resistors are in *parallel,* and the equivalent resistance between c and b is

$$\frac{1}{R_{cb}} = \frac{1}{3} + \frac{1}{6} = \frac{3}{6}$$

or,

$$R_{cb} = 2\Omega$$

The circuit can be reduced pictorially as follows:

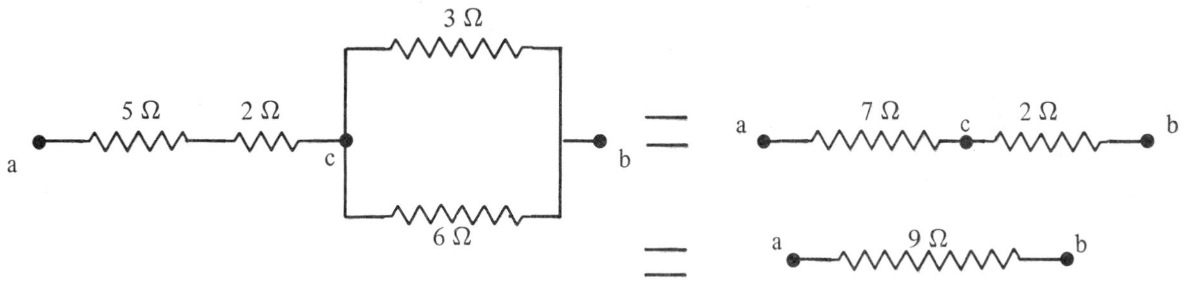

The equivalent resistance is 9Ω.

(b) A potential difference of 18 V is maintained between a and b, where $V_b > V_a$. Find the current through each resistor.

Since the equivalent resistance is 9Ω, the current I through the 5Ω and 2Ω resistors is given by

(1)
$$I = \frac{V_{ab}}{R_{ab}} = \frac{18\text{ V}}{9\Omega} = 2\text{ A}$$

The current that enters the junction at c equals the current that leaves the junction, so we have

(2)
$$I = I_1 + I_2 = 2\text{ A}$$

The potential differences across the 3Ω and 6Ω resistors are *equal,* since they are connected in parallel. Therefore, since $V_{cb} = 3I_1 = 6I_2$, we get

$$3I_1 = 6I_2$$

or,

(3)
$$I_1 = 2I_2$$

Substituting (3) into (2) gives

(4)
$$I_2 = \frac{2}{3}\text{ A}$$

(5)

$$I_1 = \frac{4}{3} A$$

Since the potential at b is greater than the potential at a, the direction of the currents are as shown in Figure 4–8.

(c) Find the potential differences V_{ac} and V_{cb}, and show that they add up to V_{ab}.

$$V_{ac} = I\,R_{ac} = 2\text{ A} \times 7\Omega = 14\text{ V}$$

$$V_{bc} = 3I_1 = 3\Omega \times \frac{4}{3}\text{A} = 4\text{ V}$$

$$[\text{or,}\quad V_{bc} = 6I_2 = 6\Omega \times \frac{2}{3}\text{A} = 4\text{ V}]$$

Therefore,

$$V_{ac} + V_{bc} = 18\text{ V} = V_{ab}$$

4.5 POWER DISSIPATION IN CIRCUITS

When a charge dq passes between two points in a circuit differing in potential by an amount V, the loss in potential energy of the charge is given by

$$dU = dq\ V$$

If the potential difference V remains constant in time, then the time rate of change of the loss in energy is

$$\frac{dU}{dt} = \frac{dq}{dt}V = IV$$

The rate at which energy is lost is the power P, given by

$$P = IV \tag{4.19}$$

When I is in amperes and V is in volts, P is in watts (W). More basically, 1 watt is 1 joule per second.

$$1\text{ W} = 1\text{ A} \times 1\text{ V} = 1\frac{\text{C}}{\text{sec}} \times \frac{\text{J}}{\text{C}}$$

or,

$$1\text{ W} = 1\frac{\text{J}}{\text{sec}}$$

Since the voltage across a resistor is given by V = IR, we can write Equation (4.19) as

$$P = I^2 R \quad \text{Power Dissipated by a Resistor} \tag{4.20}$$

An equivalent form of Equation (4.20) for a resistor is

$$P = \frac{V^2}{R} \tag{4.21}$$

Example 4.4

Three resistors are connected in parallel, as in Figure 4-9. A potential difference of 18 V is maintained across *each* resistor.

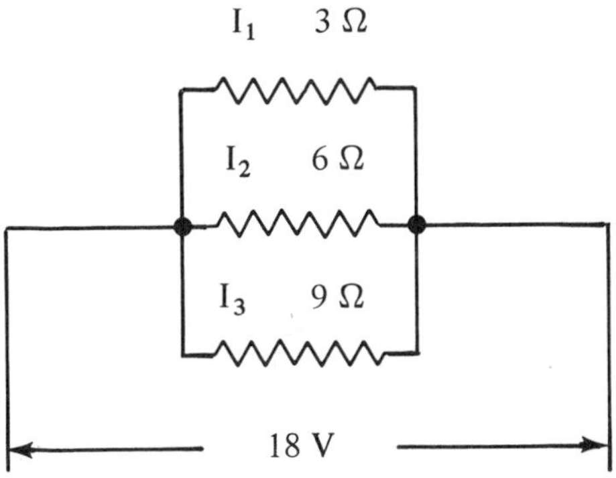

Figure 4–9

(a) Find the current in each resistor.

$$V = I_1 R_1 = 18\ \text{V}$$

$$I_1 = \frac{18\ \text{V}}{3\Omega} = 6\ \text{A}$$

Likewise,

$$I_2 = \frac{18\ \text{V}}{6\Omega} = 3\ \text{A}$$

$$I_3 = \frac{18\ \text{V}}{9\Omega} = 2\ \text{A}$$

(b) Calculate the power dissipated by each resistor. Applying Equation (4.20) to each resistor gives

3Ω: $P_1 = I_1{}^2 R_1 = 6^2 \times 3 = 108\ \text{W}$

6Ω: $P_2 = I_2{}^2 R_2 = 3^2 \times 6 = 54\ \text{W}$

9Ω: $P_3 = I_3{}^2 R_3 = 2^2 \times 9 = 36\ \text{W}$

Note that the smallest resistance dissipates the most power, since it carries the most current. [One can also use Equation (4.21) to obtain P for each resistor.]

(c) The total power dissipated is obviously 198 W. Show that the same result is obtained by calculating the equivalent resistance of the circuit. Use Equation (4.21) to get P.

$$\frac{1}{R} = \frac{1}{3} + \frac{1}{6} + \frac{1}{9}, \quad \text{or } R = \frac{18}{11}\,\Omega$$

$$\therefore \qquad P_{total} = \frac{V^2}{R} = \frac{18^2}{\frac{18}{11}} = 18 \times 11 = 198\text{ W}$$

4.6 KIRCHHOFF'S RULES AND DC CIRCUITS

Many circuits can be reduced step by step to series or parallel combinations. However, in the general case where reduction is not possible, there are two rules known as *Kirchhoff's rules* which make the analyses straightforward. Both rules are generalizations of results already obtained for the series and parallel combinations, and can be stated in the following manner:

1. The sum of the currents into any junction in a circuit must be *zero.* (A junction is a point where three or more wires meet.)

2. The sum of the potential differences across each element around a *closed* loop in a circuit is *zero.*

The first rule is a statement of *conservation of charge.* That is, whatever current enters a given point in a circuit must leave that point. (For example, $I = I_1 + I_2$ in Figure 4–8.) The second rule is a statement of *conservation of energy.* In other words, an electron which goes around a closed loop in a circuit (starts and ends at the same point) undergoes *no* net loss or gain of energy.

There are a few pointers that are useful in applying the *second* rule. First, the change in potential across a resistor is $-IR$ if the direction of the traversal is in the same direction as the current. If the resistor is traversed in the direction opposite to I, the change in potential is IR. Second, if a seat of emf is traversed in the direction of the emf (from − to + on a battery), the change in potential is $\mathcal{E}$; if the seat of emf is traversed in the opposite direction (from + to −), the change in potential is $-\mathcal{E}$. These "rules of thumb" are summarized in Figure 4–10.

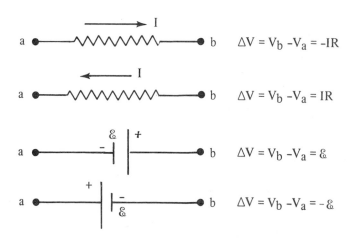

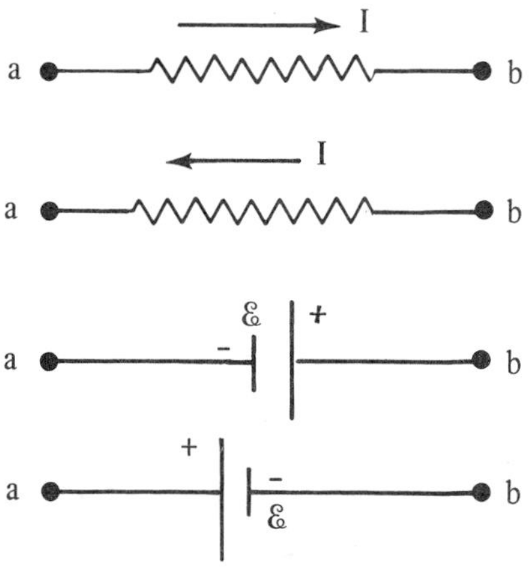

$\Delta V = V_b - V_a = -IR$

$\Delta V = V_b - V_a = IR$

$\Delta V = V_b - V_a = \mathcal{E}$

$\Delta V = V_b - V_a = -\mathcal{E}$

Figure 4–10 Changes in potential across a resistor and a seat of emf.

Example 4.5

(a) Consider the circuit in Figure 4-11, which contains two seats of emf, $\mathcal{E}_1$ and $\mathcal{E}_2$. Write the loop equation starting at a, traversing the circuit in the *clockwise* direction.

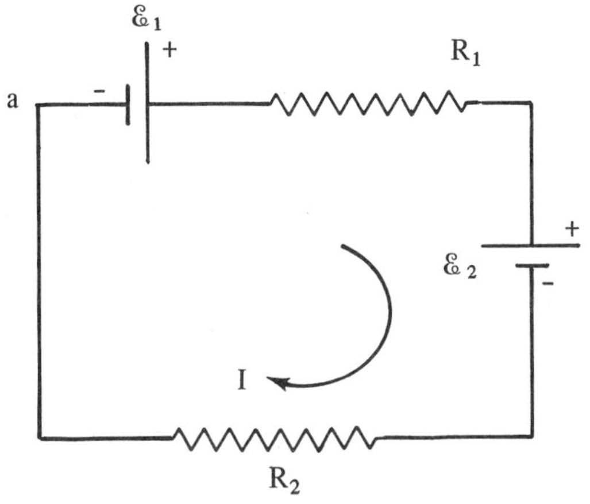

Figure 4-11

First, we assign a direction to I. If we get a negative sign for I in the numerical part, this simply means the wrong direction was chosen for I. Choosing I to be clockwise and traversing the circuit in this direction gives

$$\mathcal{E}_1 - IR_1 - \mathcal{E}_2 - IR_2 = 0$$

or,

$$I = \frac{\mathcal{E}_1 - \mathcal{E}_2}{R_1 + R_2}$$

Obviously, if $\mathcal{E}_1 > \mathcal{E}_2$, the current will be in the clockwise direction, since I will be positive. On the other hand, if $\mathcal{E}_1 < \mathcal{E}_2$, I is negative, implying a current in the counterclockwise direction.

(b) Find I if $\mathcal{E}_1 = 12$ V, $\mathcal{E}_2 = 6$ V, $R_1 = R_2 = 9\Omega$.

$$I = \frac{6\text{ V}}{18\Omega} = \frac{1}{3}\text{ A}$$

Since I is positive, the direction is clockwise, as in Figure 4-11.

(c) Calculate the power dissipated by the two resistors.

$$P = I^2R_1 + I^2R_2 = \left(\frac{1}{3}\right)^2 (18) = 2\text{ W}$$

Example 4.6

Two-loop circuit: We wish to determine the currents I_1, I_2 and I_3 in the two-loop circuit shown in Figure 4-12. First, a direction is assigned to the currents. We must adhere to these directions when writing the equations corresponding to Kirchhoff's rules.

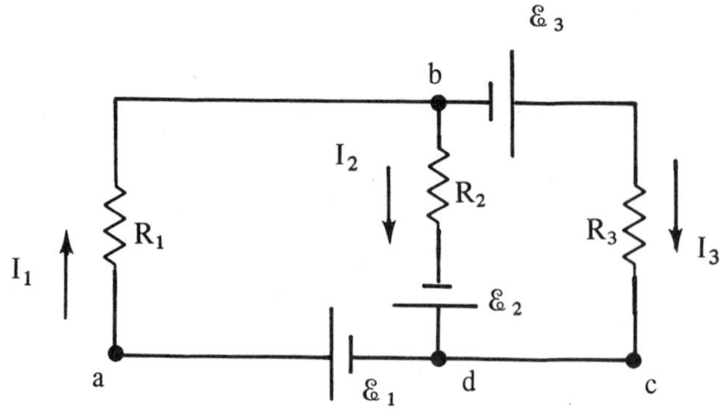

Figure 4–12

Applying Kirchhoff's first rule to the junction at b gives

(1)

$$I_1 - I_2 - I_3 = 0$$

where currents going into the junction are called positive, whereas currents leaving the junction are negative.

Now, we apply Kirchhoff's second rule to the loop abda. Starting at d, going in the clockwise direction, gives

(2)

$$\mathcal{E}_1 - I_1 R_1 - I_2 R_2 + \mathcal{E}_2 = 0$$

Likewise, Kirchhoff's second rule appled to the loop dbcd, starting at d and going clockwise, gives

(3)

$$-\mathcal{E}_2 + I_2 R_2 + \mathcal{E}_3 - I_3 R_3 = 0$$

A third loop equation could be written for the loop abcda; however, it would provide no new information. In other words, the equation would not be independent of (2) and (3). Equations (1), (2) and (3) represent three linear equations with three unknowns, namely I_1, I_2 and I_3. Solving these equations simultaneously gives (after some cumbersome algebra)

(4)

$$I_1 = \frac{(R_2 + R_3)\,\mathcal{E}_1 + R_2 \mathcal{E}_3 + R_3 \mathcal{E}_2}{R_1 R_2 + R_1 R_3 + R_2 R_3}$$

(5)

$$I_2 = \frac{(R_1 + R_3)\,\mathcal{E}_2 + R_3 \mathcal{E}_1 - R_1 \mathcal{E}_3}{R_1 R_2 + R_1 R_3 + R_2 R_3}$$

(6)

$$I_3 = \frac{(\mathcal{E}_3 - \mathcal{E}_2)\,R_1 + (\mathcal{E}_1 + \mathcal{E}_3)\,R_2}{R_1 R_2 + R_1 R_3 + R_2 R_3}$$

Note that in the *limit* $R_2 \rightarrow \infty$, the limiting values of the current are $I_2 = 0$ and

$$I_1 = I_3 = \frac{\mathcal{E}_1 + \mathcal{E}_3}{R_1 + R_3}$$

Likewise, as $R_1 \to \infty$, $I_1 = 0$ and

$$I_2 = -I_3 = \frac{\mathcal{E}_2 - \mathcal{E}_3}{R_2 + R_3}$$

Finally, as $R_3 \to \infty$, $I_3 = 0$ and

$$I_1 = I_2 = \frac{\mathcal{E}_1 + \mathcal{E}_2}{R_1 + R_2}$$

The student should verify these limiting results.

(b) Find the currents if $R_1 = 5\Omega$, $R_2 = 8\Omega$, $R_3 = 6\Omega$, $\mathcal{E}_1 = 4$ V, $\mathcal{E}_2 = 6$ V and $\mathcal{E}_3 = 8$ V.

Direct substitution of these values into (4), (5) and (6) gives $I_1 = 1.32$ A, $I_2 = 0.42$ A and $I_3 = 0.90$ A. Since the values are all positive, the correct directions were chosen for the currents. Note also that the junction rule $I_1 = I_2 + I_3$ is verified.

Example 4.7

The multiloop circuit in Figure 4–13 contains resistors, a capacitor and three seats of emf. (a) Find the unknown currents, labeled as in Figure 4–13.

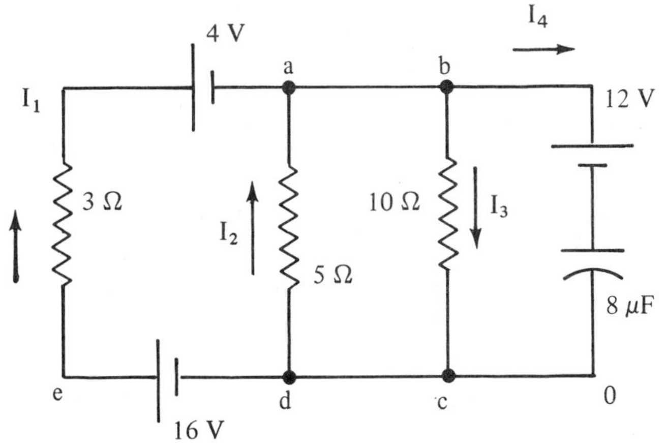

Figure 4–13

Since a capacitor is an *open* circuit, no current will be carried by it under steady-state conditions; therefore, $I_4 = 0$.

Applying Kirchhoff's first rule to the junction at a and assuming the directions of currents indicated gives

(1)

$$I_1 + I_2 - I_3 = 0$$

Kirchhoff's second rule applied to the loops adea and abcda gives

(2)

$$5\,I_2 + 16 - 3\,I_1 - 4 = 0 \qquad \text{(adea)}$$

(3)

$$-10\,I_3 - 5\,I_2 = 0 \qquad \text{(abcda)}$$

Again, (1), (2) and (3) represent three equations with three unknowns. Solving them simultaneously gives

$$I_1 = 1.89\text{ A}, \quad I_2 = -1.26\text{ A} \text{ and } I_3 = 0.63\text{ A}$$

The negative result for I_2 simply means we picked the wrong direction for I_2 in the figure.

(b) Find the potential difference across the capacitor and the charge on it. We can apply Kirchhoff's second rule to the loop b0cb. Calling the potential difference across the capacitor V_c and traversing the loop in the *clockwise* direction gives

$$-12 + V_c + 10\, I_3 = 0$$

Since $I_3 = 0.63$ A,

$$V_c = 5.7\text{ V}$$

The charge on a capacitor is given by Q = CV, so

$$Q = 8\mu F \times 5.7\text{ V} = 45.6\mu C$$

4.7 RC CIRCUITS

In this section, we will show that an uncharged capacitor does not become charged instantaneously when a potential difference is maintained across it. In fact, the rate at which the capacitor charges up depends on its capacitance and the resistance of the circuit. Consider the series circuit shown in Figure 4–14. When the switch is in position 1, a current will flow until the capacitor becomes fully charged. Once the maximum charge is reached (which depends, of course, on the emf of the battery), there will be no current.

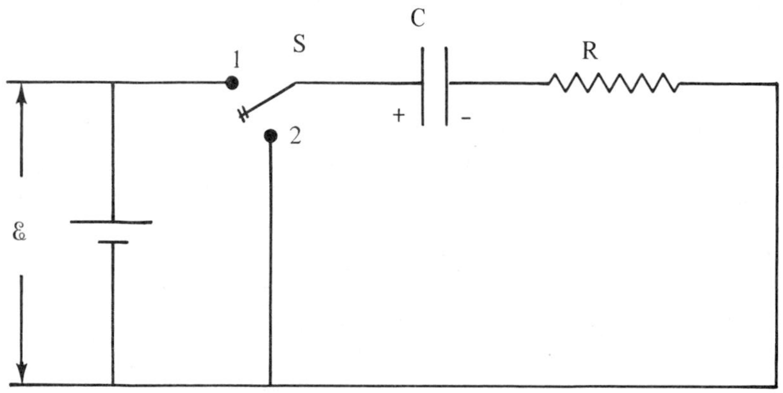

Figure 4–14 In this circuit the capacitor is charged when the switch is in position 1, and then discharged when the switch is in position 2.

Kirchhoff's second rule applied to the outer loop when the switch is in position 1 gives

$$\mathcal{E} - \frac{q}{C} - IR = 0 \tag{4.22}$$

where $\frac{q}{C}$ is the voltage drop across the capacitor and IR is the voltage drop across the resistor. Both q and I change as the capacitor is charging up. When the switch is first closed, $q = 0$ and I is a maximum. As the charge increases, the current decreases. Both changes occur in an exponential fashion. To demonstrate this, we differentiate Equation (4.22) with respect to time. Since $\mathcal{E}$ is a constant and $I = \frac{dq}{dt}$, differentiation gives

$$\frac{I}{C} + R\frac{dI}{dt} = 0$$

Rearranging terms, we can write this as

$$\frac{dI}{I} = -\frac{dt}{RC} \tag{4.23}$$

Since R and C are constants, direct integration of Equation (4.23) is possible. It is left as an exercise to show that the solution is

$$I = I_0 e^{-t/RC} \tag{4.24}$$

where $I_0 = \frac{\mathcal{E}}{R}$ is the *initial* current when the switch is first thrown and "e" is the base of natural logarithms (e = 2.718 . . .), not to be confused with electronic charge. Now, we can use the fact that $I = \frac{dq}{dt}$ in Equation (4.24) and integrate once more to get q as a function of time. Again, it is left as an exercise to show that

$$q = q_0 \left[1 - e^{-t/RC}\right] \tag{4.25}$$

where $q_0 = C\mathcal{E}$ is the *maximum* charge on the capacitor which results as $t \to \infty$.

Plots of I *vs* t and q *vs* t are shown in Figure 4–15. Note the exponential decrease in I and the corresponding exponential increase in q. At $t = 0$, the current is simply $\frac{\mathcal{E}}{R}$, that is, it is independent of C since there is no charge initially on the capacitor ($q = 0$ at $t = 0$). The quantity RC which appears in the exponential is called the *time constant,* τ, of the circuit. It is a quantity which is the measure of the time it takes the current to fall to $\frac{1}{e}$ of its initial value.

$\tau = RC$ **Time Constant of an RC Circuit**

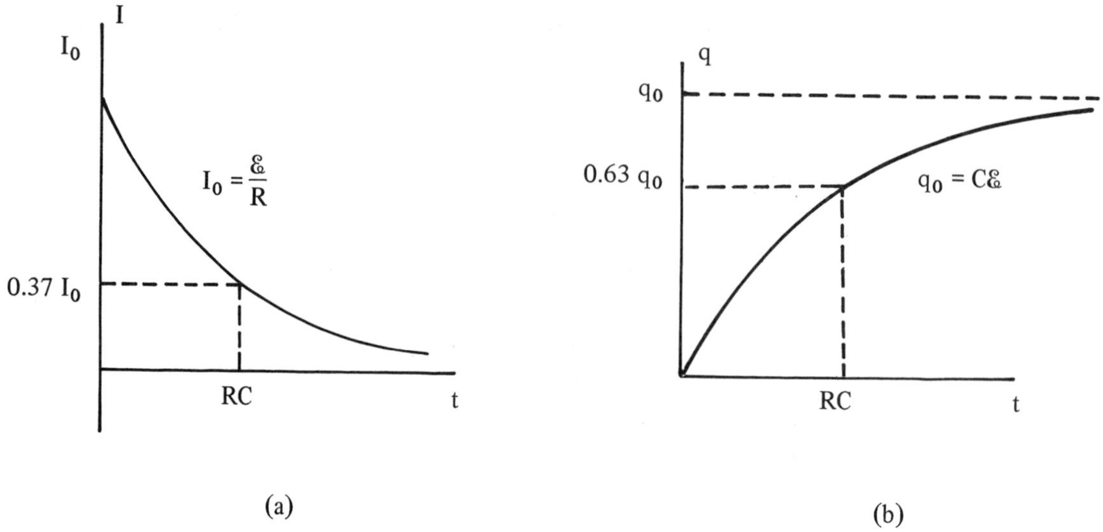

Figure 4–15 (*a*) Current *vs* time in the RC circuit in Figure 4–14 when the switch is in position 1. (*b*) Charge *vs* time in the same RC circuit.

To obtain the value of I after a time τ has elapsed, we substitute $t = \tau = RC$ into Equation (4.24). This gives

$$I(\tau) = I_0 e^{-1} \cong 0.37\ I_0$$

Likewise, in the time τ, we see from Equation (4.25) that the charge will increase to

$$q(\tau) = q_0\ (1 - e^{-1}) \cong 0.63\ q_0$$

Now suppose the switch in the circuit shown in Figure 4–14 is thrown to position 2. The battery is now *out* of the circuit and the capacitor will discharge through the resistor. We will assume the capacitor is fully charged initially. Kirchhoff's second rule applied to the circuit now gives

$$-\frac{q}{C} - IR = 0$$

Since $I = \frac{dq}{dt}$, this expression can be written as

$$-\frac{q}{C} - \frac{dq}{dt} R = 0$$

or,

$$\frac{dq}{q} = -\frac{dt}{RC} \tag{4.26}$$

We can integrate this expression, using the condition that $q = q_0$ at $t = 0$. This gives

$$q = q_0 e^{-t/RC} \tag{4.27}$$

Differentiating this with respect to time gives the current. The result is

$$I = -I_0 e^{-t/RC} \tag{4.28}$$

The negative sign signifies the *reversal* in the direction of the current when the capacitor discharges. Therefore, we see that the charge on both the capacitor and the current decreases exponentially in time, with a time constant of RC.

Example 4.8

A capacitor and resistor are connected in series to a battery, as in Figure 4-14. Using the values $\mathcal{E} = 6$ V, $C = 8 \mu F$, and $R = 5 \times 10^5 \Omega$, (a) calculate the time constant of the circuit.

$$\tau = RC = 5 \times 10^5 \Omega \times 8 \times 10^{-6} F = 4 \text{ sec}$$

(b) Show that the units of RC are time units.

$$[RC] = \frac{V}{I} \times \frac{Q}{V} = \frac{Q}{I} = \frac{Q}{\frac{Q}{T}} = T$$

(c) Determine the maximum charge on the capacitor, and the maximum current in the circuit.

$$q_0 = C\mathcal{E} = 8 \times 10^{-6} F \times 6 V = 4.8 \times 10^{-5} C$$

$$I_0 = \frac{\mathcal{E}}{R} = \frac{6 V}{5 \times 10^5 \Omega} = 1.2 \times 10^{-5} A$$

(d) Write expressions for the charge and current as functions of time when the capacitor is charging. Using Equations (4.24) and (4.25) gives

$$q = 4.8 \times 10^{-5} [1 - e^{-t/4}] C$$

$$I = 1.2 \times 10^{-5} e^{-t/4} A$$

4.8 PROGRAMMED EXERCISES

1 A conducting wire has a length l and uniform cross-sectional area A. The wire carries a *constant* current I.

1.A

Define the current in the wire.

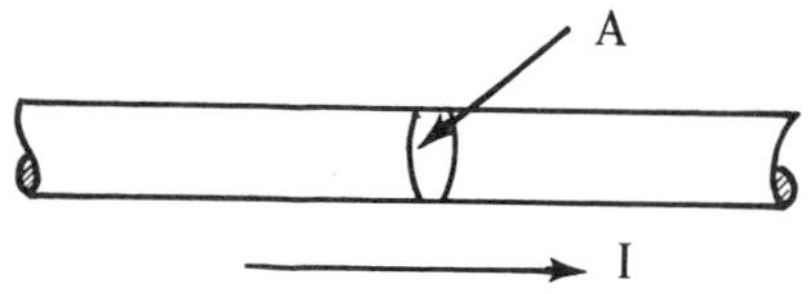

Current is the charge, Δq, that passes through a given cross-section per unit time, Δt.

$$I = \frac{\Delta q}{\Delta t} \qquad (1)$$

1.B

What is the unit of current in the SI system?

$$[I] = \frac{\text{coulomb}}{\text{second}} = 1 \text{ ampere},$$

$$1\text{A} = 1\,\frac{\text{C}}{\text{sec}}$$

1.C

Define the magnitude of the current density, J, in the wire. What is the unit of J in the SI system?

Current density is the current per unit area.

$$J = \frac{I}{A} \qquad (2)$$

where A in (2) is the cross-sectional area. J has units of $\frac{\text{A}}{\text{m}^2}$.

1.D

If I = 5 A, find the charge that passes through a given cross-section in a time interval of 8 sec.

From (1), we have

$\Delta q = I\Delta t$

$\Delta q = 5 \text{ A} \times 8 \text{ sec} = 5\,\frac{\text{C}}{\text{sec}} \times 8 \text{ sec}$

$\Delta q = 40 \text{ C}$

1.E

Suppose the same wire, which carries a current of 5 A, has a circular cross-section and a radius of 2 mm. Find the current density.

$$\text{Area} = \pi r^2 = \pi\,(2 \times 10^{-3})^2\,\text{m}^2$$
$$= 4\pi \times 10^{-6} \text{ m}^2$$

From (2), we have

$$J = \frac{I}{A} = \frac{5 \text{ A}}{4\pi \times 10^{-6}\,\text{m}^2} \cong 4 \times 10^5\,\frac{\text{A}}{\text{m}^2}$$

1.F

What condition must exist in the wire in order to maintain a current?

An electric field must exist within the conductor.

1.G

Does the fact that an electric field exists contradict our statement that $E = 0$ in a conductor in Chapter 2? Explain.

No. $E = 0$ in a conductor only in the *static* case, where there is no net motion of charges; hence, $I = 0$. In the present situation, charges *are* in motion and have a preferred direction, namely, the direction of the internal E field.

1.H

How can an E field be maintained in the wire?

The opposite ends of the wire must be at *different* electric potentials. The potential difference can be established by means of a battery.

1.I

The charge carriers in a conductor move with an average *drift speed,* v_d. Write an expression for the current density in terms of v_d, and the concentration of charge carriers, n. Assume the charge carriers are electrons.

$$J = nev_d \qquad (3)$$

where n is the number of electrons per unit volume and e is the electronic charge.

1.J

What is the direction of **J** compared to the direction of motion of electrons? (Recall that $\mathbf{J} = \sigma\mathbf{E}$.)

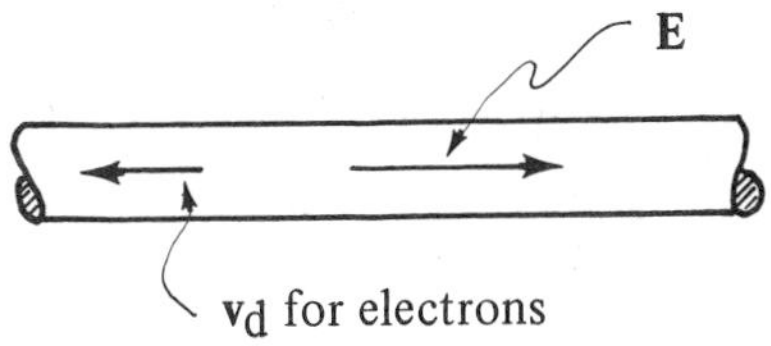

J is in the direction of **E**, while the electron motion is in the direction $-\mathbf{E}$. Therefore, the current density is opposite to the direction of motion of negative charges.

1.K

In a monatomic metal, such as silver or copper, how can you determine the value of n? Assume there is one conduction electron per atom, the density is ρ and the atomic weight M.

The number of atoms per mole is N_0, Avogadro's number. Therefore, the number of conduction electrons per unit volume is

$$n = \frac{N_0 \rho}{M}$$

Note that M has units of $\frac{g}{mole}$, so

$$n \sim \frac{\frac{atoms}{mole} \times \frac{g}{cm^3}}{\frac{g}{mole}} \times 1 \frac{electron}{atom} \sim \frac{electrons}{cm^3}$$

1.L

For copper, $n \cong 8.3 \times 10^{28}$ electrons per cubic meter. Use this value together with (3) and the numerical result for J in frame 1.E to find v_d for the conduction electrons in copper.

$$v_d = \frac{J}{ne}$$

$$v_d = \frac{4 \times 10^5 \frac{A}{m^2}}{8.3 \times 10^{28} \frac{electrons}{m^3} \times 1.6 \times 10^{-19} C}$$

$$v_d \cong 3 \times 10^{-5} \frac{m}{sec}$$

1.M

If the length of the conductor is 10 m, calculate the time it takes a conduction electron to travel this distance.

$$t = \frac{l}{v_d}$$

$$t = \frac{10 \text{ m}}{3 \times 10^{-5} \frac{m}{sec}} \cong 3.3 \times 10^5 \text{ sec}$$

This is about 100 hours!

2 *Ohm's Law.* Consider a cylindrical conductor of length l, uniform cross-sectional area A. The potential difference across the ends is V.

2.A

Define the resistance of the conductor if it carries a current I.

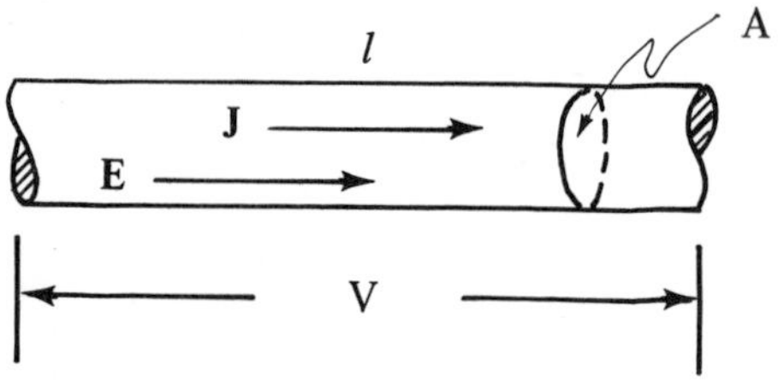

$$R = \frac{V}{I} \qquad (1)$$

Equation (1) is a *general* definition of resistance for any conductor. *If* the ratio $\frac{V}{I}$ is constant, the conductor is said to obey Ohm's law.

2.B

What is the unit of resistance in the SI system?

When V is in volts, I in amperes, R has units of ohms (Ω).

$$1\Omega = 1\,\frac{V}{A}$$

2.C

Write an expression which relates the electric field in the conductor to V and l.

$$E = \frac{V}{l} \qquad (2)$$

2.D

We can substitute (2) into (1) and use the fact that I = JA to obtain an expression for J. What is this expression?

$$R = \frac{El}{JA}$$

$$J = \left(\frac{l}{RA}\right) E = \sigma E \qquad (3)$$

This is known as Ohm's law.

2.E

What is the constant σ called? What are its units?

This constant is the *electrical conductivity*, σ, of the conductor.

$$\sigma = \frac{l}{RA} \quad (\Omega m)^{-1} \qquad (4)$$

2.F

What is the *reciprocal* of σ called? What are its units?

The reciprocal of σ is the *electrical resistivity*, ρ.

$$\rho = \frac{1}{\sigma} = \frac{RA}{l} \quad \Omega m \qquad (5)$$

2.G

Give a qualitative description of the size of σ for good conductors and poor conductors.

σ is *large* for good conductors (since R is small), whereas σ is *small* for poor conductors.

2.H

In Exercise 1, we found that $J = nqv_d$, where v_d is the drift velocity of the charge carrier. Use this in (3) to obtain an expression for v_d in terms of σ and E.

$$J = \sigma E$$

$$nqv_d = \sigma E$$

$$v_d = \frac{\sigma}{nq} E \qquad (5)$$

Note: v_d is in the direction of **E** if q is positive, opposite to **E** if q is negative.

2.I

Do all conductors obey Ohm's law? That is, is the ratio $\frac{V}{I}$ constant for all conductors? Give examples.

No. Many conductors do not obey Ohm's law; for example, vacuum tubes and various semiconducting devices have *nonlinear* V *vs* I relations.

2.J

A very special conductor is a material called a *superconductor.* What is the resistance of such a metal in the superconducting state?

R = 0. That is, if a current is established in a superconducting loop, it will persist *without* an applied voltage.

2.K

What is the power dissipated by a conductor whose resistance is R, carrying a current I? What are the units of power?

$$P = I^2 R \qquad (6)$$

or,

$$P = IV \qquad (7)$$

P has units of watts (W).

$$1\,W = 1 \frac{J}{sec}$$

2.L

A *silver* wire has a length of 200 m and a cross-sectional area of 2 mm². What is its resistance? $\sigma = 6.1 \times 10^7 (\Omega m)^{-1}$ for silver.

Using (4), we have

$$R = \frac{l}{A\sigma} = \frac{2 \times 10^2}{2 \times 10^{-6} \times 6.1 \times 10^7}$$

$$R \cong 1.6\Omega$$

2.M

The same wire carries a current of 3 A. Find the current density and electric field in the wire.

$$J = \frac{I}{A} = \frac{3\text{ A}}{2 \times 10^{-6}\,m^2} = 1.5 \times 10^6 \frac{A}{m^2}$$

$$E = \frac{J}{\sigma} = \frac{1.5 \times 10^6}{6.1 \times 10^7} \cong 2.5 \times 10^{-2} \frac{N}{C}$$

2.N

What is the potential difference across the ends of the wire for a current of 3 A?

From (1), we have

$$V = IR = 3A \times 1.6\Omega$$

$$V = 4.8\text{ V}$$

2.0

How much power is dissipated by the wire for a current of 3 A?

We can use (6) or (7) to get P.

$$P = I^2 R = 3^2 \times 1.6 = 9 \times 1.6 = 14.4 \text{ W}$$

or,

$$P = IV = 3 \times 4.8 = 14.4 \text{ W}$$

3 Four resistors are connected as shown in the figure below. All resistances are in ohms. A potential difference of 48 V is maintained between points a and b.

3.A

What is the equivalent resistance of the 20Ω and 30Ω combination?

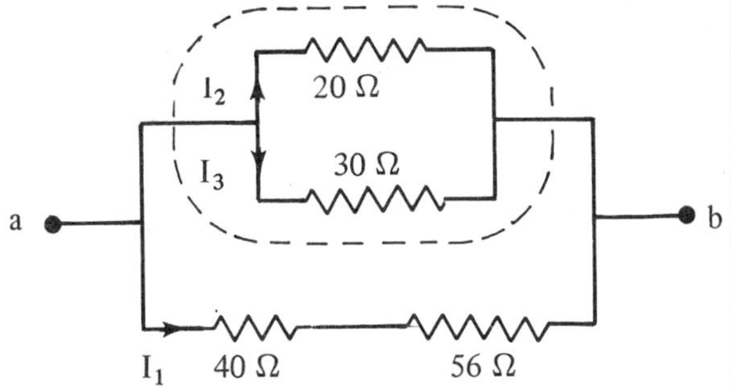

These are connected in *parallel,* so

$$\frac{1}{R} = \frac{1}{R_1} + \frac{1}{R_2} = \frac{1}{20} + \frac{1}{30}$$

$$R = 12\Omega \qquad (1)$$

3.B

Find the equivalent resistance of the 40Ω and 56Ω resistors.

These are connected in *series;* therefore,

$$R = R_1 + R_2 = 96\Omega \qquad (2)$$

3.C

The circuit is now reduced to

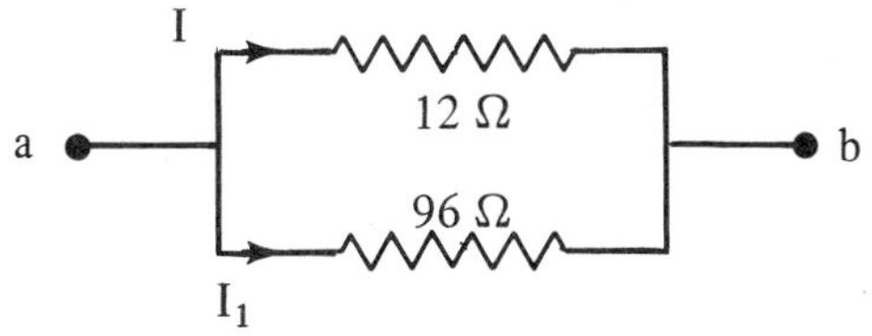

What is the equivalent resistance of the circuit?

$$\frac{1}{R_{eq}} = \frac{1}{12} + \frac{1}{96} = \frac{9}{96}$$

$$R_{eq} \cong 10.7\Omega$$

3.D

Determine the current through the 40Ω and 56Ω resistors.

The potential across this combination is 48 V and the equivalent resistance of the combination is 96Ω.

$$\therefore \quad I_1 = \frac{V}{R} = \frac{48}{96} = 0.5 \text{ A} \qquad (3)$$

3.E

Determine the current, I, through the equivalent 12Ω resistor of the upper branch.

From frame 3.C, we see that the potential across this equivalent resistor is also 48 V; ∴

$$I = \frac{48 \text{ V}}{12\Omega} = 4 \text{ A} \qquad (4)$$

3.F

Find the currents I_2 and I_3 through the 20Ω and 30Ω resistors. Note that V = 48 V across each.

$$20\Omega: I_2 = \frac{48}{20} = 2.4 \text{ A} \qquad (5)$$

$$30\Omega: I_3 = \frac{48}{30} = 1.6 \text{ A} \qquad (6)$$

3.G

Are the results to frame 3.F consistent with (4)? Explain.

Yes. $I = I_2 + I_3$ is verified. That is, the current through the equivalent resistance of 12Ω must equal the sum of the currents through the 20Ω and 30Ω resistors.

3.H

Find the power dissipated by each resistor. Use results (3), (5) and (6).

$40\Omega: P = I_1{}^2 R = (0.5)^2 40 = 10$ W

$56\Omega: P = (0.5)^2 (56) = 14$ W

$20\Omega: P = I_2{}^2 R = (2.4)^2 20 = 115$ W

$30\Omega: P = I_3{}^2 R = (1.6)^2 30 = 77$ W

3.I

What is the total power dissipated by the circuit?

Adding the figures in frame 3.H gives

$$P_{total} = 216 \text{ W} \qquad (7)$$

3.J

From (3) and (4) we see that the total current delivered by the battery is 4.5 A. (Or, $I = \frac{V}{R_{eq}}$.) Show that the total power dissipated by the circuit can be obtained using $P = I^2 R_{eq}$. Compare this with (7).

$$P_{total} = I^2 R_{eq}$$

$$P_{total} = (4.5)^2 \times 10.7 = 216 \text{ W}$$

or,

$$P_{total} = IV = 4.5 \times 48 = 216 \text{ W}$$

The result is the same as (7).

3.K

If the potential difference between a and b is halved, what happens to the resistance of the circuit?

It remains the same. Resistance is independent of applied voltage for devices which obey Ohm's law.

3.L

What is the *total* current delivered by the battery if the potential difference between a and b is halved?

Since $I = \frac{V}{R}$, reducing the voltage by one-half would also halve the current. $\therefore I = 2.25$ A.

4 *Kirchhoff's Rules.* In the analyses of complex networks, or those involving more than one loop, the application of Kirchhoff's rules is extremely useful.

4.A

Describe Kirchhoff's first rule. What conservation law does it represent?

Kirchhoff's first rule states that the net current into a junction is zero, or the total current entering must equal the total current leaving the junction. This is a statement of *conservation of charge.*

4.B

Apply Kirchhoff's first rule to the junction below.

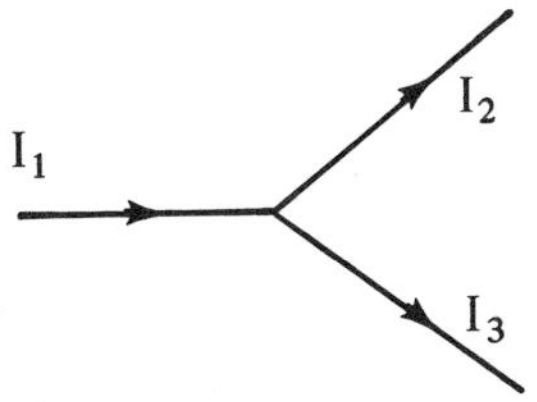

$$I_1 - I_2 - I_3 = 0$$

or,

$$I_1 = I_2 + I_3$$

4.C

Describe Kirchhoff's second rule. What conservation law does it represent?

Kirchhoff's second rule states that the sum of the *changes* in potential in any *complete* loop of a circuit must be zero. This represents a statement of *conservation of energy.*

4.D

The symbol below represents a seat of emf. If we traverse this circuit element from left to right, the change in potential is ____________, whereas if we traverse it from right to left, the change in potential is ____________.

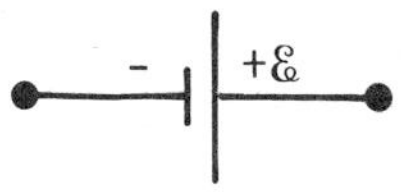

$\underline{\mathcal{E}}$, $-\underline{\mathcal{E}}$

The positive side of an isolated battery is always at a higher potential than the negative side.

4.E

The symbol below represents a resistor. If the resistor is traversed from left to right, the change in potential is ____________, whereas if the resistor is traversed from right to left, the change in potential is ____________.

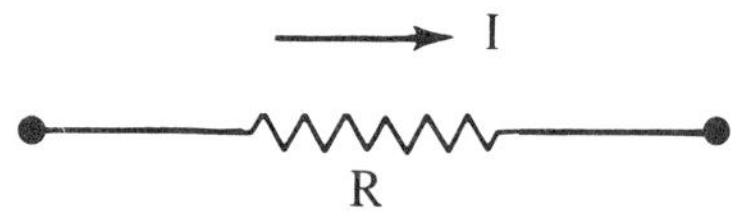

$\underline{-IR}$, $\underline{IR}$

The change in potential is $-IR$ if the resistor is traversed in the same direction as I, whereas the change is IR if it is traversed in the opposite direction to I.

4.F

With these rules in mind, apply Kirchhoff's second rule to the circuit below. Traverse the circuit in the clockwise direction, starting at a.

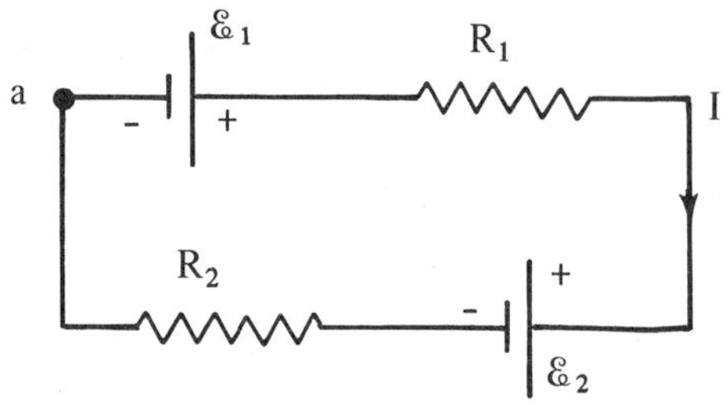

$$\mathcal{E}_1 - IR_1 - \mathcal{E}_2 - IR_2 = 0$$

or,

$$\mathcal{E}_1 - \mathcal{E}_2 = I(R_1 + R_2) \qquad (1)$$

Note: We have *assumed* that I is in the clockwise direction. This may or may not be true, depending on the values of $\mathcal{E}_1$ and $\mathcal{E}_2$.

4.G

In the circuit shown in frame 4.F, let $\mathcal{E}_1 = 6$ V, $\mathcal{E}_2 = 18$ V, $R_1 = 10\ \Omega$ and $R_2 = 14\ \Omega$. Find I.

Substituting these values into (1) gives

$$I = \frac{6 - 18}{10 + 14} = -\ 0.5\ \text{A}$$

4.H

What does the negative sign in the answer for I indicate?

This means we picked the *wrong* direction for I in the circuit. For these values, I is in the *counterclockwise* direction and its magnitude is 0.5 A.

4.I

Determine the potential difference across each resistor for this value of I.

$V_1 = IR_1 = 0.5 \times 10 = 5$ V

$V_2 = IR_2 = 0.5 \times 14 = 7$ V

4.J

Note that $V_1 + V_2 = 12$ V. Is this result surprising? Explain.

The result should not surprise you. The sum of the potential drops across the resistors simply adds up to the *net emf* in the circuit; that is, Equation (1) is satisfied:

$$\mathcal{E}_2 - \mathcal{E}_1 = 18\ \text{V} - 6\ \text{V} = 12\ \text{V}$$

$$I\,(R_1 + R_2) = 5 + 7 = 12\ \text{V}$$

4.K

Calculate the power dissipated by each resistor and the total power delivered by the batteries.

$P_1 = I^2 R_1 = (0.5)^2 \times 10 = 2.5$ W

$P_2 = I^2 R_2 = (0.5)^2 \times 14 = 3.5$ W

$P_{total} = P_1 + P_2 = 6$ W

[or $P = I\,(\mathcal{E}_2 - \mathcal{E}_1) = 0.5\,(12) = 6$ W]

5 *Multiloop Circuit.* Let us apply Kirchhoff's rules to the circuit below. We wish to solve for the currents in the various branches.

5.A

We will assume the direction of the currents are as shown below. Write Kirchhoff's first rule applied to junction d.

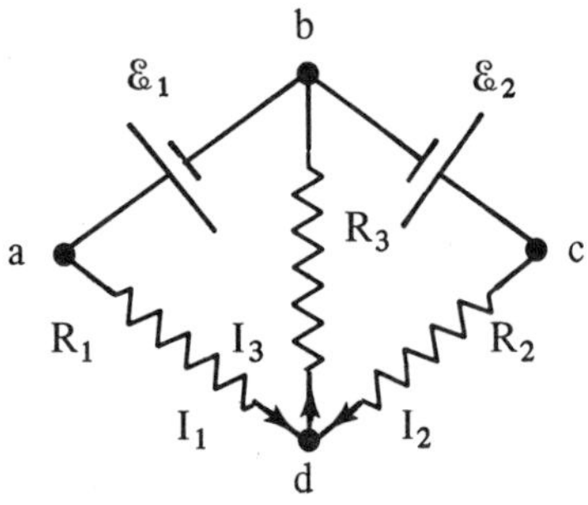

$$I_1 + I_2 - I_3 = 0 \qquad (1)$$

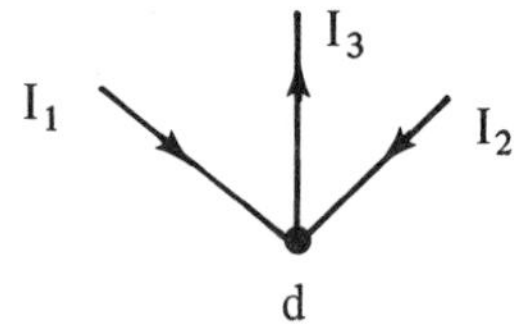

5.B

Apply Kirchhoff's second rule to the loop abda. Traverse the loop in the clockwise direction, starting at a.

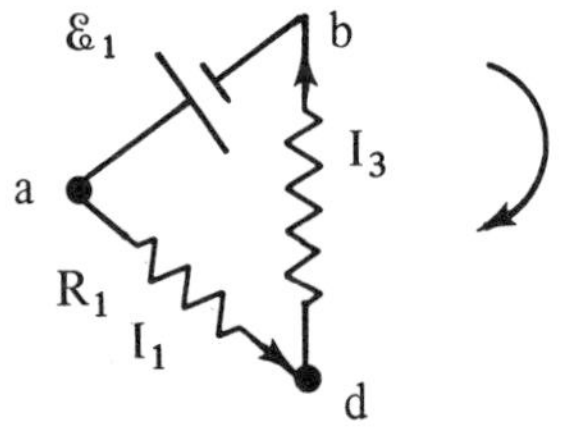

$$-\mathcal{E}_1 + I_3 R_3 + I_1 R_1 = 0 \qquad (2)$$

5.C

Now apply Kirchhoff's second rule to the loop bcdb, going in the clockwise direction from b.

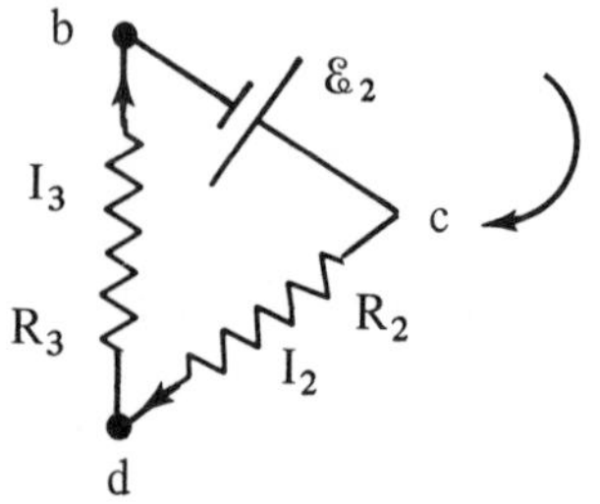

$$\mathcal{E}_2 - I_2 R_2 - I_3 R_3 = 0 \qquad (3)$$

5.D

Write a loop equation for the path abcda. Does this equation present any "new" information? Explain.

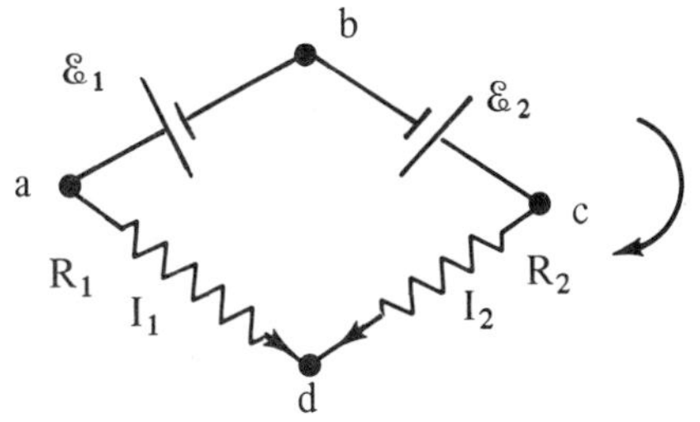

$$-\mathcal{E}_1 + \mathcal{E}_2 - I_2R_2 + I_1R_1 = 0 \quad (4)$$

No. Equation (4) is simply the *sum* of Equations (2) and (3); therefore, it gives no additional information. Equations (1), (2) and (3) are sufficient to solve the problem. They represent three linear equations with three unknown currents.

5.E

Substitute the value of I from (1) into (2) to get another expression involving I_2 and I_3.

$$-\mathcal{E}_1 + I_3R_3 + I_3R_1 - I_2R_1 = 0$$

or,

$$-\mathcal{E}_1 + I_2R_1 + I_3(R_1 + R_3) = 0 \quad (5)$$

5.F

We now have reduced the problem to *two* equations with two unknowns, I_2 and I_3. Solve (3) and (5) simultaneously and obtain an expression for I_3.

From (3), $I_2 = \dfrac{\mathcal{E}_2 - I_3R_3}{R_2}$. Substitute this into (5).

$$-\mathcal{E}_1 - \left(\frac{\mathcal{E}_2 - I_3R_3}{R_2}\right)R_1 + I_3(R_1 + R_3) = 0$$

$$I_3 = \frac{\mathcal{E}_1R_2 + \mathcal{E}_2R_1}{D} \quad (6)$$

where $D = R_1R_2 + R_1R_3 + R_2R_3$. After some algebra, we also find that

$$I_2 = \frac{\mathcal{E}_2(R_1 + R_3) - \mathcal{E}_1R_3}{D} \quad (7)$$

and

$$I_1 = \frac{\mathcal{E}_1(R_2 + R_3) - \mathcal{E}_2R_3}{D} \quad (8)$$

5.G

In the limit of $R_3 \to \infty$, does the solution (6) for I_3 make sense? Explain.

Yes. As $R_3 \to \infty$, we see from (6) that $I_3 \to 0$. That is, no current will pass through an infinitely large R.

5.H

Suppose $\mathcal{E}_1 = 3$ V, $\mathcal{E}_2 = 5$ V, $R_1 = R_2 = 4\Omega$ and $R_3 = 2\Omega$. Find numerical values for I_1, I_2 and I_3.

Using (6), (7) and (8), we get

$$I_1 = \frac{1}{4} \text{ A}$$

$$I_2 = \frac{3}{4} \text{ A}$$

$$I_3 = 1 \text{ A}$$

Note: These values are consistent with (1).

5.I

Calculate the power dissipated by each resistor and the total power dissipated. Use the numerical data in frame 5.H.

$$P_1 = I_1{}^2 R_1 = \left(\frac{1}{4}\right)^2 \times 4 = 0.25 \text{ W}$$

$$P_2 = I_2{}^2 R_2 = \left(\frac{3}{4}\right)^2 \times 4 = 2.25 \text{ W}$$

$$P_3 = I_3{}^2 R_3 = (1)^2 \times 2 = 2 \text{ W}$$

$$P_{total} = 4.5 \text{ W} \tag{9}$$

5.J

Calculate the power delivered by the 3 V battery, $\mathcal{E}_1$. Do the same for the 5 V battery, $\mathcal{E}_2$. What is the total power delivered by the batteries?

$$P_1 = I_1 \mathcal{E}_1 = \frac{1}{4} \times 3 = \frac{3}{4} \text{ W}$$

$$P_2 = I_2 \mathcal{E}_2 = \frac{3}{4} \times 5 = \frac{15}{4} \text{ W}$$

$$P_{total} = \frac{18}{4} = 4.5 \text{ W} \tag{10}$$

5.K

We see that the power delivered by the batteries is exactly equal to the power dissipated by the resistors. Is the result surprising? Explain.

No. Energy must be conserved. The energy supplied by the battery in a given time interval is *lost* in the form of heat in the resistors.

6 This is similar to Exercise 5, except we have included a capacitor in the circuit. The symbol (A) represents an ammeter, a *low* resistance device (ideally zero) which measures current. The symbol (V) represents a voltmeter, a high resistance device (ideally ∞) which measures potential differences.

6.A

What is the current through the capacitor? Explain.

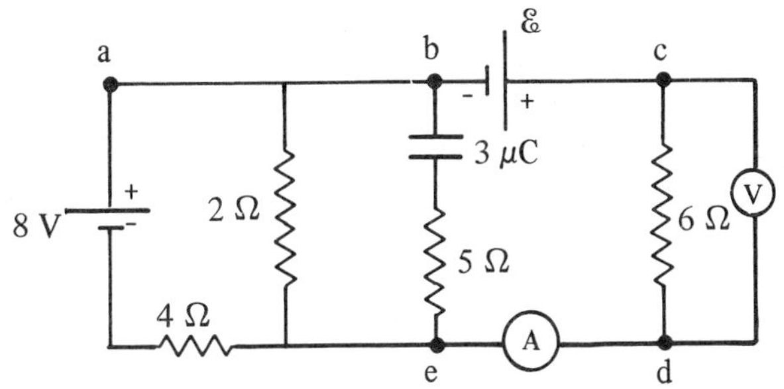

Zero. The capacitor represents an *open* circuit for a DC current.

6.B

Is the potential difference across the capacitor zero? Explain.

No. For an arbitrary value of &, there will be a potential difference across the capacitor; hence, it will have a charge.

6.C

Is there a potential difference across the 5Ω resistor? Explain.

No. Since there is *zero* current in this branch, then $V = IR = 0$ across the resistor.

6.D

We can now eliminate the branch containing the capacitor and solve for the unknown emf and currents. Assume the voltmeter reads 18 V. Redraw the circuit and label the unknowns.

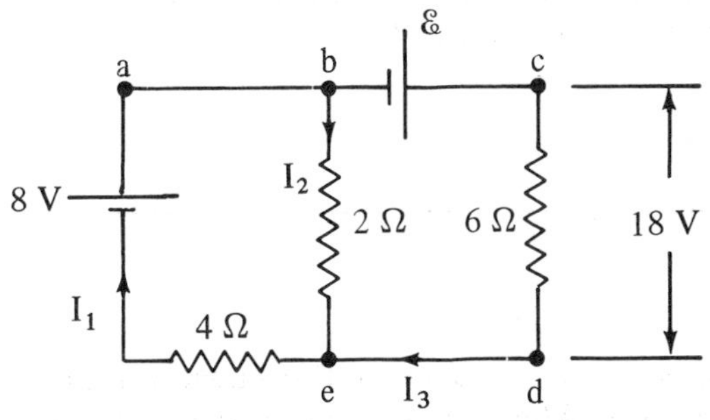

6.E

We have simplified the circuit to two loops. There are four unknowns, I_1, I_2, I_3 and &. Apply Kirchhoff's first rule to junction e. Assume the directions of currents shown in frame 6.D.

$$I_2 + I_3 - I_1 = 0$$

or,

$$I_1 = I_2 + I_3 \qquad (1)$$

6.F

Now apply Kirchhoff's second rule to the loops abea and bcdeb. Traverse the loops in the clockwise direction.

abea: $-2I_2 - 4I_1 + 8 = 0 \qquad (2)$

bcdeb: $\mathcal{E} - 6I_3 + 2I_2 = 0 \qquad (3)$

6.G

Note that the potential difference across the 6Ω resistor is 18 V. From this information find I_3.

$$6I_3 = 18$$

$$I_3 = 3\text{ A} \qquad (4)$$

6.H

Substitute (1) into (2), with $I_3 = 3$ A, and solve for I_1 and I_2.

(1) gives $\quad I_1 = I_2 + 3$

(2) becomes $-2I_2 - 4(I_2 + 3) + 8 = 0$

$$I_2 = -\frac{2}{3}\text{A} \qquad (5)$$

and

$$I_1 = \frac{7}{3}\text{A} \qquad (6)$$

Since I_2 is negative, we picked the *wrong* direction for it in frame 6.D.

6.I

Calculate the unknown emf, $\mathcal{E}$, using (3) and the values of I_2 and I_3.

$$\mathcal{E} = 6I_3 - 2I_2$$

$$\mathcal{E} = 6(3) - 2\left(-\frac{2}{3}\right) = 19.3\text{ V} \qquad (7)$$

6.J

Now let us return to the capacitor in the circuit. What is the potential difference across the capacitor?

By inspection of the circuit in frame 6.A, we see that the potential difference across the capacitor is *equal* to the potential difference across the 2Ω resistor. This potential difference is $I_2 R$, or

$$V_{be} = \frac{2}{3} \times 2 = \frac{4}{3}\text{V}$$

6.K

Find the charge on the 3μF capacitor. Recall that $q = CV$.

$$q = CV_{be} = 3\mu\text{F} \times \frac{4}{3}\text{V}$$

$$q = 4\mu\text{C}$$

6.L

What is the energy stored in the capacitor? Recall that

$$U = \frac{1}{2}qV = \frac{1}{2}CV^2$$

$$U = \frac{1}{2}qV_{be} = \frac{1}{2} \times 4\mu C \left(\frac{4}{3}V\right)$$

or,

$$U = \frac{8}{3}\mu J$$

†7 *RC Circuits.* In this exercise, the energy balance in an RC circuit is demonstrated.

7.A

Before the switch is thrown, we will assume C is *uncharged.* What is the energy stored in the capacitor after S is closed and its charge is q?

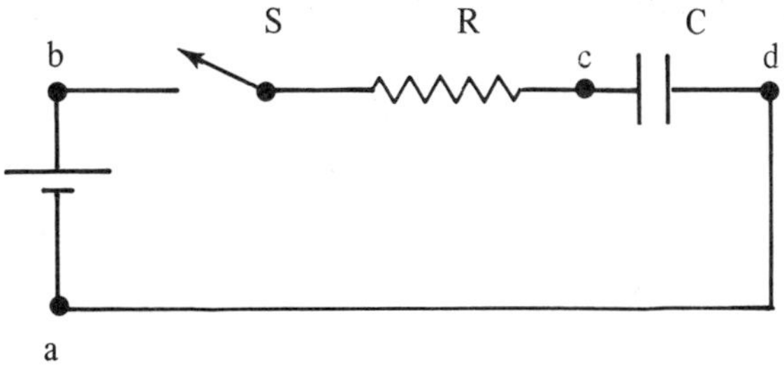

$$U = \frac{q^2}{2C} \qquad (1)$$

7.B

Where does this charge come from?

The *battery* displaces charges from one plate of the capacitor to the other, so that one side of C has a charge +q, the other −q.

7.C

What is the *change* in energy stored in the capacitor if a small charge dq is added to its plates?

From (1), we have

$$dU = \frac{q}{C}\,dq \qquad (2)$$

7.D

How much work does the battery do in moving the charge dq from a to b?

$$dW = \mathcal{E}\,dq \qquad (3)$$

7.E

As the charge dq passes through the resistor from b to c, power is dissipated in the form of heat. What is this power in terms of I and R?

$$P = I^2R \qquad (4)$$

7.F

Since power is defined as $\frac{dW'}{dt}$, how much energy is lost in the form of heat in the time dt?

$$\frac{dW'}{dt} = I^2 R$$

$$dW' = I^2 R dt \qquad (5)$$

7.G

Write an expression for energy conservation, showing how all the energy supplied by the battery is accounted for. Your result should reduce to Equation (4.22), corresponding to Kirchhoff's second rule for the loop.

The energy supplied by the battery is converted into energy stored in the capacitor plus energy lost in the resistor. So from (2), (3) and (5) we get

$$\mathcal{E} dq = \frac{q}{C} dq + I^2 R dt \qquad (6)$$

Dividing by dq gives

$$\mathcal{E} = \frac{q}{C} + IR \qquad (7)$$

†8 In the RC circuit shown below, the capacitor is initially uncharged when S is opened.

8.A

Write an expression for Kirchhoff's second rule applied to the loop when S is closed.

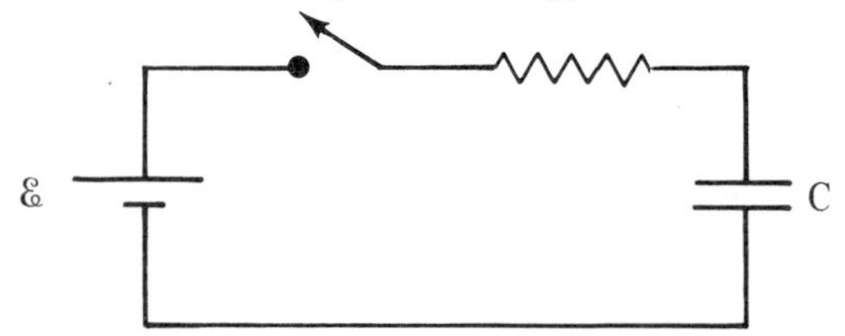

$$\mathcal{E} - IR - \frac{q}{C} = 0 \qquad (1)$$

This is equivalent to (7) of Exercise 7.

8.B

Take the time derivative of (1) and obtain an expression for the change in current dI in the time interval dt.

Since $\mathcal{E}$ = constant, differentiating (1) with respect to time gives

$$R\frac{dI}{dt} + \frac{1}{C}\frac{dq}{dt} = 0 \qquad (2)$$

8.C

Recall that $I = \frac{dq}{dt}$, and rewrite (2) in a separable form.

$$R\frac{dI}{dt} = -\frac{1}{C}I$$

or,

$$\frac{dI}{I} = -\frac{1}{RC}dt \qquad (3)$$

8.D

Integrate the left and right sides of (3). Let $I = I_0$ at $t = 0$ (when S is closed). Note that

$$\int \frac{dx}{x} = \ln x$$

$$\int_{I_0}^{I} \frac{dI}{I} = -\frac{1}{RC}\int_0^t dt$$

$$\ln\left(\frac{I}{I_0}\right) = -\frac{t}{RC} \qquad (4)$$

8.E

Now use (4) and write an exponential expression for I (t).

$$I(t) = I_0 e^{-t/RC} \qquad (5)$$

This is equivalent to (4), which can be seen by taking *ln* of both sides.

8.F

What is the product RC called and what are its units?

RC is the *time constant,* τ, of the circuit and has units of *time.* It is a measure of the time it takes the current to decay to $\frac{1}{e}$ of its initial value, I_0.

8.G

Make a rough sketch of I *vs* t. Note that $I_0 = \frac{\mathcal{E}}{R}$.

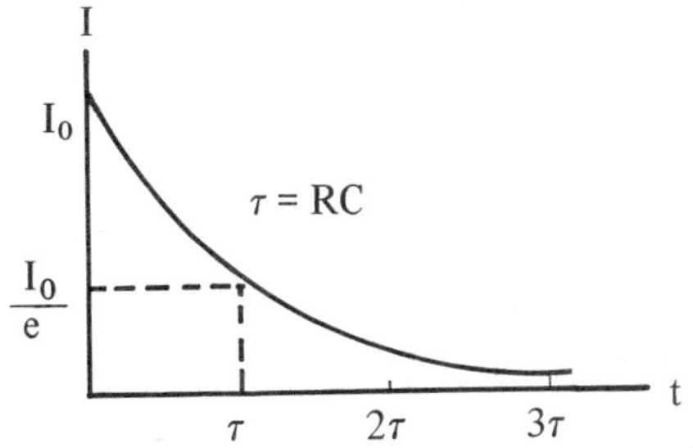

8.H

Now we wish to determine the charge on the capacitor as a function of time. Substitute $I = \frac{dq}{dt}$ into (5) and write a differential expression relating dq to dt.

$$\frac{dq}{dt} = I_0 e^{-t/\tau}$$

$$dq = I_0 e^{-t/\tau} dt \qquad (6)$$

8.I

Since (6) represents an equation with *two variables,* q and t, we can integrate it directly. Perform the integration, letting q = 0 at t = 0. Again, t = 0 represents the instant S is closed. Recall that

$$\int e^{ax}\,dx = \frac{e^{ax}}{a}$$

$$\int_0^q dq = I_0 \int_0^t e^{-t/\tau}\,dt$$

$$q = -I_0\tau\, e^{-t/\tau}\Big]_0^t$$

$$q = I_0\tau\,[1 - e^{-t/\tau}] \qquad (7)$$

8.J

Show that the coefficient $I_0\tau$ in (7) is the *maximum* charge that can be obtained on the capacitor, $q_0 = C\mathcal{E}$.

$$I_0\tau = I_0 RC = \frac{\mathcal{E}}{R} RC$$

or,

$$I_0\tau = \mathcal{E}C = q_0$$

$$\therefore \quad q = q_0\,[1 - e^{-t/\tau}] \qquad (8)$$

8.K

Make a rough sketch of q *vs* t. Note that the limiting value of q at t = ∞ is q_0, the maximum charge on the capacitor.

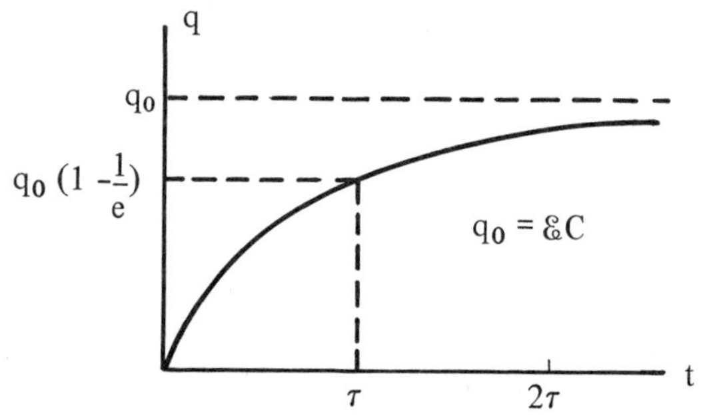

8.L

Suppose C = 5μF, R = 4 × 10⁴Ω and $\mathcal{E}$ = 20 V. Find the time constant of the circuit.

$$\tau = RC = 4 \times 10^4\,\Omega \times 5 \times 10^{-6}\,\text{F}$$

$$\tau = 0.2\ \text{sec} \qquad (9)$$

8.M

Find the *maximum* charge on the capacitor and the *maximum* current in the circuit. Use the numerical values in frame 8.L.

$$q_0 = C\mathcal{E} = 5 \times 10^{-6}\,\text{F} \times 20\ \text{V}$$

$$q_0 = 10^{-4}\ \text{C} \qquad (10)$$

$$I_0 = \frac{\mathcal{E}}{R} = \frac{20\ \text{V}}{4 \times 10^4\,\Omega}$$

$$I_0 = 5 \times 10^{-4}\ \text{A} \qquad (11)$$

8.N

Now write expressions for I(t) and q(t) using (5), (8) and the numerical values for $\mathcal{E}$, q_0 and I_0. Note that these expressions predict the current and charge for any value of $t \geqslant 0$.

$$I(t) = 5 \times 10^{-4}\ e^{-5t}\ \text{A} \qquad (12)$$

$$q(t) = 10^{-4}\ [1 - e^{-5t}]\ \text{C} \qquad (13)$$

8.O

Determine the energy stored in the capacitor as a function of time.

In general, $U = \frac{q^2}{2C}$. (14)

Since C = 5μF and q(t) is given by (13), we can write (14) as

$$U(t) = 10^{-3}\ [1 - e^{-5t}]^2\ \text{J} \qquad (15)$$

4.9 SUMMARY

Current is the rate of flow of charge in a conductor and is given by

$$I = \frac{dq}{dt} \tag{4.2}$$

Current density in a conductor of uniform cross-sectional area A carrying a current I has a magnitude given by

$$J = \frac{I}{A} \tag{4.6}$$

The current density in a conductor of electrical conductivity, σ, is given by Ohm's law:

$$J = \sigma E \tag{4.7}$$

The potential difference across a conductor of resistance R carrying a current I is

$$V = IR \tag{4.11}$$

Two or more resistors R_1, R_2, R_3 . . . connected in *series* have an equivalent resistance given by

$$R = R_1 + R_2 + R_3 + \ldots \tag{4.16}$$

Two or more resistors connected in *parallel* have an equivalent resistance R which can be obtained from the expression

$$\frac{1}{R} = \frac{1}{R_1} + \frac{1}{R_2} + \frac{1}{R_3} + \ldots \tag{4.18}$$

Kirchhoff's rules when used to solve DC circuits are as follows: (1) the sum of the currents into any junction of the circuit is zero; (2) the sum of the potential differences across each element around a closed loop is zero.

4.10 PROBLEMS

1. The current of an electron beam in a cathode ray tube is measured to be $70\mu A$. How many electrons hit the screen in 5 sec?

2. A gold ribbon has a length of 21 m, a width of 2 cm and a thickness of 0.1 mm. (a) Find the resistance of the ribbon from one end to the other. (b) What potential difference between the ends would produce a current of 30 A? (c) What is the current density in the wire if the current is 30 A?

3. A silver wire 50 m in length has a circular cross-section and radius 0.3 mm. The wire carries a current of 15 A. (a) Calculate the resistance of the wire. (b) What is the electric field intensity in the wire? (c) Determine the potential difference between the ends of the wire.

4. A tungsten filament in a certain incandescent lamp has a resistance of 10Ω when it is at room temperature (20°C). If the temperature coefficient of resistivity for tungsten is taken to be 4.5×10^{-3} $(°C)^{-1}$, calculate (a) the resistance of the filament if its temperature is raised to 520°C, (b) the reduction in current through the filament for a constant potential difference of 110 V and (c) the power dissipated by the filament at 520°C. Assume the resistance change is due only to a change in resistivity, according to Equation (4.14).

5. (a) Find the equivalent resistance of the circuit shown in Figure 4-16. (b) If the potential difference between b and a is 25 V, find the currents in each resistor. (c) What is the total power dissipated by the circuit?

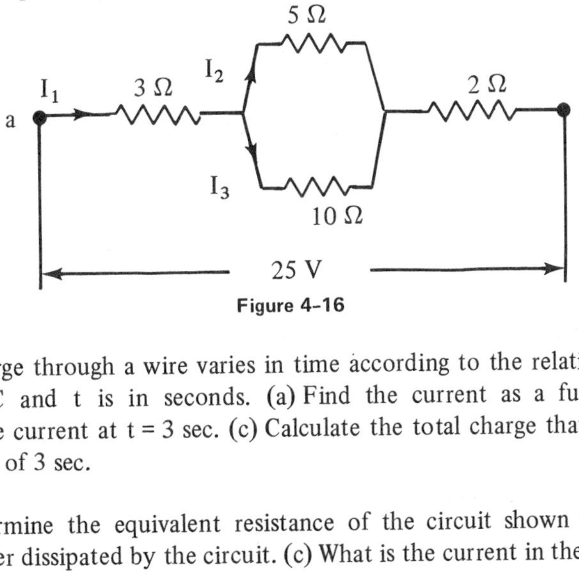

Figure 4-16

6. The charge through a wire varies in time according to the relation $q = 3t + 5t^2$, where q is in C and t is in seconds. (a) Find the current as a function of time. (b) Determine the current at $t = 3$ sec. (c) Calculate the total charge that passes through the wire in a time of 3 sec.

7. (a) Determine the equivalent resistance of the circuit shown in Figure 4-17. (b) Find the power dissipated by the circuit. (c) What is the current in the 5Ω resistor?

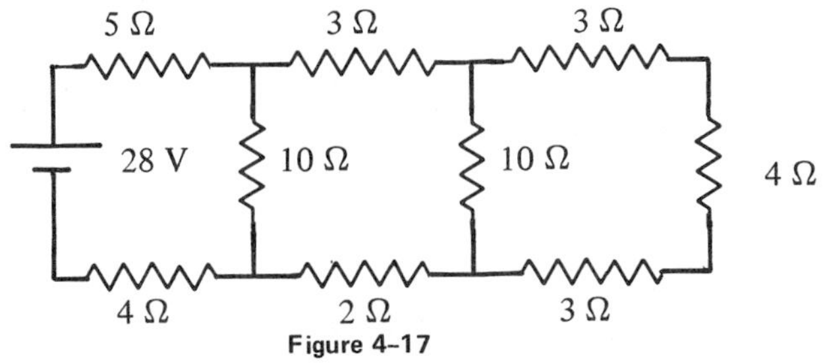

Figure 4-17

8. A 60 W and 100 W lamp are connected in parallel to a 110 V source of emf. Find (a) the current in each lamp and (b) the resistance of each lamp.

9. An electric heater operating at full power draws a current of 15 A when connected to a 220 V circuit. (a) What is the resistance of the heater? (b) How much current should it draw if it is to dissipate only 1200 W? Assume R is constant.

10. *Terminal Voltage of a Battery.* A seat of emf is represented by the symbols within the dashed rectangle of Figure 4-18, where r is the internal resistance of the battery and $\mathcal{E}$ represents the ideal emf, or *open circuit* voltage. (a) If a "load" resistor R is connected across the battery, show that the *terminal voltage* is given by

$$V_t = V_b - V_a = \mathcal{E}\left(\frac{r}{r+R}\right)$$

This shows that the terminal voltage of a seat of emf *varies* with the load resistance. (b) Suppose the open circuit voltage of a battery is 22 V, but a load resistance of 5Ω reduces the terminal voltage to 12 V. Find the internal resistance of the battery.

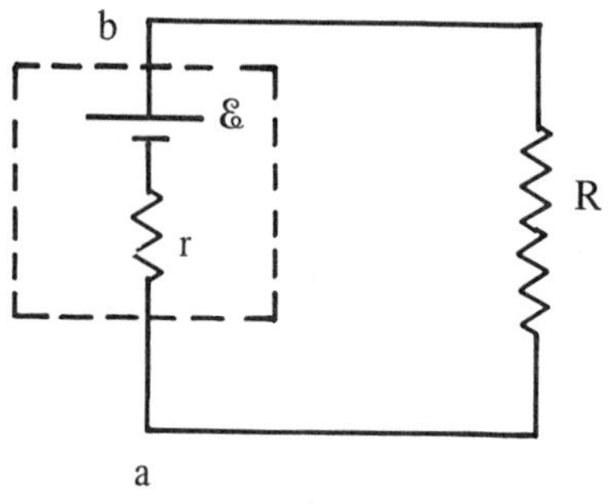

Figure 4-18

11. Find the unknown currents for the circuit shown in Figure 4-19. Assume the directions of currents shown.

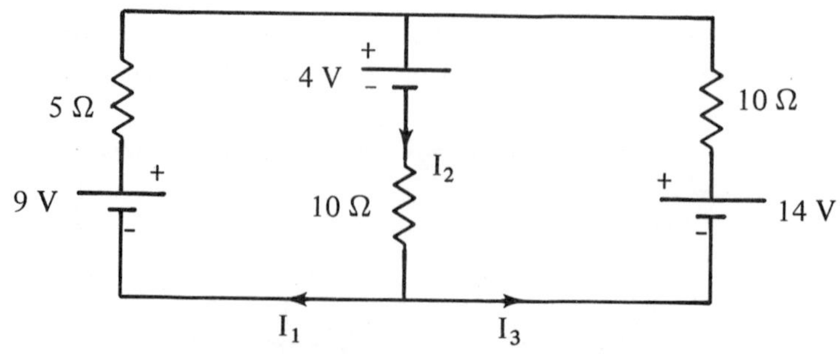

Figure 4-19

12. The ammeter (a zero resistance device) in the circuit shown in Figure 4–20 reads 1 A. Find the unknown currents I_1 and I_2 and the unknown emf.

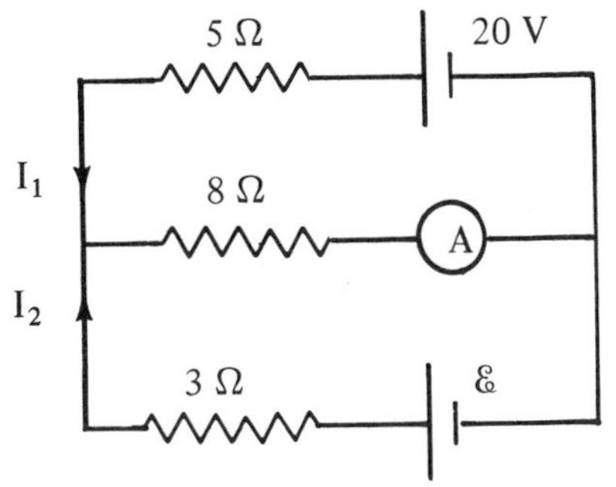

Figure 4–20

13. In the circuit shown in Figure 4–21, find (a) the unknown currents, (b) the voltage drop across each capacitor, (c) the charge on each capacitor and (d) the energy stored in each capacitor. (Assume the circuit has reached steady-state, so the capacitors are fully charged.)

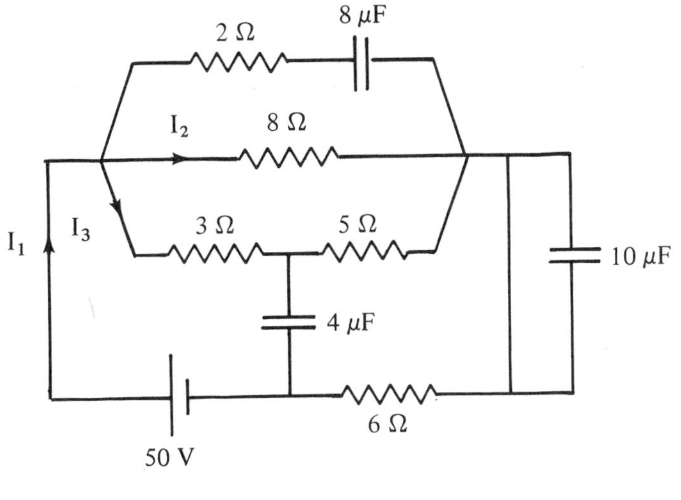

Figure 4–21

14. In a "pinch," an experimenter wishes to use an ammeter as a voltmeter. Suppose the ammeter normally reads 0 to 10 mA on a certain scale, and a full scale of 500 V is required. (a) Determine the value of a series resistor that must be placed in the circuit to give the 500 V on full scale. (b) What is the maximum power which can be dissipated by this resistor? *Hint:* An ammeter can be considered to be a zero resistance device.

15. A galvanometer can be used as an ammeter by using a "shunt" resistor R_S as shown in Figure 4–22. Suppose the galvanometer has a resistance $R_g = 200\Omega$, and a $50\mu A$ current produces a full scale deflection when there is no shunt resistance. Determine the value of R_S if the galvanometer is to read 10 mA at full scale. *Hint:* The potential differences across R_S and R_g must be equal.

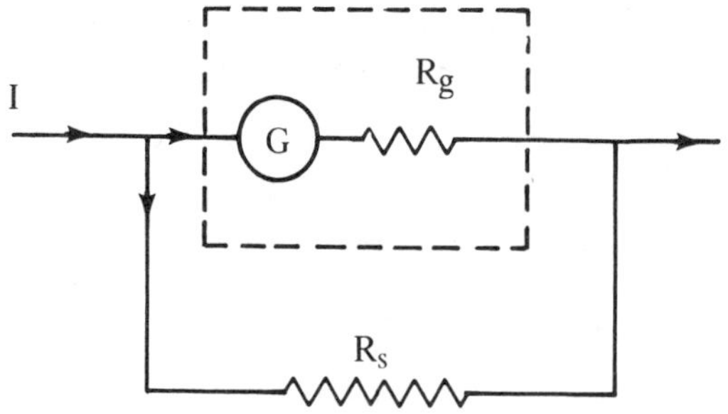

Figure 4-22

16. *Voltage Divider.* The circuit shown in Figure 4-23 is known as a voltage divider. With this circuit, a voltage V_1 is applied across two resistors.

A voltage V_2 is obtained across one of the resistors, where $V_2 < V_1$. Show that the voltage across R_2 is given by

$$V_2 = V_1 \frac{R_2}{R_1 + R_2}$$

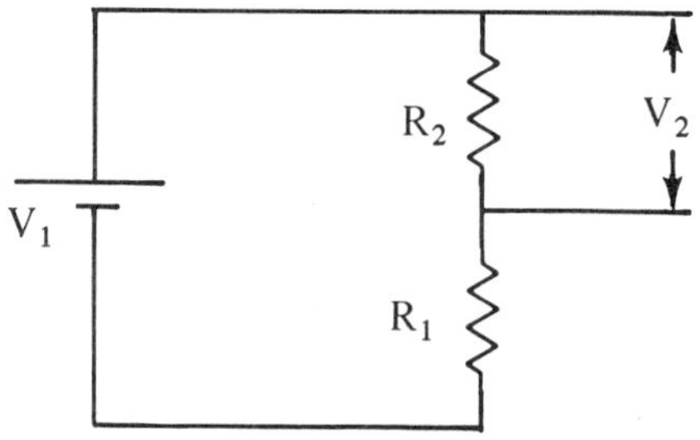

Figure 4-23 A voltage divider.

17. *The Wheatstone Bridge.* The circuit shown in Figure 4-24 is known as a Wheatstone Bridge. It is a common circuit used for measuring unknown resistances. R_1 represents a variable resistance, R_2 and R_3 are fixed, and R_x is the unknown resistance. The symbol (G) represents a galvanometer, which is a sensitive current-measuring instrument. The variable resistor is adjusted until *no* current goes through the galvanometer. At that point, $V_a = V_b$, that is, the potential drop across R_x equals the potential drop across R_1. Use this condition, and show that

$$R_x = \frac{R_1 R_2}{R_3}$$

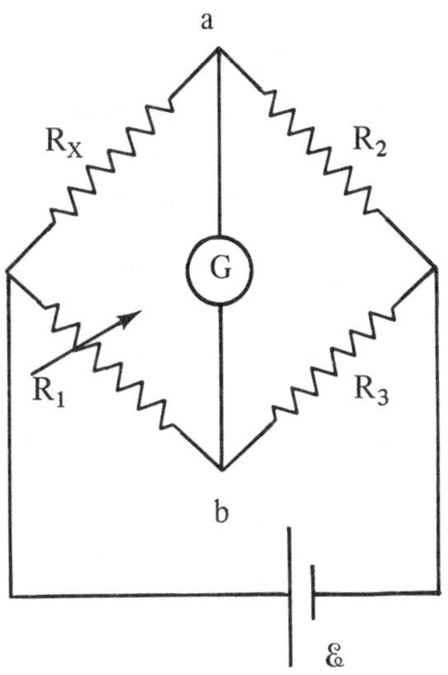

Figure 4-24 The Wheatstone Bridge Circuit

18. *The Potentiometer.* The potentiometer is an electrical circuit used to measure an unknown emf, $\mathcal{E}_X$. The basic circuit shown in Figure 4-25 consists of a battery, $\mathcal{E}$ (where $\mathcal{E} > \mathcal{E}_X$ or $\mathcal{E}_0$), a standard cell, $\mathcal{E}_0$, a galvanometer and a variable resistance from a to b. The switch S is placed in position 1, putting $\mathcal{E}_0$ in the circuit. Then the sliding contact at b is varied until the galvanometer reads zero (corresponding to $I = 0$). For this situation, call the resistance from a to b, R_0. Now the process is repeated with the switch in position 2, so that $\mathcal{E}_X$ is in the circuit. Call the "new" resistance from a to b, R_X, in the "balanced" condition. Comparing these two steps, show that the unknown emf is given by

$$\mathcal{E}_X = \mathcal{E}_0 \frac{R_X}{R_0}$$

Note that $\frac{R_X}{R_0}$ is the ratio of two lengths measured along the slide wire from a to c.

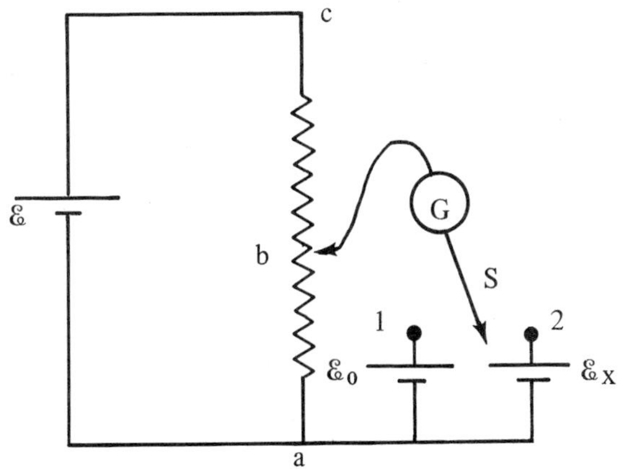

Figure 4-25 A potentiometer circuit.

19. An RC circuit is wired as shown in Figure 4-14, with $R = 8 \times 10^5\,\Omega$, $C = 0.5\mu F$ and $\mathcal{E} = 80$ V. The switch is thrown to position 1. Find (a) the maximum current, (b) the maximum charge on the capacitor, (c) the time constant of the circuit, (d) the current at $t = 2$ RC and (e) the charge at $t = 2$ RC. (f) Find the energy stored in the capacitor at $t = 2$ RC. (g) Make plots of q *vs* t and I *vs* t for this circuit.

20. For the RC circuit shown in Figure 4-14, the switch is in position 1 long enough to give a maximum charge to the capacitor, at which time $I \approx 0$.

Now suppose the switch is suddenly thrown to position 2. Using the numerical values given in Problem 19, and calling $t = 0$ the time the switch is thrown to 2, find (a) the charge on the capacitor after one time constant has elapsed, that is, at $t =$ RC, (b) the voltage drop across the capacitor at $t =$ RC and (c) the voltage drop across the resistor at $t =$ RC. (d) Do the results to (b) and (c) violate Kirchhoff's second rule? Explain. (e) Make plots of q *vs* t and I *vs* t for the discharging capacitor.

5

MAGNETIC FIELDS

Up until now we have worked with two types of fields, gravitational and electric. A third type of field which arises from the motion of electrical charges is the *magnetic field.* The science of magnetism to the novice perhaps means the "whys and wherefores" of the behavior of an ordinary bar magnet. For example, why does a magnet attract iron nails, or why does it rotate a compass needle? These effects were known for many centuries, and it was not until 1820 that a link was discovered between electrical and magnetic phenomena. In that year, the Danish scientist, Hans Christian Oersted, showed that a compass needle is deflected when placed in the vicinity of a current-carrying conductor. Several years later, Michael Faraday and Joseph Henry showed that an electrical current could be produced in a conductor through the motion of a magnet near the conductor. In this chapter, we will be concerned with the fundamental aspects of magnetic fields. We will show that the origin of *all* magnetic fields can be attributed to some form of motion of electrical charges.

5.1 DEFINITION AND PROPERTIES OF THE MAGNETIC FIELD

Magnetic forces arise due to the relative motion of charged particles. We describe the effect by saying that a charge in motion produces both an electric and a magnetic field at the position of the second charge. Therefore, two types of forces act on the second charge, one due to the magnetic field, the other due to the electric field. We will concern ourselves only with the effects of the magnetic field set up by the moving charge. The important point to remember is the association of a magnetic field with a *moving* charged particle. A stationary charge produces *no* magnetic field relative to a stationary observer. In addition, there is *no* magnetic force on a stationary charge situated in a magnetic field. *A magnetic field exists at a given point only if a magnetic force acts on a moving charged particle at that point.*

Suppose that a particle whose charge is q moves with a velocity **v** in a region where the *magnetic field* is **B** (sometimes referred to as the magnetic induction). Experiments on the deflection of a beam of charged particles moving in a magnetic field give the following results: (1) the magnetic force on the charged particle is *always perpendicular* to the *plane formed by the vectors* **v** *and* **B**; (2) the magnetic force is proportional to **v**, **B** and q; (3) the magnetic force on a positive charge is *opposite* in direction to the force on a negative charge moving in the same direction;

and (4) the magnetic force is *zero* when **v** is *parallel* to **B**. These observations can be interpreted with the expression

$$\mathbf{F} = q\mathbf{v} \times \mathbf{B} \tag{5.1}$$

where the direction of the magnetic force **F** is determined by the right-hand rule. If q is *positive*, the direction of **F** is in the direction of **v** × **B**, as shown in Figure 5–1. If q is *negative*, the direction of **F** is in the direction of −**v** × **B**. If θ is the angle between **v** and **B**, the magnitude of **F** can be written as

$$F = qvB\sin\theta \tag{5.2}$$

From this expression, note that F is *zero* when v is parallel or antiparallel to B ($\theta = 0$ or π), while F = qvB, which is a maximum value, when **v** is ⊥ to **B** $\left(\theta = \frac{\pi}{2}\right)$.

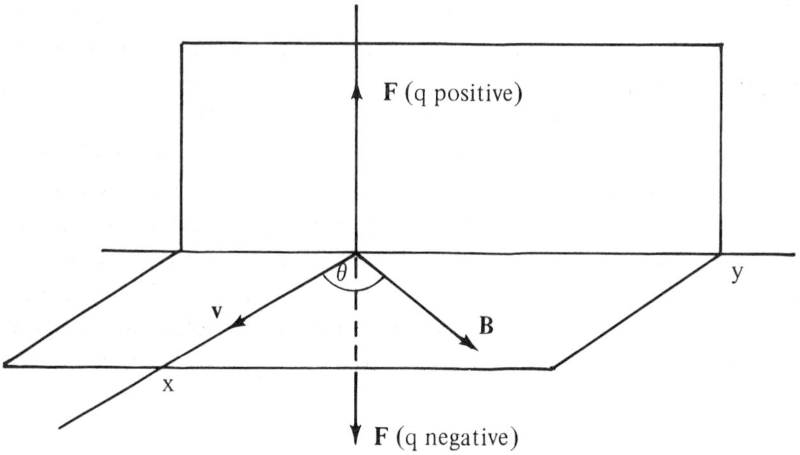

Figure 5–1 A vector diagram showing the relative directions of **v**, **B** and **F** for a charged particle moving in a magnetic field. The direction of **F** is determined from the law **F** = q**v** × **B**. If q is positive, **F** points in the +z direction of this figure; if q is negative, **F** is in the −z direction.

We can regard Equation (5.2) as the definition of B at a given point. That is, the magnetic field is defined by the force on a moving charged particle. This can be compared with the electric field, defined by the force per unit charge. There are several important differences, however, between the two forces. First, the direction of the electric force is parallel to the **E** field, whereas the direction of the magnetic force is perpendicular to the **B** field. Second, the magnetic force is proportional to **v**, whereas the electric force is independent of **v**. Finally, the units of **E** and **B** differ.

The SI unit of B is the weber per square meter $\frac{Wb}{m^2}$, which is given the name tesla, T. This unit can be reduced to more fundamental units by inspecting Equation (5.1).

$$[B] = T = \frac{Wb}{m^2} = \frac{N}{C\,\frac{m}{sec}} = \frac{N}{A\,m} \tag{5.3}$$

where $1\ A = 1\ \frac{C}{sec}$. The cgs unit of magnetic field is the gauss (G), which is related to the tesla by the conversion

$$1\ T = 10^4\ G \tag{5.4}$$

Laboratory magnets can be constructed to produce magnetic fields ranging from about 10^{-2} T to 1 T, but more recently constructed superconducting magnets can generate fields as high as about 25 T (or 250,000 G). This can be compared with the earth's magnetic field near its surface, which is about 5×10^{-5} T or 0.5 G.

Example 5.1

A proton moves along the +x axis with a speed of $5 \times 10^6\ \frac{m}{sec}$. A magnetic field whose magnitude is $2\ \frac{Wb}{m^2}$ lies in the xy plane, and makes an angle of 45° with the +x axis, as in Figure 5-1. Find the magnetic force on the proton.

Solution

$$F = qvB\sin\theta$$

$$F = (1.6 \times 10^{-19}\ C) \times \left(5 \times 10^6\ \frac{m}{sec}\right) \times 2\ \frac{Wb}{m^2} \times \frac{\sqrt{2}}{2}$$

$$F = 1.13 \times 10^{-12}\ N$$

The direction of F is in the +z direction. The student should verify that the unit of F in the expression above reduces to newtons.

The *flux* Φ_B associated with a magnetic field is defined in a manner identical to the electric flux:

$$\Phi_B = \int \mathbf{B} \cdot d\mathbf{A} \tag{5.5}$$

where the integral is over a *surface,* and Φ_B has units of webers. If the integral is taken over a *closed* surface, the result gives *zero,* since *magnetic monopoles do not exist* as far as we know today. Therefore, the magnetic equivalent of Gauss's law can be written as

$$\oint \mathbf{B} \cdot d\mathbf{A} = 0 \tag{5.6}$$

Since magnetic monopoles do not exist, magnetic field lines do not terminate at any point, but form loops that never end.

Another important property of a static magnetic field stems from the fact that the velocity of the charged particle is always perpendicular to **B**. Consequently, the work done by the magnetic force is *zero* for *any* displacement of the particle. (The mathematical proof of this statement is left as an exercise.) Therefore, the *kinetic energy* of a charged particle *cannot* be altered by a static magnetic field, but the *direction* of **v** can change.

5.2 MOTION OF A CHARGED PARTICLE IN A UNIFORM MAGNETIC FIELD

Consider a positively charged particle moving in a uniform magnetic field with its velocity vector perpendicular to the field, as in Figure 5-2. We see that the magnetic force on the charge is *always* at right angles to **v** and **B**. Therefore, **F** is always changing its direction, and its magnitude is qvB. Again, **F** does no work on the charge, since it is perpendicular to **v**. Therefore, the speed of the particle is constant, and we see that **F** provides the centripetal force. The orbit of the particle is a circle of radius r, since we are assuming that the magnetic force is the only force acting on the particle. From Newton's second law, we have

$$F = qvB = \frac{mv^2}{r} \qquad \text{or} \qquad r = \frac{mv}{qB} \tag{5.7}$$

If q is a *negative* charge, then the "sense" of its orbit would be counterclockwise in Figure 5-2.

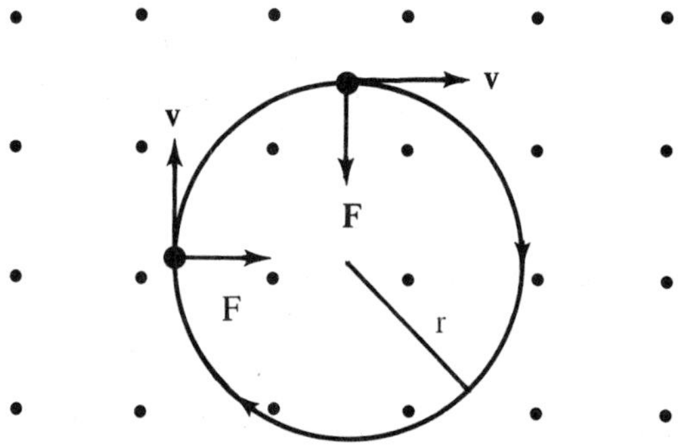

Figure 5-2 Circular orbit of a positively charged particle moving in a uniform magnetic field with $\mathbf{v} \perp \mathbf{B}$. The dots represent the field coming *out* of the paper.

When **v** makes an angle of θ with **B**, the charged particle will move in a *helix*, whose *cross-section* has a radius of $r\sin\theta$. The axis of the helix will coincide with the **B** field.

Example 5.2

A proton moves in a circular orbit of radius 6.0 cm when placed in a uniform magnetic field of $0.50 \frac{Wb}{m^2}$. (a) What is the speed of the proton?

$$v = \frac{rqB}{m} = \frac{6 \times 10^{-2} m \times 1.6 \times 10^{-19} C \times 0.5 \frac{Wb}{m^2}}{1.67 \times 10^{-27} kg} = 2.9 \times 10^6 \frac{m}{sec}$$

(b) What is the angular frequency of the proton (sometimes called the cyclotron frequency)?

$$\omega = \frac{v}{r} = \frac{qB}{m} = \frac{2.9 \times 10^6 \frac{m}{sec}}{6 \times 10^{-2}\ m} = 4.8 \times 10^7 \frac{rad}{sec}$$

(c) Find the period of revolution for the proton.

$$T = \frac{2\pi}{\omega} = \frac{2\pi}{4.8 \times 10^7} = 1.3 \times 10^{-7}\ sec$$

The circular motion of a charged particle in a uniform magnetic field is the basis of such important devices as the *cyclotron* (a particle accelerator) and the *mass spectrograph* (used in the search of isotopes).

5.3 MAGNETIC FORCE ON A CURRENT-CARRYING CONDUCTOR

If a conductor carries a current, I, and is placed in an external magnetic field, a magnetic force will be exerted on the conductor. This follows from the fact that the conductor represents a collection of many charges in motion, with a magnetic force acting on each charge. Consider an element whose charge is dq and whose velocity is **v**, as in Figure 5–3. The magnetic force on this element is $d\mathbf{F} = dq\ \mathbf{v} \times \mathbf{B}$. Since the velocity of the charge dq is given by $\mathbf{v} = \frac{ds}{dt}$, and $I = \frac{dq}{dt}$, the force reduces to

$$d\mathbf{F} = dq\frac{d\mathbf{s}}{dt} \times \mathbf{B} = I\ d\mathbf{s} \times \mathbf{B} \tag{5.8}$$

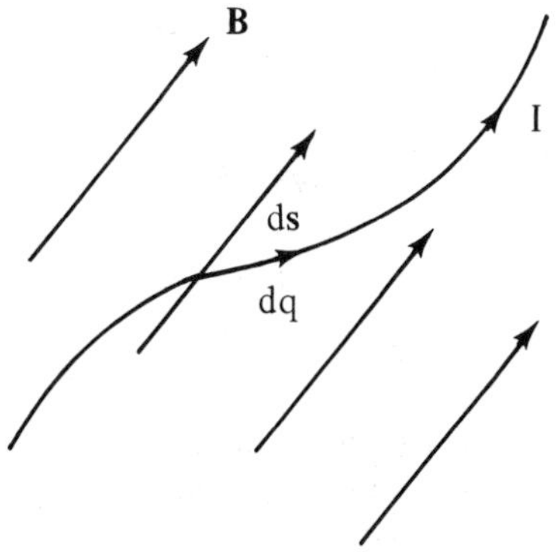

Figure 5–3 A current-carrying conductor in an external magnetic field.

If the magnetic field is *uniform*, Equation (5.8) can be integrated directly to give

$$\mathbf{F} = I\boldsymbol{l} \times \mathbf{B} \tag{5.9}$$

where $\boldsymbol{l}$ is a displacement measured *in the direction* of I.

Example 5.3

A straight wire, 2 m in length, carries a current of 5 A. The wire is placed in a uniform magnetic field whose magnitude is 3×10^{-2}T, which makes an angle of 53° with the wire, as in Figure 5-4. Find the magnetic force on the wire.

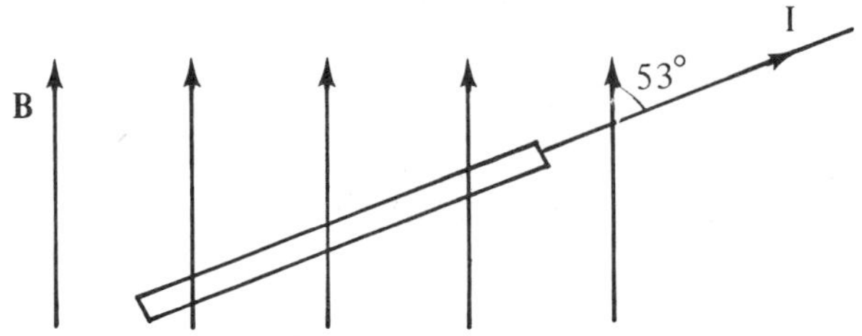

Figure 5-4

Solution

$$F = IlB\sin\theta = 5 \times 2 \times 3 \times 10^{-2} \sin(53)$$

$$F = 0.24 \text{ N}$$

From the right-hand rule, we see that **F** is *out* of the paper.

Example 5.4

A wire bent into the shape of a semicircle of radius R forms a closed circuit and carries a current I. A uniform magnetic field is present along the +y direction, as shown in Figure 5-5. Find the forces on the straight part of the wire and on the curved portion.

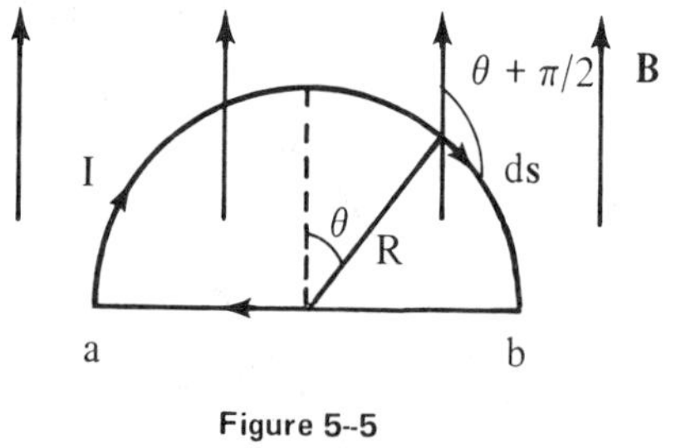

Figure 5-5

Solution

The force on the straight portion is simply $F_1 = IlB$, since the wire is ⊥ to **B**. But $l = 2R$; ∴ $F_1 = 2IRB$. The direction of F_1 is *into* the paper, since $l \times \mathbf{B}$ is inwards.

To find the force on the curved part, we first write an expression for the force $d\mathbf{F}_2$ on the element whose length is ds. Since the angle between ds and **B** is $\theta + \frac{\pi}{2}$ we have

$$d\mathbf{F}_2 = I\, ds \times \mathbf{B}$$

or,

$$dF_2 = IB\sin\left(\theta + \frac{\pi}{2}\right) ds = IB\cos\theta\, ds$$

To get the total force on the curved portion, we can integrate this expression, noting that the direction of F_2 is *out* of the paper. That is, each element of the arc will have a force on it coming out of the paper, so the total force on the arc is a superposition of these forces. Since $ds = Rd\theta$, we have

$$F_2 = IRB \int_{-\pi/2}^{\pi/2} \cos\theta \, d\theta = IRB [\sin\theta]_{-\pi/2}^{\pi/2} = 2IRB$$

If we call $\mathbf{k}$ a unit vector out of the paper, then $\mathbf{F}_2 = 2IRB\,\mathbf{k}$ and $\mathbf{F}_1 = -2IRB\,\mathbf{k}$. Hence, the *total* force on the closed circuit is *zero.* In fact, the *net force on any closed loop* in a magnetic field is *zero.*

5.4 THE LORENTZ FORCE

When a charge q moves through a region where both **E** and **B** fields are present, the force on the charge is given by the *Lorentz force,*

$$\mathbf{F} = q\mathbf{E} + q\mathbf{v} \times \mathbf{B} \tag{5.10}$$

It is possible to obtain a Lorentz force of *zero* by adjusting **E** and **B** such that the relation $\mathbf{E} = -\mathbf{v} \times \mathbf{B}$ is satisfied. That is, **E** must be perpendicular to the plane formed by **v** and **B**, and opposite in direction to the magnetic force.

5.5 AMPERE'S LAW

The experiment performed by Oersted in 1820 clearly demonstrates that a *current-carrying conductor* is a *source of a magnetic field.* If a compass is placed in a horizontal plane near a long vertical wire carrying a large current (~10 A), the compass needle will deflect, as in Figure 5–6. By convention, the North pole of the compass needle points in the direction of **B**. Therefore, from the experiment, we conclude that the **B** field lines are concentric circles whose sense is determined by the right-hand rule. That is, we grasp the wire with our right hand with the thumb pointing in the direction of I. The remaining four fingers are wrapped around the wire in the direction of **B**.

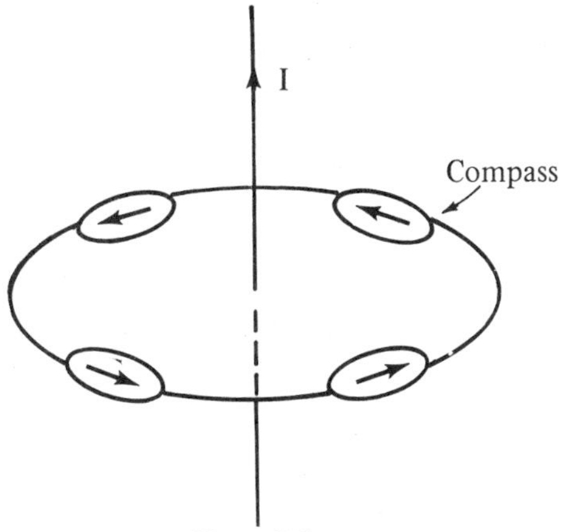

Figure 5–6

Ampere's law, which relates the magnetic field to the current producing it, is given by

$$\oint \mathbf{B} \cdot d\mathbf{s} = \mu_0 I \tag{5.11}$$

This represents a *line integral* around a *closed path,* where I is the *net* current *linked* by the path. The constant μ_0 is called the *permeability of free space,** and has the value

$$\mu_0 = 4\pi \times 10^{-7} \frac{\text{Wb}}{\text{Am}} \tag{5.12}$$

The meaning of Ampere's law may be readily understood with a few examples.

Example 5.5

(a) We wish to calculate the magnetic field at a distance r from a long straight wire carrying a current I, as in Figure 5-7.

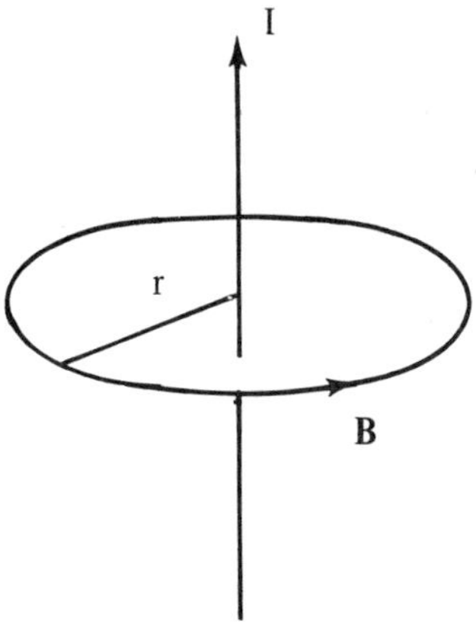

Figure 5-7.

The simplest path to choose is a circle of radius r, centered at the wire. Along this path, B is constant in magnitude and parallel everywhere to ds. Therefore, Ampere's law gives

$$\oint \mathbf{B} \cdot d\mathbf{s} = B\oint ds = B2\pi r = \mu_0 I$$

or,

$$B = \frac{\mu_0 I}{2\pi r} \qquad (\text{for } r \geqslant R) \tag{5.13}$$

*The constants μ_0 and ϵ_0 are related to the *speed of light* through the relation $c = (\epsilon_0 \mu_0)^{-1/2}$.

Note the B varies as $\frac{1}{r}$ just as E varies as $\frac{1}{r}$ for a long charged filament.

(b) What is the field of a long wire carrying a current of 8 A, at a distance of 4 cm from the wire?

$$B = \frac{\mu_0 I}{2\pi r} = \frac{4\pi \times 10^{-7} \frac{Wb}{Am} \times 8\ A}{2\pi \times 4 \times 10^{-2}\ m} = 4 \times 10^{-5} \frac{Wb}{m^2}$$

Example 5.6

A long straight wire of radius R carries a current I, uniformly distributed across its cross-section, as in Figure 5-8. Find the magnetic field at a point *within* the wire, a distance r from its center.

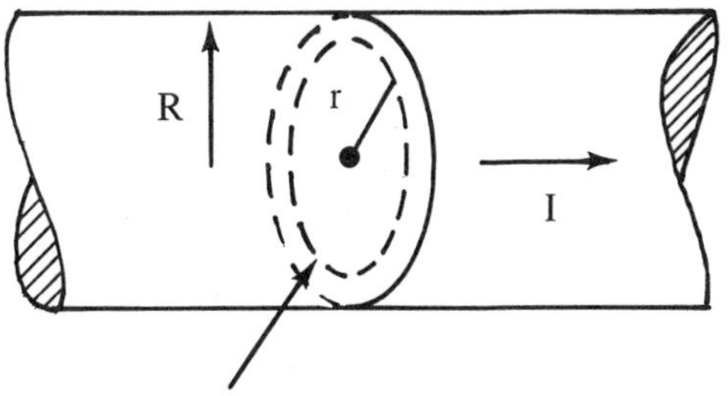

Figure 5-8

The field at an interior point of the wire can be obtained with Ampere's law, using a path of radius $r < R$. However, note that the current linked by this path is *less* than I. To get this current, we write Ampere's law in a more general way as

$$\oint \mathbf{B} \cdot ds = \mu_0 \int \mathbf{J} \cdot d\mathbf{A}$$

where the right side represents the μ_0 times the current *enclosed* by the path. In this form, the right side is a *surface integral.* In our problem, **J** is parallel to d**A**, and the magnitude of the current density $J = \frac{I}{\pi R^2}$. Also, the area in question is the area *enclosed* by the path of the line integral. Since **J** is uniform over the cross-section of the wire, we get

$$B 2\pi r = \mu_0 \frac{I}{\pi R^2} \pi r^2 = \frac{\mu_0 I}{R^2} r^2$$

or,

$$B = \frac{\mu_0 I}{2\pi R^2} r \qquad (\text{for } r \leqslant R) \qquad (5.14)$$

We see that $B \rightarrow 0$ as $r \rightarrow 0$. This result is similar to the case where $E \rightarrow 0$ as $r \rightarrow 0$ for a uniformly charged filament.

Example 5.7

The toroid consists of N turns of wire wrapped around a doughnut shaped air core, as in Figure 5-9. (a) Let us calculate the magnetic field inside the coil, a distance r from the center.

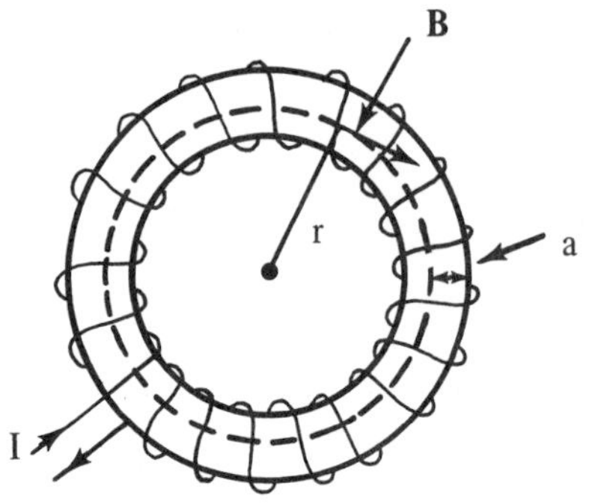

Figure 5-9

We let the path of our line integral be a circle of radius r. Note that the path links N loops of wire over the total path, each of which carries a current I. Therefore, the right side of Ampere's law is $\mu_0 NI$. The path integral reduces to a simple form, since **B** is along the circular path as shown in the figure. (Convince yourself with the right-hand rule that the sense of **B** is as shown in Figure 5-9.) Ampere's law then gives

$$\oint \mathbf{B}\cdot ds = B\oint ds = B2\pi r = \mu_0 NI$$

or,

$$B = \frac{\mu_0 NI}{2\pi r} \tag{5.15}$$

We see that B goes at $\frac{1}{r}$. However, if the inner and outer radii of the toroid are *large* compared to r, then the field will be *approximately* constant in the toroid.

(b) Suppose r is *large* compared to a, the cross-sectional radius of the toroid. In that case, a small section of the toroid approximates a *solenoid,* that is, a long straight coil, whose number of turns per unit length is $n = \frac{N}{2\pi r}$. Hence, the B field of a large radius toroid can be written as

$$B = \mu_0 nI \tag{5.16}$$

(c) Find the magnetic flux through the toroid, assuming $r \gg a$ and the cross-section is circular.

$$\Phi = \int \mathbf{B}\cdot d\mathbf{A} = BA = \mu_0 nI\pi a^2$$

(The field of a solenoid is also given by $B = \mu_0 nI$, and can be obtained by applying Ampere's law. This is a common procedure in most textbooks.)

5.6 BIOT-SAVART LAW

Ampere's law is useful only when the field has a high symmetry and B can be extracted from the line integral. In a more general situation where the current distribution does not have high symmetry, we break the current distribution up into small segments. Then, we write an expression for the field dB due to a given current element, and integrate to get **B** for the entire current-carrying conductor.

Consider a segment ds of a length of conductor carrying a current I, as in Figure 5-10. The field dB due to this segment at the point P according to the *Biot-Savart law* is given by

$$dB = \frac{\mu_0 I}{4\pi} \frac{ds \times \hat{r}}{r^2} \tag{5.17}$$

where $\hat{r}$ is a *unit vector directed* from ds to P, and r is the distance from ds to P. It should be recognized that the Biot-Savart law *is not* a new principle, but simply a useful method for calculating **B** fields. Note that d**B** points *out* of the paper at P, but *into* the paper at P′ (assuming the points P and P′ are in the plane of the wire). If θ is the angle between ds and $\hat{r}$, then the *magnitude* of d**B** is given by

$$dB = \frac{\mu_0 I}{4\pi} \frac{\sin\theta \, ds}{r^2} \tag{5.18}$$

It is important to pick ds *in the direction* of I to get the correct direction for **B**. Also, when the total field is calculated, it must be remembered that d**B** is a *vector* quantity, and the integral must be handled properly. This is a common point of misunderstanding in such calculations.

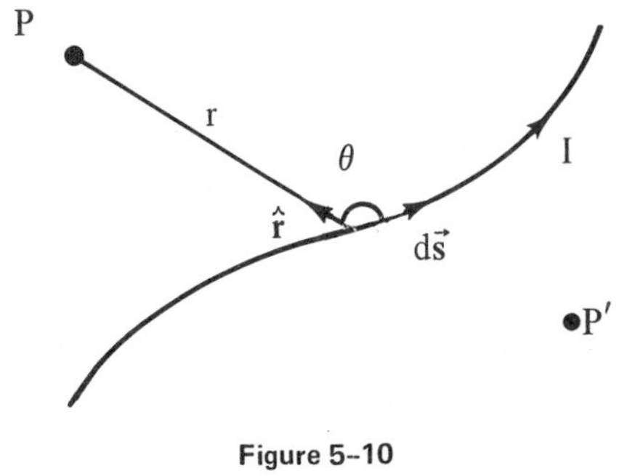

Figure 5-10

Example 5.8

Calculate the magnetic field at the point 0 for the closed circuit shown in Figure 5-11. The curved portion is a circular arc of radius R, and the wire carries a current I.

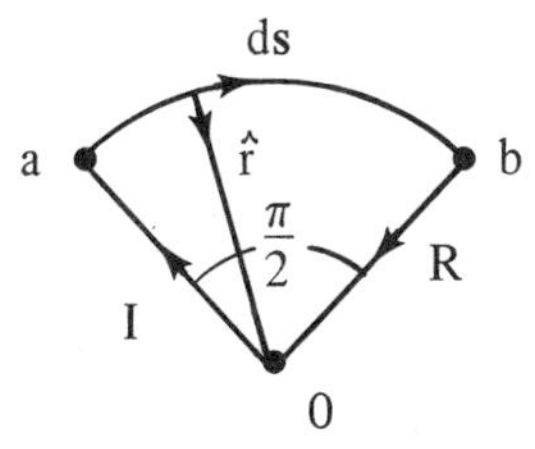

Figure 5-11

First, note that the magnetic field at 0 due to the straight segments 0a and 0b is *identically* zero, since ds is parallel to $\hat{r}$; ∴ ds × $\hat{r}$ = 0. Therefore, we only have to calculate **B** at 0 due to the curved portion ab. Along the segment ab, note that ds is ⊥ $\hat{r}$, and each segment is at a distance R from 0. The field at the point 0 due to the segment ds can be obtained from the Biot-Savart law. This gives

$$dB = \frac{\mu_0 I}{4\pi} \frac{ds}{R^2}$$

But ds = Rdθ, and R and I are constants, so we get

$$B = \frac{\mu_0 I}{4\pi R} \int_{-\pi/4}^{\pi/4} d\theta = \frac{\mu_0 I}{8R}$$

Also, note that **B** at 0 is *into* the paper, since ds × **r** is into the paper for every element along ab. We can also see that **B** is into the paper at 0 by the right-hand method.

If the circuit consisted of a *full* circular loop of radius R, the integral would have limits of 0 to 2π, and the result for B at the center would be $B = \frac{\mu_0 I}{2R}$.

5.7 MAGNETIC DIPOLE MOMENT

In section 5.3 we showed that the resultant magnetic *force* on a closed loop carrying a current is *zero* when placed in a *uniform* magnetic field. However, the resultant *torque* on the loop is generally not zero. Consider the rectangular loop shown in Figure 5–12. In Programmed Exercise 3, we show that the *torque* on such a loop is given by

$$\tau = \mu \times B \tag{5.19}$$

where μ is the *magnetic dipole moment* of the current loop, defined by

$$\mu = IA \tag{5.20}$$

If the loop has N turns, its magnetic dipole moment is NIA. The magnitude of **A** represents the area of the loop, and the direction of **A** is the same as the direction of the magnetic field produced *by* the loop at its center. Consequently, one can use the right-hand rule to get the direction of μ, as in Figure 5–12. Equations (5.20) and (5.21) are valid for a current loop of *any* shape.

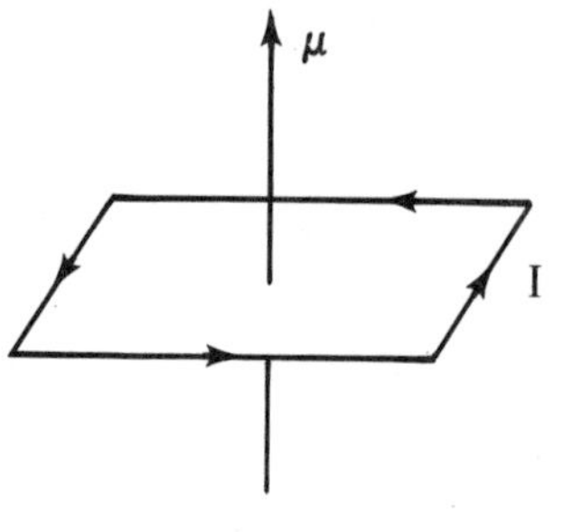

Figure 5–12

The *potential energy* of a magnetic dipole in an external field **B** is given as

$$U = -\boldsymbol{\mu} \cdot \mathbf{B} \tag{5.21}$$

Therefore, U is a maximum (μB) when $\boldsymbol{\mu}$ is aligned *against* the field ($\theta = \pi$), whereas U is a minimum ($-\mu$B) when $\boldsymbol{\mu}$ is aligned with the field ($\theta = 0$). Also, U = 0 when $\theta = \frac{\pi}{2}$.

These expressions have an electrical analog. When an *electric* dipole **p** is placed in an electric field, the torque on the dipole is given by $\boldsymbol{\tau} = \mathbf{p} \times \mathbf{E}$. and its potential energy is given by $U = -\mathbf{p} \cdot \mathbf{E}$.

5.8 MAGNETIC PROPERTIES OF MATTER

There are three types of substances which are important in describing the magnetic properties of materials. These are diamagnetic, paramagnetic and ferromagnetic materials.

Diamagnetic materials are weakly repelled by an ordinary magnet, which is a result of an induced magnetic dipole moment by the external field. Such materials exhibit a very feeble magnetism and, therefore, are not useful for magnetic applications. Water, glass, helium, neon and other rare gases are examples of diamagnetic substances.

Paramagnetic materials are substances made up of atoms having a permanent magnetic dipole moment. The dipole moment can be due to both the orbital and spin motions of the electrons. However, the orientation of the moments is random in the absence of an external field. When a magnetic field is applied, the dipoles tend to align themselves with the field and thereby increase the internal field. The degree of alignment depends on the strength of the external field, as well as on temperature. In fact, the magnetic moment of the sample is generally proportional to B and inversely proportional to T. Paramagnetic materials are weakly attracted to a permanent magnet. Examples of paramagnetic substances are oxygen, transition elements and rare-earth elements.

Ferromagnetic materials are a class of substances in which the magnetic moments of *many* atoms are self-aligned within a small region called a *domain*. The magnetic moments arise *primarily* from the spin motion of the electrons. Each domain behaves as a permanent magnet, but the domains are randomly oriental in the absence of an applied field. The domains become partially aligned when a magnetic field is applied; therefore, the internal field is greatly enhanced. Consequently, ferromagnetic substances are strongly attracted to an external field and are vital to the development of magnetic devices. Examples are iron, nickel, cobalt and a host of alloys.

If we assume that the magnetic field in a certain region is B_0 in the *absence* of a magnetic substance, the magnetic field increases by a factor κ_m (the relative permeability) when a ferromagnetic or paramagnetic material fills the region. We can represent this increase by the expression

$$B = \kappa_m B_0 \,. \tag{5.22}$$

The relative permeability, κ_m, is a dimensionless quantity and ranges from about 1 (for paramagnetic materials) to 10^5 (for supermalloy). For some experiments, it is useful to define the magnetic susceptibility as $\chi = \kappa_m - 1$.

Example 5.9

A toroid has 300 turns of wire and carries a current of 5 A, as in Figure 5-9. The inner and outer radii of the toroid are 10 cm and 15 cm, respectively. (a) What is the magnetic field in the toroid at r = 12 cm if the core is filled with air?

Using Equation (5.15), we have

$$B_0 = \frac{\mu_0 NI}{2\pi r} = \frac{4\pi \times 10^{-7}\,\frac{\text{Wb}}{\text{Am}} \times 300 \times 5\text{A}}{2\pi \times 12 \times 10^{-2}\ \text{m}} = 2.5 \times 10^{-3}\,\frac{\text{Wb}}{\text{m}^2}$$

(b) Suppose the core is filled with annealed iron, which has a relative permeability of 400 at 20°C. What is the new value of B at r = 12 cm?

$$B = \kappa_m B_0 = 400 \times 2.5 \times 10^{-3}\,\frac{\text{Wb}}{\text{m}^2} = 1\,\frac{\text{Wb}}{\text{m}^2}$$

That is, the field within the toroid increases from 25 G to 10,000 G!

5.9 PROGRAMMED EXERCISES

1.A

What is the magnetic force on a charge q moving with a velocity **v** in a magnetic field **B**?

$$\mathbf{F} = q\mathbf{v} \times \mathbf{B} \qquad (1)$$

This expression defines **B**.

1.B

What is the magnitude of **F** if **v** makes an angle θ with **B**?

$$F = qvB\sin\theta \qquad (2)$$

1.C

What are the SI units of B?

By definition, B has units of $\frac{F}{qv}$ or, in SI units,

$$[B] = \frac{N}{C\frac{m}{sec}} = \frac{N}{A\,m} = T = \frac{Wb}{m^2}$$

where T = tesla and Wb = weber.

1.D

Determine the direction of **F** if q is positive in the figure.

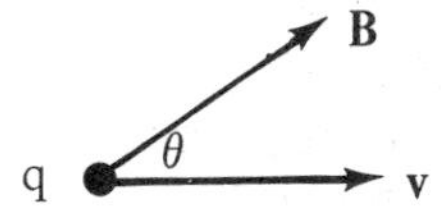

By the right-hand rule, **v** × **B** is out of the paper. Since q is positive, **F** is *out* of the paper. **F** is always ⊥ to the plane formed by **v** and **B**.

1.E

Suppose q is *negative* in the figure. What is the direction of **F**?

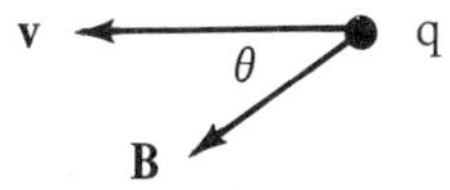

v × **B** is out of the paper, but since q is negative, **F** = q**v** × **B** is *into* the paper.

1.F

What is the magnetic force on q if it is stationary?

Zero. From (1) we see that **F** = 0 when **v** = 0.

1.G

Can the magnetic force on q be zero if $\mathbf{v} \neq 0$? Explain.

Yes. When **v** is *parallel* to **B**, $\mathbf{v} \times \mathbf{B} = 0$, so $\mathbf{F} = 0$.

1.H

Determine the angle between **v** and **B** such that **F** is a maximum.

From (2) we see that **F** is a maximum of $\theta = \frac{\pi}{2}$, that is, when **v** is $\perp$ to **B**. This maximum value is qvB.

1.I

Find the magnetic force on a proton at the instant its velocity is $(5 \times 10^6\ \mathbf{i})\ \frac{\text{m}}{\text{sec}}$, moving in a magnetic field of $2\,\mathbf{j}\ \frac{\text{Wb}}{\text{m}^2}$.

$\mathbf{F} = e\mathbf{v} \times \mathbf{B}$

$\mathbf{F} = 1.6 \times 10^{-19}\ (5 \times 10^6\ \mathbf{i}) \times 2\mathbf{j}\ \text{N}$

$\mathbf{F} = (1.6 \times 10^{-12}\ \mathbf{k})\ \text{N}$

2 The motion of charged particles in a uniform magnetic field when v is perpendicular to B.

2.A

A *positive* charge q is given an initial horizontal velocity **v**. Suddenly, a magnetic field is introduced, directed into the paper as shown below. Describe the motion of the charge when the field is present.

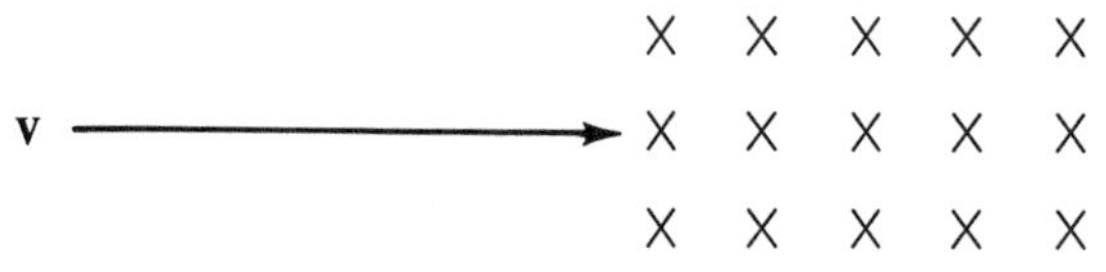

B *is into the paper.*

As soon as the field is turned on, the charge experiences an *upwards* deflection, since $q\mathbf{v} \times \mathbf{B}$ is up. As it proceeds in its motion, the magnetic force will be constant in magnitude (qvB), but its direction will vary in time. If this is the only force on q, the motion will be *circular.*

2.B

Make a sketch of the orbit of the charge while it moves in the magnetic field.

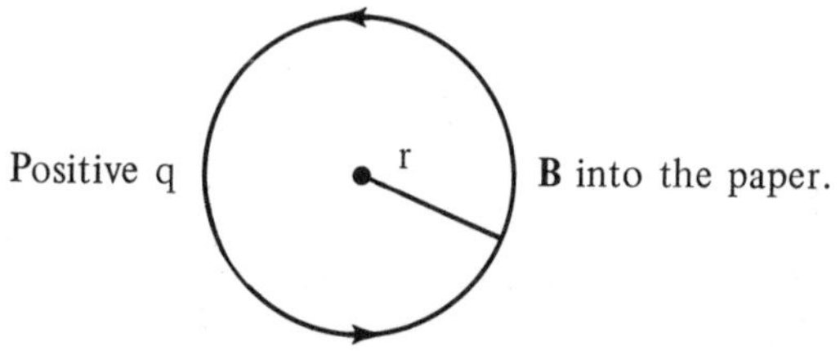

2.C

Sketch the orbit if q is assumed to be *negative.*

Negative q

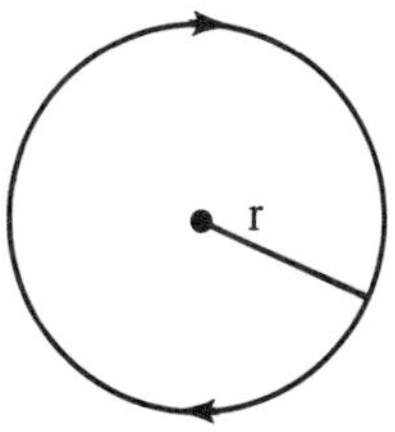

B into the paper.

2.D

Write Newton's second law for the circular motion of the positive charge. Assume its mass and the radius of the orbit is r. Calculate the speed of the particle.

The magnetic force is always *radial,* so we have

$$F_r = qvB = ma_r = \frac{mv^2}{r}$$

$$\therefore \qquad v = \frac{qrB}{m} \qquad (1)$$

2.E

Show that the *period* of the circular motion can be obtained from (1). Your result should be *independent* of r.

The period T is the time for one revolution. Since the particle moves a distance $2\pi r$ in a time T,

$$v = \frac{2\pi r}{T} = \frac{qrB}{m}$$

$$\therefore \qquad T = \frac{2\pi m}{qB} \qquad (2)$$

2.F

Determine the *cyclotron frequency,* ω, for the motion. Use (2) and recall that $\omega = \frac{2\pi}{T}$.

$$\omega = \frac{2\pi}{T} = \frac{qB}{m} \qquad (3)$$

2.G

If the proton is projected into the field with **v** at an angle to **B**, describe the path of its motion.

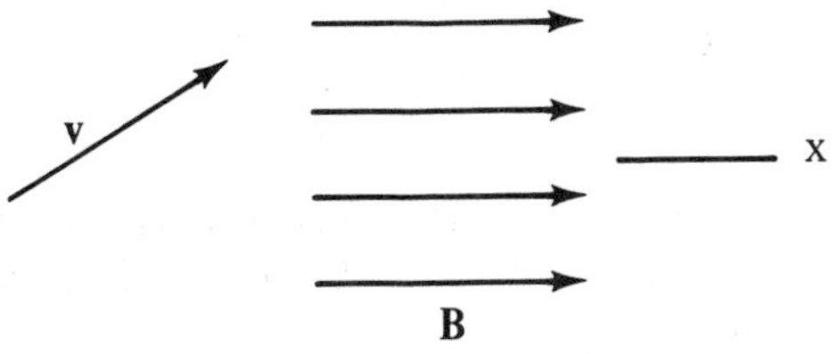

The path will be a *helix.* The component v_x will remain *constant,* since $a_x = 0$, but v_y and v_z *change* in time.

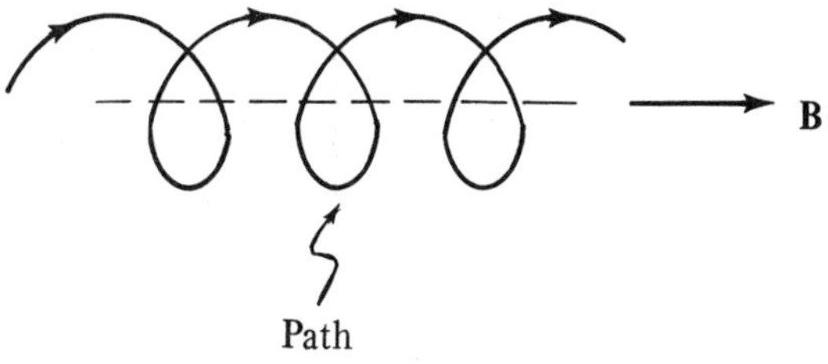

†3 When a current-carrying conductor is placed in an external magnetic field, a force can be exerted on it.

3.A

Suppose the conductor is a straight wire of length l carrying a current I. The wire makes an angle θ with **B**. What is the magnetic force on an element dl?

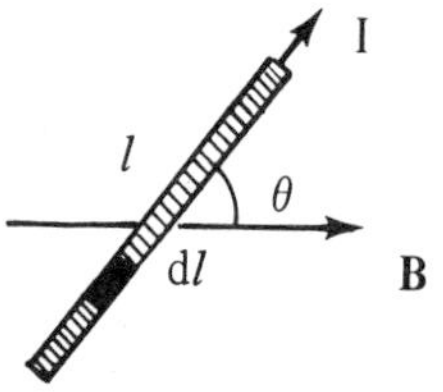

In general,

$$d\mathbf{F} = I\, d\mathbf{l} \times \mathbf{B} \qquad (1)$$

where dl *points along* I. Since dl × **B** is into the paper (−**k**)

$$d\mathbf{F} = -I\, dlB\sin\theta\ \mathbf{k} \qquad (2)$$

3.B

Find the *total* magnetic force on the wire.

Since I, B and θ are constants,

$$\mathbf{F} = \int d\mathbf{F} = -I\, B\sin\theta\ \mathbf{k}\int_0^l dl$$

or

$$\mathbf{F} = -IlB\sin\theta\ \mathbf{k} \qquad (3)$$

3.C

What is the physical basis of the magnetic force on the wire?

Since current in a wire represents charges in motion, the force **F** is the resultant magnetic force on the moving charges.

3.D

Suppose we complicate the situation somewhat and form a *closed* rectangular loop as shown below, with B *out* of the paper, ⊥ to the *plane* of the loop. Draw a diagram showing the forces on the various sides.

B out of the paper

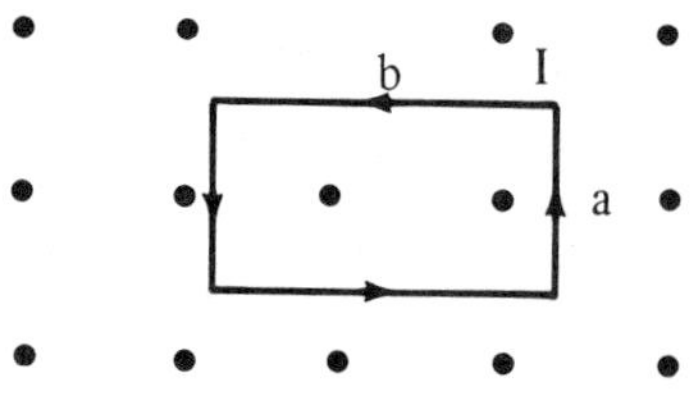

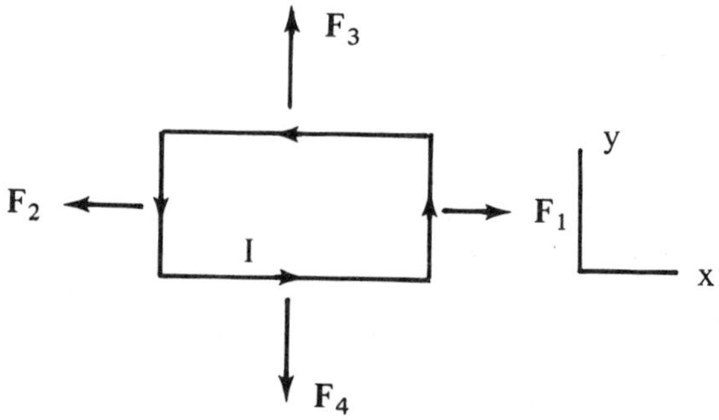

The directions of the forces correspond to the expression $\mathbf{F} = I\mathbf{l} \times \mathbf{B}$, where l is in the direction of I.

3.E

Write *vector* expressions for the forces on the various sides in terms of I, B and the loop dimensions a and b.

$F_1 = Il \times B = IaBi$

$F_2 = -IaBi$

$F_3 = IbBj$

$F_4 = -IbBj$

3.F

What is the *resultant* force on the loop?

$$F = F_1 + F_2 + F_3 + F_4 = 0$$

The resultant magnetic force on *any* closed current-carrying loop is *zero.*

3.G

If the normal to the loop makes an angle θ with B as shown below, the resultant force is *still* zero. What is the *torque* acting on the loop about 0?

Left end view of loop

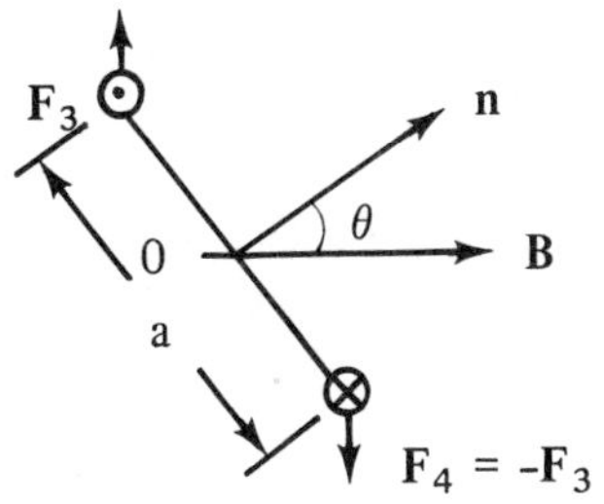

The forces F_1 and F_2 do *not* contribute to the torque. Since $F_3 = F_4 = IbB$, and the moment arm of each of these forces is $\frac{a}{2}\sin\theta$,

$$\tau = F_3 \frac{a}{2}\sin\theta + F_4 \frac{a}{2}\sin\theta$$

$$\tau = Iab\ B\sin\theta \qquad (4)$$

The "sense" of the rotation is *clockwise.*

3.H

If the loop has N turns of wire, write an expression for τ with this addition.

$$\tau = NIabB\sin\theta$$

Since the area of the loop A = ab,

$$\tau = NIAB\sin\theta \qquad (5)$$

3.I

If **A** is vector whose magnitude is the area, pointing in the direction of **n**, write a vector expression for τ.

$$\tau = NI\ A \times B \qquad (6)$$

Note: **A** points in the direction ⊥ to the plane of the loop, in the direction determined by the right-hand rule.

3.J

For what orientation is the torque on the loop a maximum? When is the torque zero?

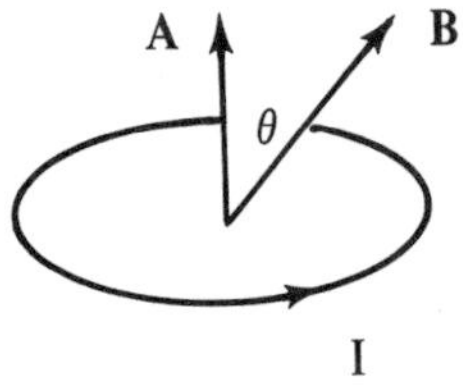

τ is a *maximum* when **A** is ⊥ to **B**, or when $\theta = \frac{\pi}{2}$ (that is, **B** coincides with the plane of the loop). τ is *zero* when **A** is || to **B**, or when $\theta = 0$ (that is, **B** is ⊥ to the plane of the loop).

3.K

The vector NIA defines the *magnetic moment* $\boldsymbol{\mu}$. Write $\boldsymbol{\tau}$ in terms of $\boldsymbol{\mu}$ and **B**.

$$\boldsymbol{\tau} = \boldsymbol{\mu} \times \mathbf{B} \qquad (7)$$

where

$$\boldsymbol{\mu} = \text{NIA}$$

3.L

Show the direction of μ for the circular loop shown where the loop is in the xy plane.

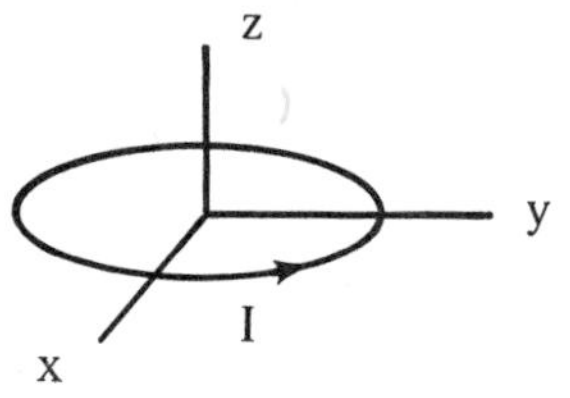

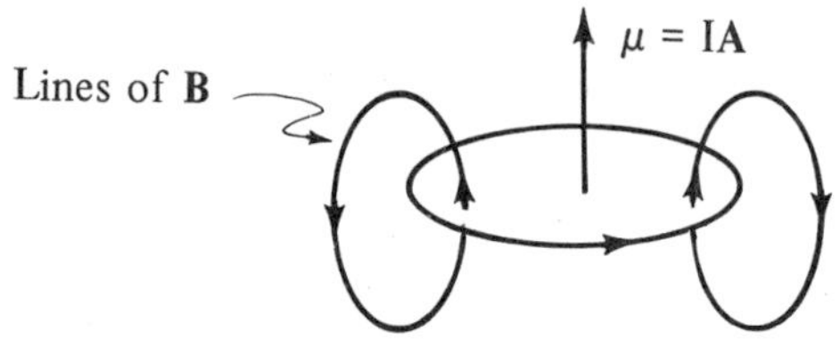

μ is in the +z direction and points in the direction of **B** produced by the current loop along the axis.

4.A

State *Ampere's law*, defining the terms in the expression.

$$\oint \mathbf{B} \cdot d\mathbf{s} = \mu_0 I \qquad (1)$$

where $\mu_0 = 4\pi \times 10^{-7} \frac{\text{Wb}}{\text{Am}}$, I is the net current linked by the *closed path* of the line integral and ds is a vector displacement along that path.

4.B

Is it possible to use Ampere's law to find **B** for *any* current-carrying conductor? Explain.

In principle, yes. However, the law is only useful when there is a high degree of symmetry in the field. One generally looks for a closed path for which **B** is constant in magnitude. This will reduce the line integral to a simple expression.

4.C

A long, straight wire carries a current I. Use Ampere's law to find the magnetic field at the point P.

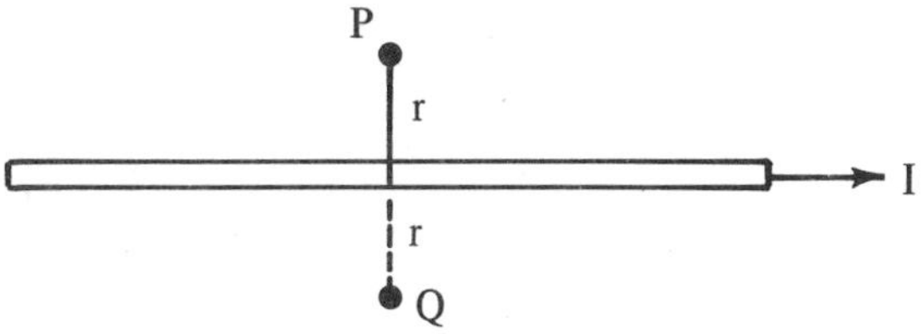

We construct a circular path of radius r concentric with the wire. Since **B** is constant in magnitude along this path and || to ds, we get

$$\oint \mathbf{B} \cdot d\mathbf{s} = B 2\pi r = \mu_0 I$$

$$B = \frac{\mu_0 I}{2\pi r} \qquad (2)$$

4.D

What are the directions of **B** at the points P and Q shown below? Show these in an end view sketch.

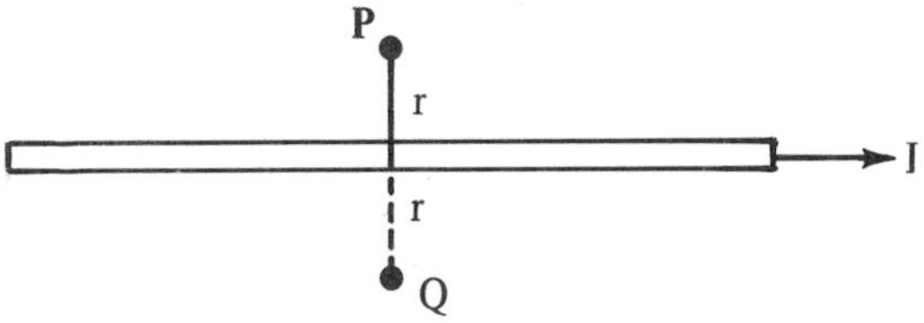

We "grasp" the wire with our right hand, with the thumb in the direction of I. The four fingers "wrap" in the direction of **B**. The fields at P and Q are shown in an end view picture.

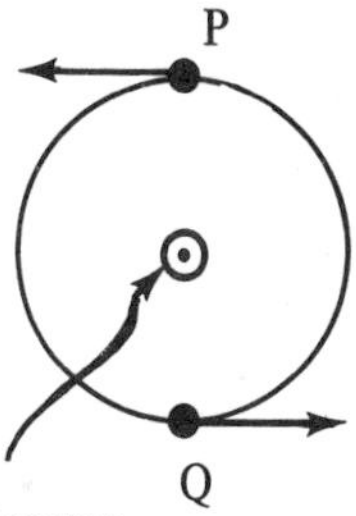

I out of paper

4.E

A *positive* charge q has a horizontal velocity **v** at the point P, as shown below. What is the magnitude and direction of the magnetic force on q at this instant?

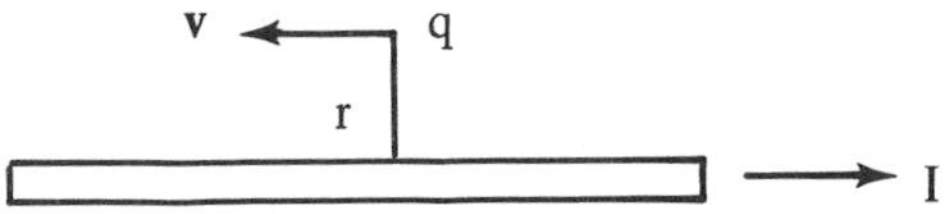

Since **v** is ⊥ to **B**,

$$F = qvB = qv\frac{\mu_0 I}{2\pi r} \qquad (3)$$

The direction of **F** is *upwards,* since **v** × **B** is up and q is positive.

4.F

Will the motion of the charge be circular? Explain.

No. The charge will initially deflect upwards, but since **B** is *not* uniform, **F** will vary in magnitude and direction.

4.G

Suppose a second wire of length l and current I_2 is placed parallel to the first, as shown below. What is the field B_1 at the second wire due to the first?

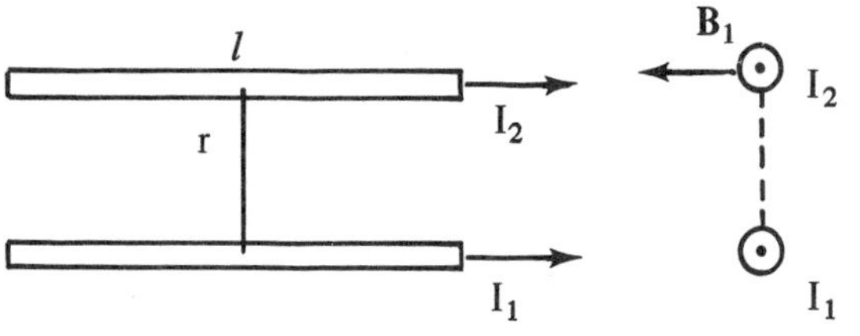

End view

From (2) we see that the field due to the first wire whose current is I is

$$B_1 = \frac{\mu_0 I_1}{2\pi r} \tag{4}$$

The direction of $\mathbf{B}_1$ is as shown in the right end view, or *out* of the paper in the first figure.

4.H

Find the magnitude and direction of the magnetic force on the second wire whose current is I_2.

$$F_2 = I_2 l \times B_1 \tag{5}$$

But l is to the right, and $\mathbf{B}_1$ is out of the paper; $\therefore \mathbf{F}_2$ is *down*. That is, the force is *attractive*. Since l is $\perp \mathbf{B}_1$, (4) and (5) give

$$F_2 = \frac{\mu_0 I_1 I_2 l}{2\pi r} \tag{6}$$

4.I

What is the magnetic force F_1 on the wire whose current is I_1? Show the forces $\mathbf{F}_1$ and $\mathbf{F}_2$ in a diagram.

From Newton's third law,

$$F_1 = -F_2; \quad \therefore F_1 = \frac{\mu_0 I_1 I_2 l}{2\pi r}$$

and its direction is *up*.

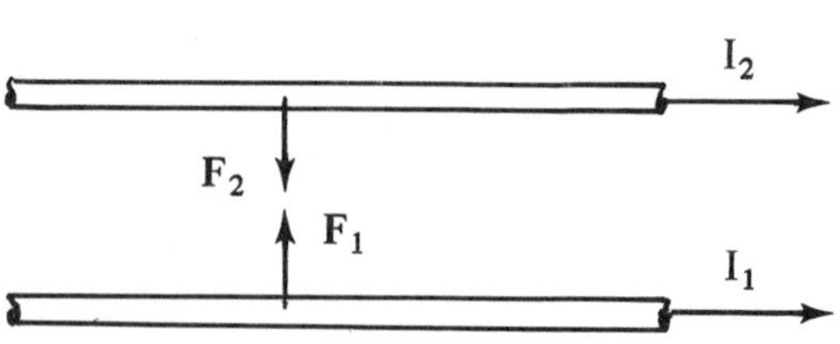

4.J

If the direction of I_2 were reversed, that is, to the left, how would the results differ?

The magnitudes of F_1 and F_2 would still be given by (6), but their directions would be *reversed*. That is, the wires would *repel* one another if the currents are in opposite directions.

5 Two long parallel wires carry currents of I_1 and I_2 moving in the same direction. The wires are separated by a distance 2d. An *end view* is shown below.

5.A

Draw vector diagrams showing the magnetic fields $\mathbf{B}_1$ and $\mathbf{B}_2$ at P due to the two currents.

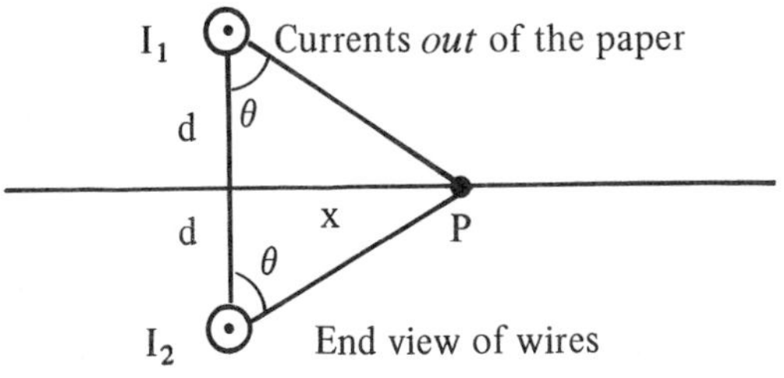

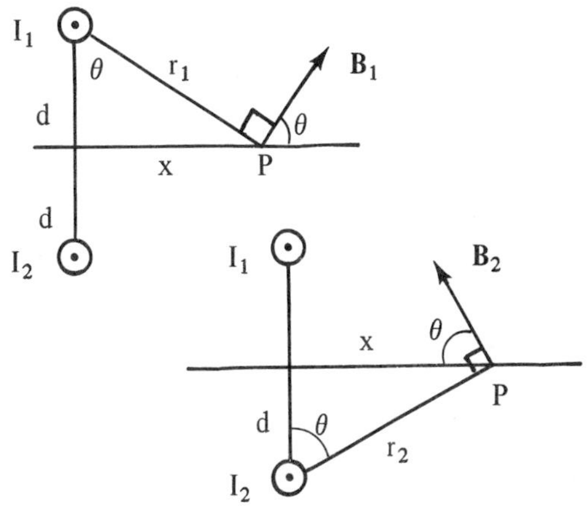

5.B

What are the magnitudes of $\mathbf{B}_1$ and $\mathbf{B}_2$ at P in terms of d and x?

$$B_1 = \frac{\mu_0 I_1}{2\pi r_1} = \frac{\mu_0 I_1}{2\pi (x^2 + d^2)^{1/2}} \qquad (1)$$

$$B_2 = \frac{\mu_0 I_2}{2\pi r_2} = \frac{\mu_0 I_2}{2\pi (x^2 + d^2)^{1/2}} \qquad (2)$$

5.C

Can you simply add B_1 and B_2 to get the resultant field at P? Explain.

No. The magnetic field is a *vector* quantity; therefore, the resultant field at P is obtained by superimposing $\mathbf{B}_1$ and $\mathbf{B}_2$ in a *vector* fashion.

5.D

Determine the x and y components of $\mathbf{B}_1$ at P. Use the geometry indicated in frame 5.A.

$$B_{1x} = B_1 \cos\theta = B_1 \frac{d}{r_1} = \frac{\mu_0 I_1 d}{2\pi (x^2 + d^2)} \qquad (3)$$

$$B_{1y} = B_1 \sin\theta = B_1 \frac{x}{r_1} = \frac{\mu_0 I_1 x}{2\pi (x^2 + d^2)} \qquad (4)$$

5.E

Determine the x and y components of $\mathbf{B}_2$ at P. Use the geometry shown in frame 5.A.

$$B_{2x} = -B_2 \cos\theta = -\frac{\mu_0 I_2 d}{2\pi (x^2 + d^2)} \qquad (5)$$

$$B_{2y} = B_2 \sin\theta = \frac{\mu_0 I_2 d}{2\pi (x^2 + d^2)} \qquad (6)$$

5.F

Calculate the resultant field in the x and y directions at P. Make use of (3), (4), (5) and (6).

$$B_x = B_{1x} + B_{2x} = \frac{\mu_0 d}{2\pi (x^2 + d^2)} (I_1 - I_2) \quad (7)$$

$$B_y = B_{1y} + B_{2y} = \frac{\mu_0 x}{2\pi (x^2 + d^2)} (I_1 + I_2) \quad (8)$$

5.G

Suppose $I_1 = I_2 = I$, that is, the currents are equal. What are the components B_x and B_y in this case?

From (7), we see that $B_x = 0$. From (8) we get

$$B_y = \frac{\mu_0 Ix}{\pi (x^2 + d^2)} \quad (9)$$

Note also that $B_y = 0$ at $x = 0$; that is, the field of the two wires annul each other.

5.H

Now consider the point Q along the y axis. Draw a vector diagram showing the fields $\mathbf{B}_1$ and $\mathbf{B}_2$ at Q.

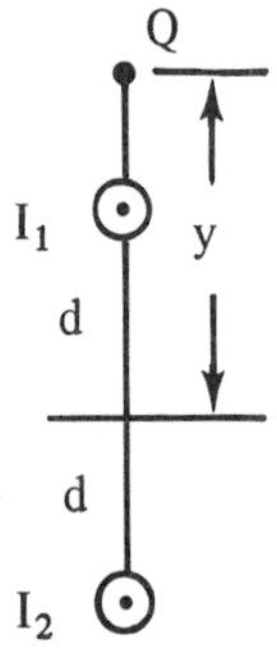

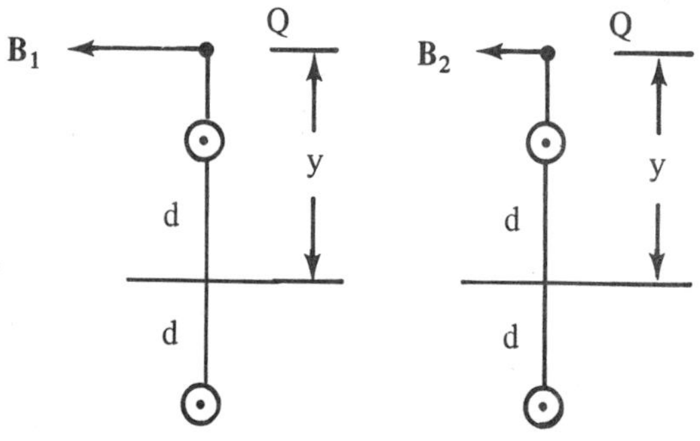

5.I

Determine the magnitudes of $\mathbf{B}_1$ and $\mathbf{B}_2$ at Q in terms of d and y.

$$B_1 = \frac{\mu_0 I_1}{2\pi (y - d)} \quad (10)$$

$$B_2 = \frac{\mu_0 I_2}{2\pi (y + d)} \quad (11)$$

5.J

Write a vector expression for the resultant field at Q. Note that $\mathbf{B}_1$ and $\mathbf{B}_2$ are both to the left at Q.

$$\mathbf{B} = \mathbf{B}_1 + \mathbf{B}_2 = -\frac{\mu_0}{2\pi}\left[\frac{I_1}{y - d} + \frac{I_2}{y + d}\right] \mathbf{i} \quad (12)$$

5.K

Are there any points along the y axis where $\mathbf{B} = 0$? If so, find them.

From (12), we see that $\mathbf{B} = 0$ *only if*

$$\frac{I_1}{y - d} + \frac{I_2}{y + d} = 0$$

or,

$$y = \left[\frac{I_1 - I_2}{I_1 + I_2}\right] d$$

Therefore, there is only *one* point along the y axis where $\mathbf{B} = 0$. If $I_2 > I_1$, the point is *above* the x axis. However, the point must always lie *between* the two wires, since $|y| < d$ for all values of I_1 and I_2.

---

†6.A

Write a general expression for the *Biot-Savart law,* which relates the magnetic field $d\mathbf{B}$ to a current element $d\mathbf{s}$.

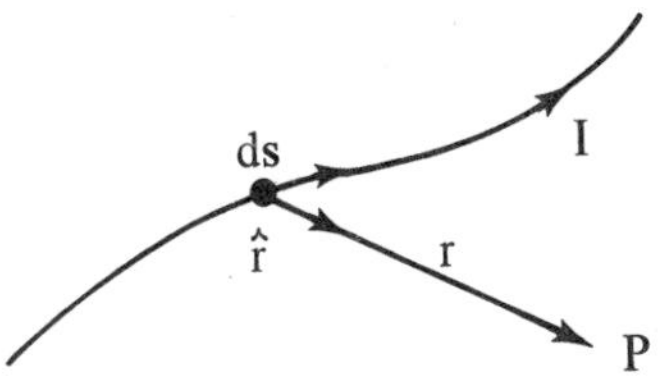

The magnetic field at P due to the element $d\mathbf{s}$ is

$$d\mathbf{B} = \frac{\mu_0 I}{4\pi} \frac{d\mathbf{s} \times \hat{\mathbf{r}}}{r^2} \qquad (1)$$

where $\hat{\mathbf{r}}$ is a unit vector directed from the element *towards* P.

6.B

Since $I = \frac{dq}{dt}$, (1) can also be written as $d\mathbf{B} = \frac{\mu_0}{4\pi} dq \frac{\mathbf{v} \times \hat{\mathbf{r}}}{r^2}$, where $\mathbf{v} = \frac{d\mathbf{s}}{dt}$. With this in mind, how can you interpret the physical meaning of (1)?

In this revised form, we can think of the Biot-Savart law as the field due to a charge dq moving with a velocity $\mathbf{v}$. The expression is correct only for speeds much less than the speed of light.

6.C

Let us apply the Biot-Savart law to a current loop carrying a current I. What is the magnitude of $d\mathbf{B}$ at P? Note that $d\mathbf{s}$ is $\perp$ to $\hat{\mathbf{r}}$ for every element on the loop.

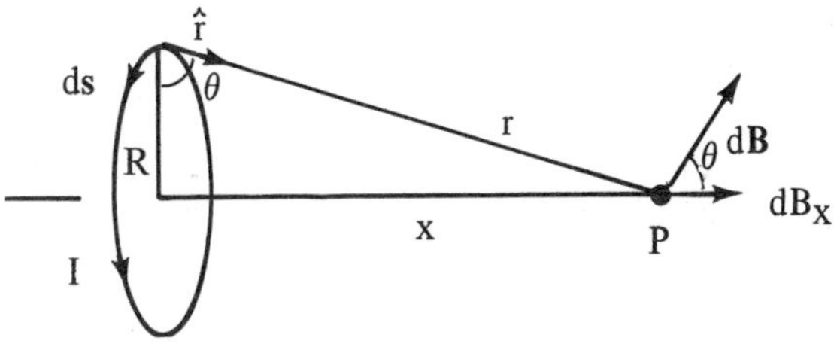

Since $d\mathbf{s}$ is $\perp \hat{\mathbf{r}}$, $|d\mathbf{s} \times \hat{\mathbf{r}}| = ds$.

$$\therefore \qquad dB = \frac{\mu_0 I}{4\pi} \frac{ds}{r^2} \qquad (2)$$

Note: $d\mathbf{s}$ is *not always* $\perp$ to $\hat{\mathbf{r}}$. When $d\mathbf{s}$ makes an angle θ with $\mathbf{r}$, then $|d\mathbf{s} \times \hat{\mathbf{r}}| = \sin\theta\, ds$.

6.D

From symmetry, we see that $B_y = 0$, so the total field is in the x-direction. Write an expression for dB_x, and note that $\cos\theta = \frac{R}{r}$.

$$dB_x = dB \cos\theta = \frac{R}{r} dB$$

$$dB_x = \frac{\mu_0 IR}{4\pi r^3} ds \qquad (3)$$

6.E

To get the total field at P, we integrate (3). Perform this integration, noting that r is the *same* for *every* element on the loop. Express the answer in terms of x and R.

$$B_x = \frac{\mu_0 IR}{4\pi r^3} \int_0^{2\pi R} ds$$

$$B_x = \frac{\mu_0 IR^2}{2r^3}$$

But $r = (x^2 + R^2)^{1/2}$; $\therefore$

$$B_x = \frac{\mu_0 IR^2}{2(x^2 + R^2)^{3/2}} \qquad (4)$$

6.F

What does the field reduce to at the *center* of the loop?

At $x = 0$, (4) reduces to

$$B_x = \frac{\mu_0 I}{2R} \qquad (5)$$

6.G

Calculate the field at the *center* of a circular loop of wire, 5 cm in radius, which carries a current of 200 A.

We use (5), with $R = 5 \times 10^{-2}$ m

$$B_x = \frac{4\pi \times 10^{-7} \times 2 \times 10^2}{2 \times 5 \times 10^{-2}}$$

$$B_x \cong 25 \times 10^{-4}\ \text{T} = 25\ \text{G}$$

6.H

Suppose the loop described in frame 6.G has *50 turns* instead of one, all of the same radius of 5 cm. What is B at the center if I = 200 A?

The field would be 50 times greater than that calculated in frame 6.G. That is,

$$B = 0.125\ \text{T} = 1250\ \text{G}$$

6.I

A wire carrying a current I is bent into a circular arc as shown below. What is the magnetic field at 0 due to the *straight portions* of the wire?

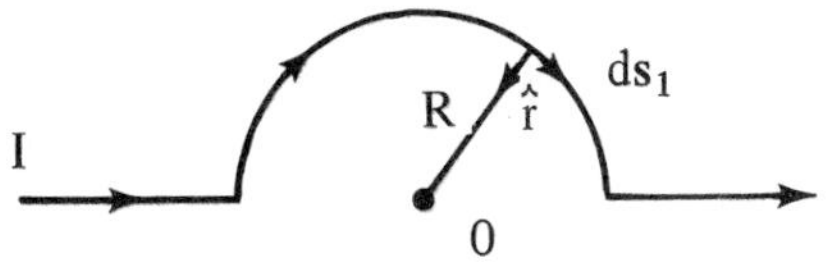

Zero. For each straight portion, ds is parallel to $\hat{r}$; therefore, $ds \times \hat{r} = 0$, and from the Biot-Savart law, $\mathbf{B} = 0$.

6.J

What is the magnitude and direction of the field at 0 due to the curved portion of the wire? Note that ds_1 is $\perp$ to $\hat{r}$ along this path, and

$$dB = \frac{\mu_0 I}{4\pi} \frac{ds_1}{R^2}$$

$ds_1 \times \hat{r}$ is *into* the paper for every segment; $\therefore$ B is *into* the paper. The magnitude of **B** is calculated in the same manner as Example 5.7, except the length of the path is πR. The result is

$$B = \frac{\mu_0 I}{4\pi R^2} \int_0^{\pi R} ds_1 = \frac{\mu_0 I}{4R} \qquad (6)$$

We could have guessed this from (5), since a semicircle will give half the field of a full circle.

5.10 SUMMARY

The force on a charge q motion with a velocity v in a magnetic field **B** is given by

$$\mathbf{F} = q\mathbf{v} \times \mathbf{B} \tag{5.1}$$

The force on a conductor of length l carrying a current I in a uniform magnetic field **B** is given by

$$\mathbf{F} = I\mathbf{l} \times \mathbf{B} \tag{5.9}$$

where l is in the direction of I.

Ampere's law states that the line integral of **B** around a closed path which links a current I is proportional to I, according to the expression

$$\oint \mathbf{B} \cdot d\mathbf{s} = \mu_0 I \tag{5.11}$$

where $\mu_0 = 4\pi \times 10^{-7} \dfrac{\text{Wb}}{\text{Am}}$ is the permeability of free space.

Biot-Savart's law states that the magnetic field d**B** due to a segment ds of a conductor carrying a current I is given by

$$d\mathbf{B} = \frac{\mu_0}{4\pi} \frac{d\mathbf{s} \times \hat{\mathbf{r}}}{r^2} \tag{5.17}$$

The magnetic field in a substance whose *relative permeability* is κ_m is given by

$$B = \kappa_m B_0 \tag{5.22}$$

where B_0 is the field in the absence of the substance and κ_m is a dimensionless constant whose value depends on the properties of the substance.

5.11 PROBLEMS

1. An electron is projected into a uniform magnetic field given by $\mathbf{B} = (0.2\,\mathbf{i} + 0.5\,\mathbf{j})$ T. Find a vector expression for the force on the electron when its velocity is $5 \times 10^6\,\mathbf{j}\,\frac{\text{m}}{\text{sec}}$.

2. A proton is accelerated from rest through a potential difference of 6000 V. It then enters a region of uniform magnetic field *perpendicular* to its velocity. (a) What is the radius of the orbit of the proton if the magnetic field is 4×10^{-3} T? (b) Determine the cyclotron frequency of the proton in $\frac{\text{rad}}{\text{sec}}$.

3. A positive charge q moves in a region where the magnetic field is $\mathbf{B} = 0.8\mathbf{k}$ T. Determine the magnitude and direction of the electric field necessary to keep the charge moving along the x axis with a velocity of $3 \times 10^5\,\mathbf{i}\,\frac{\text{m}}{\text{sec}}$.

4. Show that the work done by the magnetic force on a moving charged particle is zero for *any* displacement of the particle. *Hint:* Note that $\mathbf{v} = \frac{d\mathbf{s}}{dt}$ and $\mathbf{v} \times \mathbf{B}$ is always perpendicular to $\mathbf{v}$.

5. The wire shown in Figure 5-13 carries a current of 5 A. A uniform magnetic field of 0.2 T is applied in the x direction. (a) What is the magnetic force on the horizontal portions of the wire? (b) What is the magnitude and direction of the magnetic force on the section from a to b? (c) What is the resultant force on the conductor?

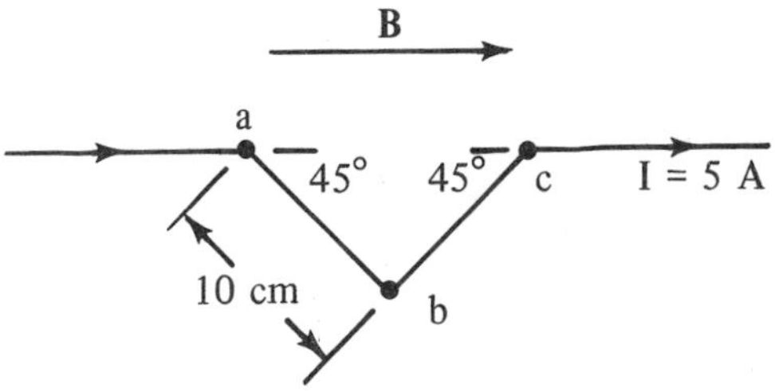

Figure 5-13

6. (a) Determine the magnetic force on the arc ab of the wire shown in Figure 5-11 if a uniform magnetic field B is present directed *into* the page. The wire carries a current I. (b) What is the *total* magnetic force on the closed loop?

7. Determine the magnitude and direction of the magnetic field at P due to the current loop shown in Figure 5-14.

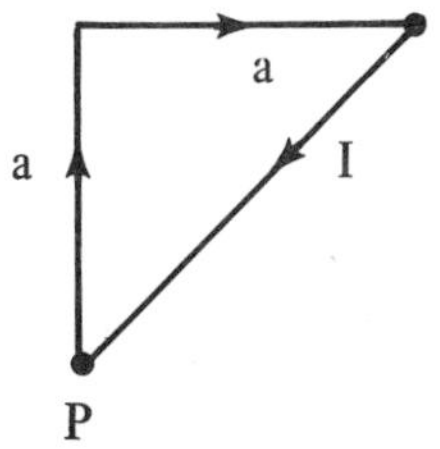

Figure 5-14

8. A superconducting solenoid is constructed using Nb_3Sn wire. The solenoid is wound in the form of a long cylinder 1 m in length and 8 cm in diameter using 5000 turns of wire. If the current is 100 A, (a) determine the intensity of the magnetic field at the center of the solenoid and (b) find the magnetic flux for a cross-section through the center. *Hint:* See Example 5.7.

9. Two long parallel wires separated by 20 cm carry currents of 5 A and 8 A in the same directions as in Figure 5-15. (a) Determine the magnitude and direction of the magnetic field at P. (b) Are there any points for which the **B** field is zero? If so, determine them. *Hint:* Superimpose the fields of the two wires, remembering that **B** is a *vector* quantity.

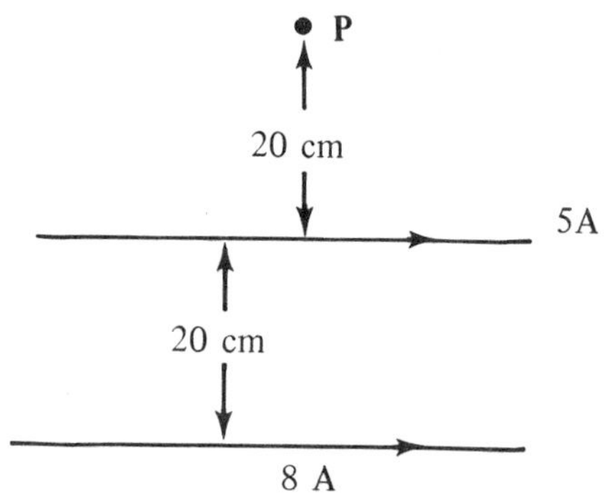

Figure 5-15

10. The loop of wire shown in Figure 5-16 carries a current I. The inner and outer circular arcs have radii of a and b, respectively. (a) Determine the magnetic field at 0 due to the straight portions of the wire. (b) Determine the magnetic field at 0 due to the arc cd. (c) Calculate the resultant magnetic field at 0.

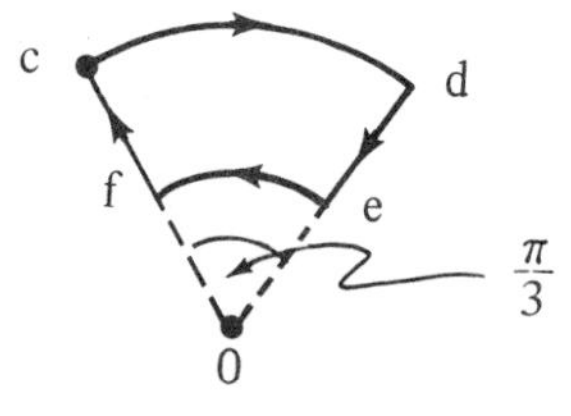

Figure 5-16.

11. *An accurate magnetic field measurement.* A rectangular coil of wire has 50 turns and a width of 5 cm. The coil is suspended from one arm of a delicate balance and is suspended between the poles of a magnet, as in Figure 5-17. The plane of the coil is perpendicular to the magnetic field. The system is balanced with no current through the coils. It is found that an additional mass of 20 g must be added to the right side to rebalance the system when a current of 2A passes through the coil. Determine the magnetic field strength.

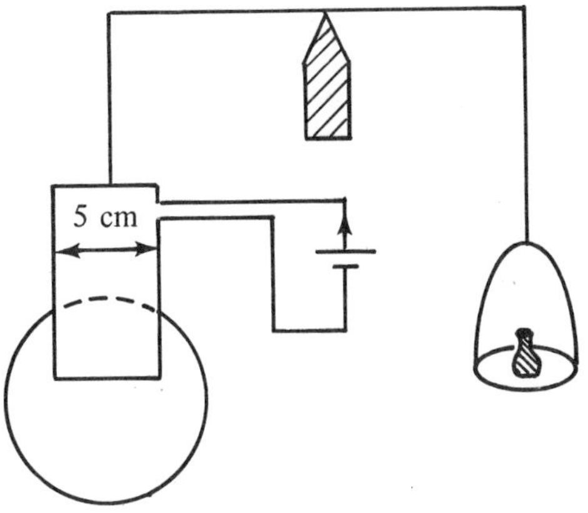

Figure 5-17

12. Assume that the electron in the hydrogen atom rotates in a circular orbit about the proton with a radius of r_0. (a) Determine the angular speed of the electron. (b) Find the equivalent current of the circulating electron and (c) calculate the magnetic field at the proton due to the orbiting electron. Obtain your answers in terms of the electronic charge e, the electron mass m, the Bohr radius r_0 and other appropriate constants. *Hint:* Recall that the Coulomb force provides the centripetal acceleration. Also, the "current" associated with the rotating charge is $I = qf$, where f is the frequency of motion.

13. A long straight wire carries a current I, as in Figure 5-18. A rectangular loop is placed a distance d from the wire. Find the *total magnetic flux* through the loop in terms of I, a, b and d. *Hint:* The B field of the wire is *not* constant, but varies with r according to Equation (5.13).

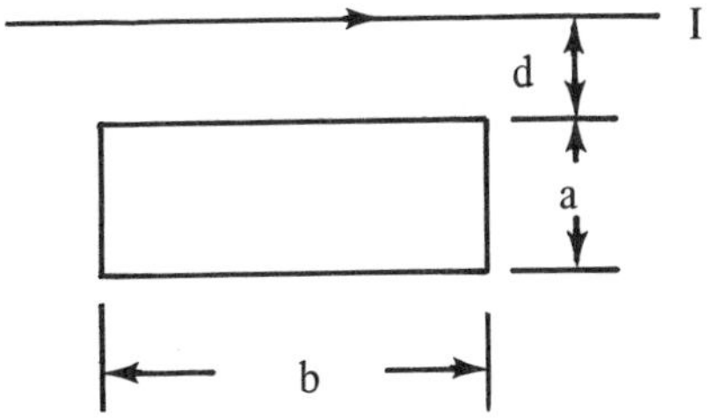

Figure 5-18

14. Two coplanar circular loops of wire carry currents of $I_1 = 5$ A and $I_2 = 3$ A in opposite directions, as in Figure 5-19. (a) If r = 10 cm, what is the resultant field at the center of the loops? (b) Determine a value for r such that the field at the center is zero.

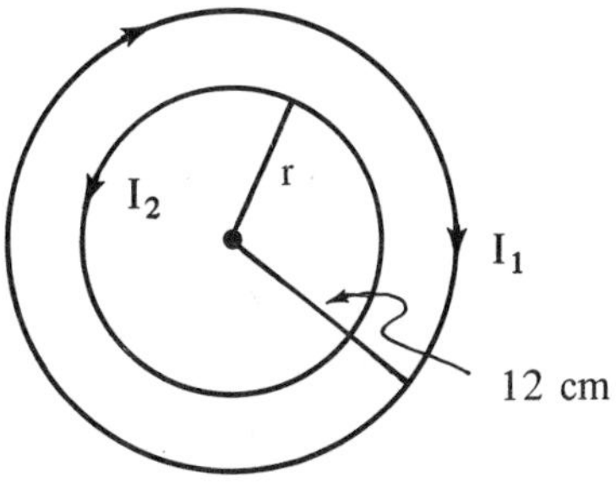

Figure 5-19

15. A straight wire of mass 10 g and length of 5 cm is suspended from two identical springs which, in turn, form a closed circuit, as in Figure 5-20. The springs are observed to stretch a distance of 0.5 cm under the weight of the wire. The circuit has a *total* resistance of 12Ω. When a magnetic field is turned on, directed *out* of the page, the springs are observed to stretch an additional 0.3 cm. What is the strength of the magnetic field? (The upper portion of the circuit is fixed.)

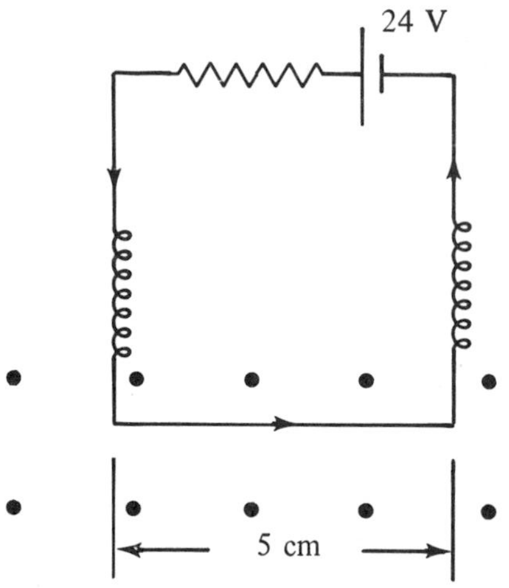

Figure 5-20

6

TIME-VARYING FIELDS AND INDUCTION

6.1 FARADAY'S LAW

If a loop of wire is connected to a galvanometer and a magnet is moved towards the loop, the galvanometer will deflect. If the magnet is moved away from the loop, the galvanometer will deflect in the opposite direction. No deflection is observed if the magnet is stationary relative to the loop. Finally, if the magnet is held stationary and the loop is moved towards or away from it, the galvanometer will deflect. These phenomena, discovered independently by Michael Faraday and Joseph Henry in 1831, demonstrate a connection between a *changing* magnetic field and an electric field. We say that a current or emf is *induced* in the coil *only* when there is relative motion between the coil and magnet.

The relation between the *induced* emf in the loop and the change in magnetic flux that passes through the loop is known as *Faraday's law of electromagnetic induction.* The law is given as

$$\mathcal{E} = -\frac{d\Phi_B}{dt} \tag{6.1}$$

where Φ_B is the magnetic flux linked by the circuit, and is given by

$$\Phi_B = \int \mathbf{B} \cdot d\mathbf{A}$$

If the coil has N turns of wire, the induced emf is given by

$$\mathcal{E} = -N\frac{d\Phi_B}{dt} \tag{6.2}$$

Faraday's law says that an emf is *induced* in a loop if the *magnetic flux* through the loop *changes* in time. Figure 6–1 illustrates a few methods of obtaining a change in the magnetic flux through the loop. The rule for determining the *direction* of the induced current is known as *Lenz's law.* This law says that the direction of the induced current in a loop will be such that the magnetic flux produced by the loop will be *opposite* to the original change in flux through the loop. The negative sign which appears in Equation (6.1) is significant for this reason.

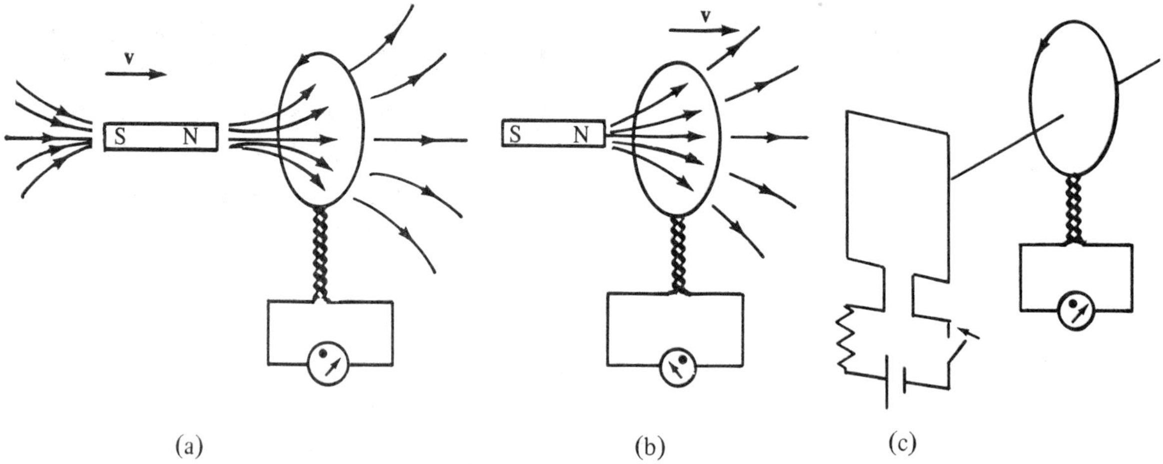

Figure 6–1 An emf is induced in the loop and the galvanometer deflects; (*a*) if a magnet is moved towards or away from the loop; (*b*) if the loop is moved towards or away from a magnet; and (*c*) momentarily when the switch is closed or opened.

The meaning of Lenz's law may be illustrated by considering the configuration in Figure 6–1. If the magnet is moved towards the fixed loop, as in Figure 6–1(*a*), the magnetic flux through the loop is *increasing* to the *right.* To counteract this increase in flux, the induced current in the loop must set up its own flux to the *left.* Therefore, by the right-hand rule, the induced current is as shown in Figure 6–1(*a*). On the other hand, if the magnet is stationary and the loop is moved to the right as in Figure 6–1(*b*), the flux through the loop is decreasing. The induced current in the loop will then be such that it produces its own flux to the *right.* This corresponds to the direction of current shown in Figure 6–1(*b*). Finally, the student should examine Figure 6–1(*c*) and show that the direction of the induced current in the loop is counterclockwise momentarily after the switch is *closed.*

Example 6.1

A circular coil of radius 0.2 m has 50 tightly wound turns, each having the same cross-sectional area. A uniform magnetic field is turned on *perpendicular* to the plane of the coil. If the field changes linearly from zero to 0.3 $\frac{\text{Wb}}{\text{m}^2}$ in a time of 0.3 sec, (a) find the induced emf in the coil while the field is changing.

The area of the loop = $\pi R^2 = 4\pi \times 10^{-2}$ m^2. The magnetic flux through the loop initially (at t = 0) is zero, since B = 0. The magnetic flux through the loop at t = 0.3 sec is $\Phi_B = BA = 0.3 \times 4\pi \times 10^{-2}$ Wb; therefore, the induced emf is

$$\mathcal{E} = -N\frac{\Delta\Phi_B}{\Delta t} = -50 \times \frac{(0.3 \times 4\pi \times 10^{-2} - 0)\ \text{Wb}}{0.3\ \text{sec}} = -2\pi\ \text{V}$$

(Note that 1 Wb = 1 V sec.)

(b) If the loop has a resistance of 4 ohms, what is the induced current while the field is changing?

$$I = \frac{|\mathcal{E}|}{R} = \frac{2\pi}{4} = \frac{\pi}{2}\ \text{A}$$

Example 6.2

A rectangular loop of area A is placed in a region where the magnetic field is *perpendicular* to the plane of the loop and varies according to the expression $B = B_0 e^{-t/\tau}$. That is, at t = 0, the field is B_0, which then decreases exponentially in time with a characteristic time constant τ. (a) Find the emf induced in the loop as a function of time.

Since **B** is ⊥ to the plane of the loop, the magnetic flux through the loop at time $t \geqslant 0$ is given by

$$\Phi_B = BA = AB_0 e^{-t/\tau}$$

But the coefficient AB_0 is constant; therefore, the induced emf can be calculated directly from Equation (6.1).

$$\mathcal{E} = -\frac{d\Phi_B}{dt} = -AB_0 \frac{d}{dt}(e^{-t/\tau}) = \frac{AB_0}{\tau} e^{-t/\tau}$$

That is, the induced emf decays exponentially in time.

(b) Suppose A = 0.1 m^2 $B_0 = 0.3 \frac{\text{Wb}}{\text{m}^2}$ and τ = 3 sec. Find the emf induced in the loop at t = 6 sec. Compare this to the maximum $\mathcal{E}$ at t = 0.

$$\mathcal{E}(3) = \frac{0.1\ \text{m}^2 \times 0.3 \frac{\text{Wb}}{\text{m}^2}}{3\ \text{sec}} e^{-6/3} = \frac{1 \times 10^{-2}}{e^2} \frac{\text{Wb}}{\text{sec}}$$

But $1 \frac{\text{Wb}}{\text{sec}} = 1$ V; therefore, $\mathcal{E}(3) = 1.35$ mV. At t = 0, $\mathcal{E}(0) = \frac{AB_0}{\tau} = 10$ mV.

6.2 MOTIONAL EMF

So far, we have considered the emf induced in a loop when the magnetic field through the loop changes in time. Now suppose the field is kept constant and a conductor is moved through the field. In particular, consider the circuit shown in Figure 6–2, where the *crossbar* moves with a constant velocity v to the right under the action of an external force. We will assume that a uniform magnetic field exists pointing perpendicular to the loop and coming *out* of the page.

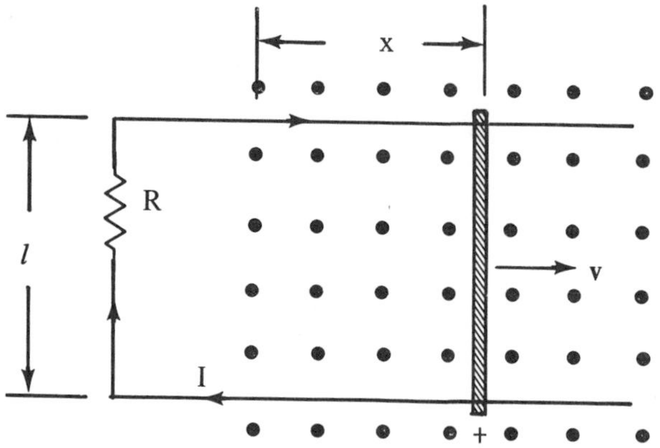

Figure 6–2 An emf is induced in the loop when the crossbar moves to the right. The **B** field is *out* of the paper. The arrows show the directions of the induced current.

Since the area of that part of the loop that is in the field is given by lx, the magnetic flux through the loop is

$$\Phi_B = BA = Blx$$

Therefore, the emf induced in the loop is given by

$$\mathcal{E} = -\frac{d\Phi_B}{dt} = -Bl\frac{dx}{dt}$$

But $v = \frac{dx}{dt}$, so $|\mathcal{E}|$ becomes

$$|\mathcal{E}| = Blv \tag{6.3}$$

If the resistance of the loop is R, the induced current is

$$I = \frac{|\mathcal{E}|}{R} = \frac{Blv}{R} \tag{6.4}$$

The direction of I is *clockwise* in Figure 6–2, since the external flux through the loop is increasing *outwards* and the induced current must counteract this and establish its own flux *inwards.*

Physically, the induced emf can be interpreted as follows: A magnetic force of $q\mathbf{v} \times \mathbf{B}$ is exerted on the conduction electrons, which forces them to move and establish a current. In Figure 6–2, the motion of free electrons is upwards. We can think of the force on the free electrons as being equivalent to an effective electric field, such that $qE = qvB$ (since v is ⊥ to B). Therefore, the equivalent electric field is given by $E = vB$, and a potential difference of $\mathcal{E} = El = Blv$ is generated by the moving bar.

Since the induced emf (that is, potential difference) has an associated electric field E in the conductor, we can relate $\mathcal{E}$ to the *line integral* of E around a closed circuit through the relation $\mathcal{E} = \oint \mathbf{E} \cdot d\mathbf{s}$. (See Chapter 3.) This line integral is zero *only* when E is a *static* field, which is conservative. However, when the fields are *time*-dependent, Equation (6.1) becomes

$$\mathcal{E} = \oint \mathbf{E} \cdot d\mathbf{s} = -\frac{d\Phi_B}{dt} = -\frac{d}{dt}\int \mathbf{B} \cdot d\mathbf{A} \tag{6.5}$$

This is another mathematical statement of *Faraday's law.* In this form, we see that an E field is generated from a time-varying B field.*

Example 6.3

A metal bar of length l is pivoted at one end and rotates with a constant angular frequency ω. A uniform magnetic field B is directed perpendicular to the plane of rotation, as in Figure 6-3. Determine the potential difference between the ends of the bar.

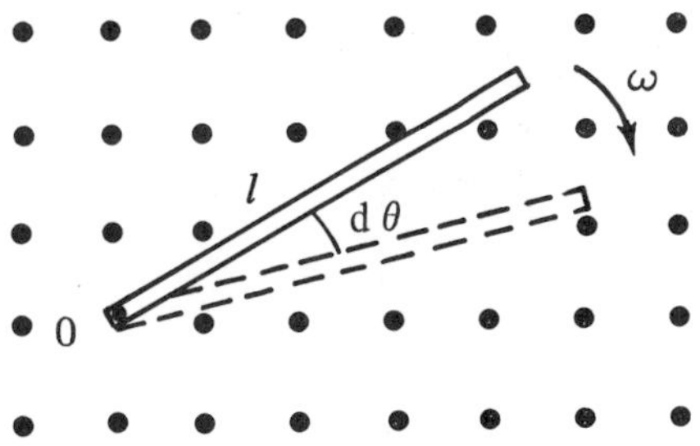

Figure 6-3

The potential difference across the ends is the same as the induced emf. Therefore, we can apply Faraday's law.

$$\mathcal{E} = -\frac{d\Phi_B}{dt} = -\frac{d}{dt}(BA) = -B\frac{dA}{dt}$$

But the area "swept out" by the bar when it rotates through an angle $d\theta$ is $\frac{1}{2}l\,(l\,d\theta)$; therefore,

$$\frac{dA}{dt} = \frac{d}{dt}\left(\frac{1}{2}l^2\,d\theta\right) = \frac{l^2}{2}\frac{d\theta}{dt} = \frac{l^2}{2}\,\omega$$

*When the fields are time-dependent, they are not conservative and we cannot use a simple scalar potential to generate E as we did in the static case, that is, in Chapter 3.

Hence, the induced emf becomes

$$\mathcal{E} = -\frac{Bl^2\omega}{2}$$

Comment: The student should verify that the *same* result is obtained by applying Equation (6.5), where $\int \mathbf{E}\cdot d\mathbf{s} = \int E_r dr$, and $E_r = vB = r\omega B$ is the electric field in the bar at a distance r from 0.

Example 6.4

Consider a coil of N turns and area A rotating in a uniform magnetic field with *constant* angular velocity ω, as in Figure 6–4. (a) Find the emf induced in the coil due to the change in magnetic flux through the coil as it rotates. (This is the mechanism responsible for the operation of generators and motors.)

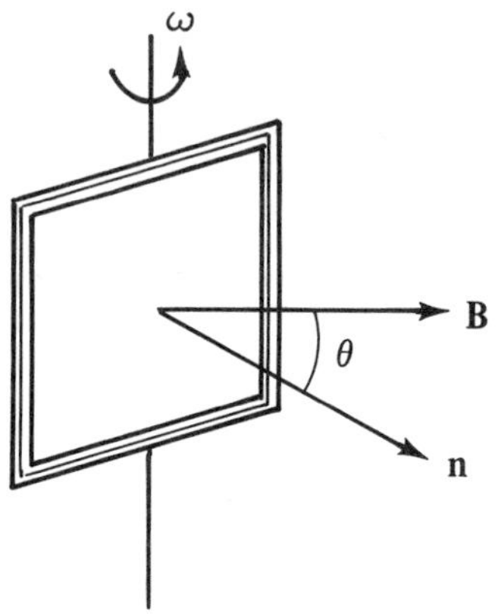

Figure 6–4

If θ is the angle between **B** and the normal **n**, then the magnetic flux through the coil is given by

$$\Phi_B = AB\cos\theta$$

We can relate θ to the angular velocity through the expression $\theta = \omega t$. Therefore, we can write Φ_B as

$$\Phi_B = AB\cos(\omega t)$$

Hence, the emf induced in the coil is

$$\mathcal{E} = -N\frac{d\Phi_B}{dt} = NAB\omega\sin(\omega t)$$

Therefore, an *alternating* (that is, sinusoidal) emf is induced in the coil if it rotates at constant angular velocity in a uniform magnetic field.

(b) Find the *maximum* emf induced in the coil if it has 80 turns, $A = 0.2\ m^2$, $B = 0.3\ T$ and the coil rotates with a frequency of $f = 6$ Hz. (Note that $\omega = 2\pi f = 12\pi\ \frac{rad}{sec}$.)

$$\mathcal{E}_{max} = NAB\omega = 80 \times 0.2 \times 0.3 \times 12\pi = 180\ V$$

(c) What is the *maximum* current induced in the coil if it has a resistance of 6Ω?

$$I_{max} = \frac{\mathcal{E}_{max}}{R} = \frac{180\ V}{6\Omega} = 30\ A$$

(d) Write expressions for variations of $\mathcal{E}$ and I with time, using the numerical data in parts (b) and (c).

$$\mathcal{E} = \mathcal{E}_{max} \sin(\omega t) = 180 \sin(12\pi t)\ V$$

$$I = \frac{\mathcal{E}}{R} = I_{max} \sin(\omega t) = 30 \sin(12\pi t)\ A$$

6.3 A MAGNETIC FIELD FORMED BY A CHANGING ELECTRIC FLUX

In the last two sections, we described the creation of an electric field from a changing magnetic flux, that is, *Faraday's law of electromagnetic induction.* Because of the symmetry that exists between electric and magnetic fields, there is also a law which relates the creation of a magnetic field by a changing electric flux. This is known as the law of *magnetoelectric induction* and is actually a *general* form of *Ampere's law.* It is given as follows:

$$\oint \mathbf{B} \cdot d\mathbf{s} = \mu_0 I + \mu_0 \epsilon_0 \frac{d\Phi_E}{dt} \qquad (6.6)$$

The left side of Equation (6.6) is sometimes called the *magnetomotive* force, but it is essentially the same line integral used in the static form of Ampere's law. (See Chapter 5.) The first term on the right of Equation (6.6) is μ_0 times the current linked by the path of integration, whereas the last term is $\mu_0 \epsilon_0$ times the time rate of change of electric flux. (Recall that $\Phi_E = \int \mathbf{E} \cdot d\mathbf{A}$.) The product $\mu_0 \epsilon_0 = \frac{1}{c^2}$, where c is the speed of light.

When Φ_E is constant in time, the last term is zero, and Equation (6.6) reduces to Ampere's law for the static case, where **B** is due only to the conduction current I. Maxwell called the term $\epsilon_0 \frac{d\Phi_E}{dt}$ a *"displacement current."* It is necessary to include this term to make the laws of electromagnetism consistent with the theory of relativity and to meet the requirement that charge is never created or destroyed. In other words, without the "displacement current" term, we would not be able to "switch" the roles of E and **B** and an asymmetric condition would prevail. Faraday did not discover this effect because the displacement current is practically zero for *slow* varying fields; therefore, it is very difficult to detect. That is, his experiments were only sensitive to the conduction current, I. However, the experiments by Hertz in 1888 using rapidly varying electromagnetic waves proved that both conduction currents *and* displacement currents are sources of the magnetic field. Hertz's experiments followed the mathematical formulation of electromagnetic theory by Maxwell.

6.4 INDUCTANCE

Suppose the switch in the circuit shown in Figure 6–5 is closed. The current in the circuit *doesn't* increase from zero to $\frac{\mathcal{E}}{R}$ immediately for the following reason. As the current starts to build up, the magnetic flux associated with the circuit changes, which in turn induces an emf. The direction of the induced emf is such that it *opposes* the magnetic flux change. This effect is called *self-induction.* For a simple circuit as shown in Figure 6–5, the effect is very small and for all practical purposes can be neglected. However, if the magnetic flux is associated with a solenoid or toroid, the effect can be very significant. In these cases, the induced emf is increased by a factor N (the number of turns in the device) over that of a single turn. We call such a device, where self-induction is important, an *inductor,* represented by the symbol . The device is characterized by a quantity called its *inductance,* L, which depends on the dimensions of the conductor, as well as the material inside it. Since $B \propto I$ for any current carrying conductor, then $\Phi_B \propto I$, so for an N turn coil we can write

$$N\Phi_B = LI \tag{6.7}$$

From Faraday's law, we can write the induced emf as

$$\mathcal{E} = -\frac{d}{dt}(N\Phi_B) = -L\frac{dI}{dt} \tag{6.8}$$

Therefore, the inductance L can also be defined as

$$L = -\frac{\mathcal{E}}{\frac{dI}{dt}} \tag{6.9}$$

Both Equations (6.7) and (6.9) can be used as definitions of inductance.*

**Note:* The inductance L (self-inductance) should not be confused with *mutual inductance,* M. M is defined by the flux Φ_2 which appears in one coil due to a current I_1 in *another* coil, through the relation $N_2\Phi_2 = MI_1$. That is, an emf $\mathcal{E}_2$ appears in coil 2 when I_1 changes in time, so $\mathcal{E}_2 = -M\frac{dI_1}{dt}$.

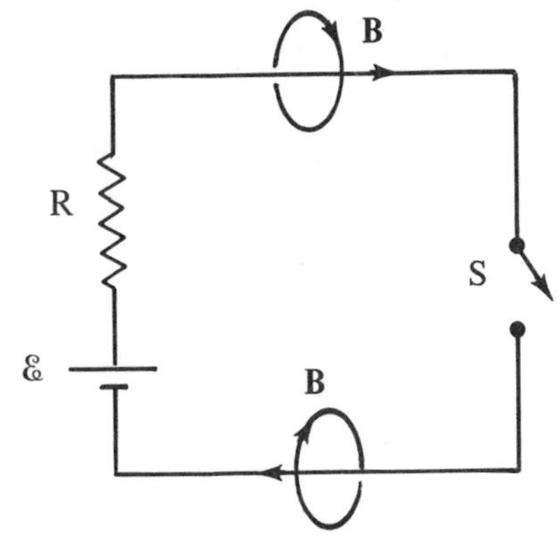

Figure 6-5

Inductance has units of volts per ampere per sec, which is given the name *henry*, H.

$$1\ \text{H} = 1\,\frac{\text{V sec}}{\text{A}} \qquad (6.10)$$

Example 6.5

(a) Calculate the inductance of a long solenoid carrying a current I, having a cross-sectional area A, length l and with a number of turns per unit length n.

For a long solenoid, recall that $B = \mu_0 nI$ if the core is air filled. Therefore,

$$L = \frac{N\Phi_B}{I} = \frac{N}{I}BA = \frac{N}{I}\mu_0 nI\ A = \mu_0 nNA$$

But $N = nl$, and the volume of the solenoid $V = Al$; therefore,

$$L = \mu_0 n^2\ Al = \mu_0 n^2\ V$$

(b) Suppose $n = 10^3\ \frac{\text{turns}}{\text{m}}$, $A = 1\ \text{cm}^2$ and $l = 10$ cm. Find the value of L for the solenoid.

$$L = 4\pi \times 10^{-7}\ \frac{\text{Wb}}{\text{Am}} \times \left(10^3\ \frac{\text{turns}}{\text{m}}\right)^2 (10^{-5}\,\text{m}^3) = 4\pi \times 10^{-6}\ \text{H}$$

6.5 ENERGY IN A MAGNETIC FIELD

If we connect a solenoid or other inductor to a battery, the battery does work in delivering current to the coil. This work goes into energy "stored" in the *magnetic field of the solenoid.* To calculate this energy, note that the power delivered by the battery is IV. But $V = |\mathcal{E}|$ for the solenoid, so

$$P = I|\mathcal{E}| = IL\frac{dI}{dt} \qquad (6.11)$$

Since $P = \frac{dW}{dt}$, we can write Equation (6.11) as $dW = ILdI$. Integrating this gives the total energy "stored" in the coil when the current is I.

$$U = L\int_0^I IdI = \frac{1}{2}LI^2 \tag{6.12}$$

This is similar to $U = \frac{1}{2}CV^2$ for a capacitor, the energy being stored in the *electric* field.

For a solenoid, we found that $B = \mu_0 nI$ and $L = \mu_0 n^2 V$. Substituting these results into Equation (6.12) gives

$$U = \frac{1}{2}\mu_0 n^2 V \frac{B^2}{\mu_0{}^2 n^2} = \frac{B^2}{2\mu_0} V \tag{6.13}$$

The energy density, or energy per unit volume, is therefore given by

$$u = \frac{B^2}{2\mu_0} \tag{6.14}$$

This expression is valid in *general.* We see that $u \sim B^2$ just as $u \sim E^2$ in the case of an electric field.

6.6 THE LR CIRCUIT

Consider the circuit shown in Figure 6–6, which includes a battery, a resistor and an inductor. When the switch is closed, a current will start to build-up, but the inductor will produce a "back" emf which effectively resists the increase in I. In other words, the inductor acts like a battery whose polarity is opposite to the battery in the circuit. Kirchhoff's loop rule applied to the circuit gives

$$\mathcal{E} - IR - L\frac{dI}{dt} = 0 \tag{6.15}$$

This can also be written as

$$\frac{\mathcal{E}}{R} - I - \frac{L}{R}\frac{dI}{dt} = 0 \tag{6.16}$$

To solve this equation, note that we can make a change of variables in the following way: Let $x = \frac{\mathcal{E}}{R} - I$, so $dx = -dI$. Then Equation (6.16) becomes

$$x + \frac{L}{R}\frac{dx}{dt} = 0 \quad \text{or} \quad \frac{dx}{x} = -\frac{R}{L}dt$$

Integrating this expression gives

$$x = x_0 e^{-\frac{R}{L}t} \tag{6.17}$$

Note that at t = 0, I = 0, so $x_0 = \frac{\mathcal{E}}{R}$. Therefore, Equation (6.17) becomes

$$\frac{\mathcal{E}}{R} - I = \frac{\mathcal{E}}{R} e^{-\frac{R}{L}t} \quad \text{or} \quad I = \frac{\mathcal{E}}{R}\left(1 - e^{-\frac{R}{L}t}\right) \tag{6.18}$$

In other words, the current increases very fast initially, but then gradually approaches an equilibrium value of $\frac{\mathcal{E}}{R}$ at $t \to \infty$. This is analogous to the RC circuit. The *time constant* of the LR circuit is given by

$$\tau = \frac{L}{R} \tag{6.19}$$

The student should verify that τ has units of time. Physically, τ is the time it takes the current to reach $(1 - e^{-1})$ of its final maximum value.

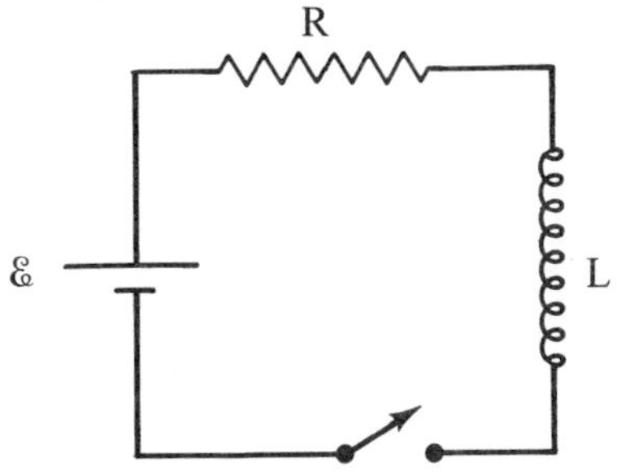

Figure 6-6

The rules for adding inductors in series or parallel are the *same* as those for adding resistances in series or parallel. This should be evident in the series case, since there is a voltage drop across one inductor given by $-L_1 \frac{dI}{dt}$, and a *series* connection of two or more gives a total voltage drop of $-(L_1 + L_2 + \ldots)\frac{dI}{dt}$, where I is the same for each inductor.* Therefore,

$$L = L_1 + L_2 + \ldots \qquad \text{Series Connection}$$

Likewise, it follows that

$$\frac{1}{L} = \frac{1}{L_1} + \frac{1}{L_2} + \ldots \qquad \text{Parallel Connection}$$

---

*Note: This is true *only* if the inductors are far apart, so that the mutual inductance can be neglected.

Finally, it is interesting to note that if we multiply Equation (6.15) by I, and rearrange the terms, we get an expression for power balance:

$$I\mathcal{E} = I^2 R + LI\frac{dI}{dt} \tag{6.20}$$

Note that $I\mathcal{E}$ is simply the power delivered by the battery, I^2R is the power dissipated by the resistor and the last term is the time rate of change of energy stored in the inductor, which can also be written as $\frac{d}{dt}\left(\frac{1}{2}LI^2\right)$.

Example 6.6

The circuit shown in Figure 6-6 has a 5Ω resistor and 30 mH inductor connected to a 60 V battery. (a) What is the time constant of the circuit?

$$\tau = \frac{L}{R} = \frac{30 \times 10^{-3}\ \text{H}}{5\Omega} = 6\ \text{msec}$$

(b) What is the current in the circuit at t = 3 msec?

$$I = \frac{\mathcal{E}}{R}(1 - e^{-t/\tau}) = \frac{60}{5}(1 - e^{-0.5}) = 4.7\ \text{A}$$

(c) Plot I *vs* t for this circuit.

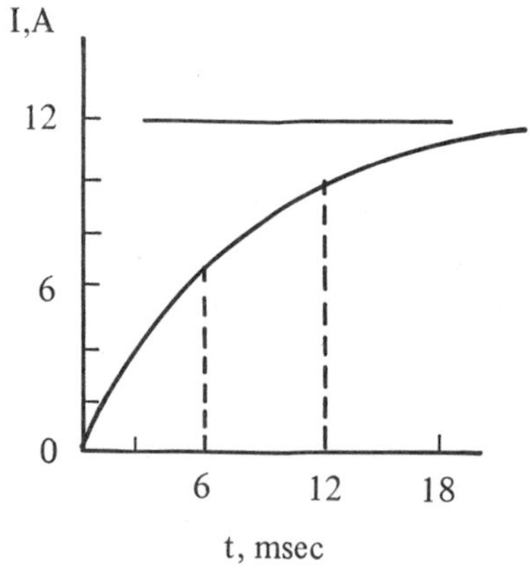

(d) How much energy is stored in the inductor at t = 3 msec?

$$U = \frac{1}{2}LI^2 = \frac{1}{2}30 \times 10^{-3}\ \text{H} \times (4.7)^2\ \text{A}^2 = 0.33\ \text{J}$$

The student should verify that $\text{H} \times \text{A}^2 = \text{J}$.

6.7 PROGRAMMED EXERCISES

†1.A

Write an expression for Faraday's law of electromagnetic induction as applied to a single loop of wire. Explain the terms involved.

$$\mathcal{E} = -\frac{d\Phi_B}{dt} \qquad (1)$$

where $\mathcal{E}$ is the induced emf in a loop and Φ_B is the magnetic flux passing through the loop.

1.B

Write an expression for the magnetic flux, Φ_B.

$$\Phi_B = \int \mathbf{B}\cdot d\mathbf{A} = \int B\cos\theta\, dA \qquad (2)$$

1.C

In view of (1) and (2), describe *three* ways that an emf can be induced in a loop.

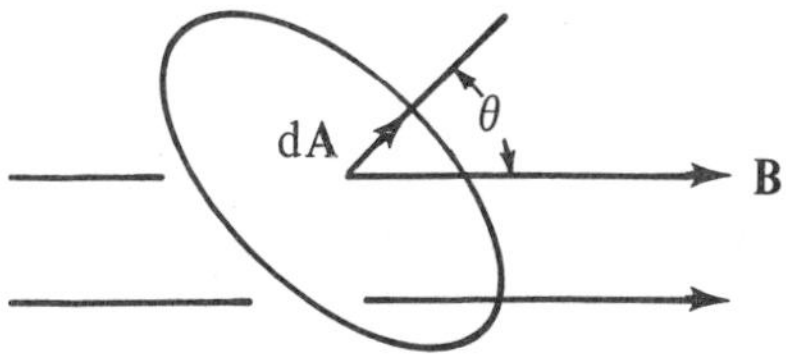

Since Φ_B depends on the external field, **B** and the area $\mathcal{E}$ can be produced: (1) by a changing field, with the area constant; (2) by a changing area and constant **B**; and (3) by varying the angle θ between **B** and d**A** (that is, rotate the loop).

1.D

Write Faraday's law as applied to a coil having N turns, each of the same area.

$$\mathcal{E} = -N\frac{d\Phi_B}{dt} \qquad (3)$$

That is, the induced emf is N times larger than that for one turn.

1.E

In SI units, $\mathcal{E}$ has units of volts (that is, a potential difference). How is the unit of volt related to the unit of magnetic flux?

From (1), we see that

$$1\text{ V} = 1\,\frac{\text{Wb}}{\text{sec}}$$

where Φ_B has units of Wb (webers).

1.F

For the *fixed* loop in the *xy plane*, what is $\mathcal{E}$ if **B** varies in time?

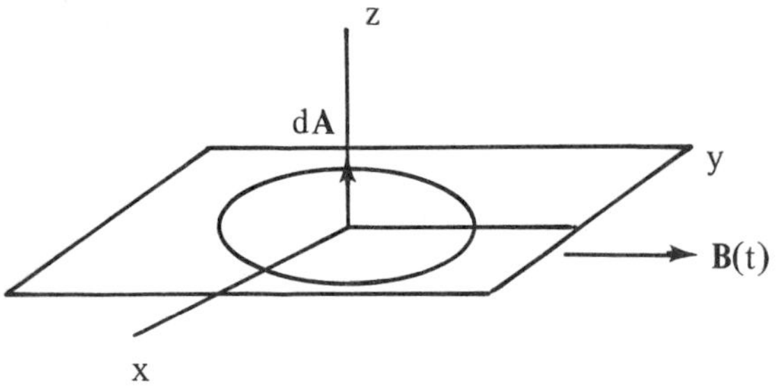

Zero. Since **B** is *parallel* to the plane of the loop, $\theta = \frac{\pi}{2}$, $\mathbf{B}\cdot d\mathbf{A} = 0$; $\therefore \Phi_B = 0$. That is, no magnetic flux penetrates the loop, hence $\mathcal{E} = 0$.

1.G

Suppose the fixed loop is in the xz plane. What is $\mathcal{E}$ if $\mathbf{B} = B_0 t\mathbf{j}$, where B_0 is a constant. Assume the area of the loop is A.

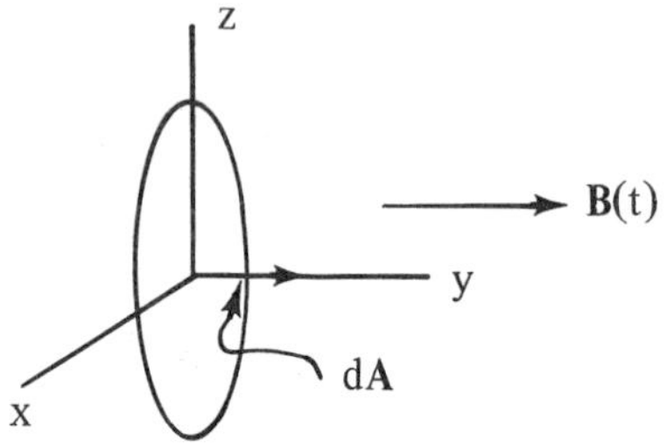

Since **B** is || to d**A**, $\theta = 0$

$$\Phi_B = \int \mathbf{B}\cdot d\mathbf{A} = \int B_0 t \, dA = B_0 A t$$

$$\therefore \mathcal{E} = -\frac{d\Phi_B}{dt} = -B_0 A$$

1.H

What does the negative sign in Faraday's law signify?

It is a consequence of Lenz's law, that is, the direction of the induced emf is such that it *opposes* the change that produced it. Alternatively, we can say that the *induced* current is such that it produces a magnetic flux which opposes the change (in flux) which produced it.

1.I

Apply Lenz's law to the loop in frame 1.G and show the direction of the induced current if $\mathbf{B} = B_0 t\mathbf{j}$, that is, it increases to the right *linearly* in time.

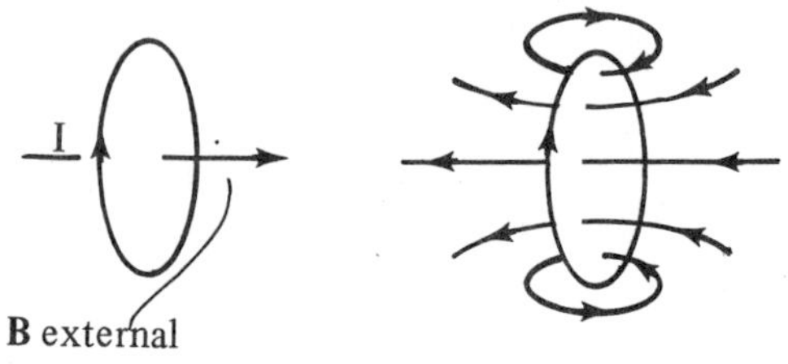

B field of induced current in loop.

1.J

Continuing along this vein, suppose the field reaches some value, say 5 B_0, and is suddenly decreased in such a way that it *decreases* to zero *exponentially,* that is,

$$B = 5B_0 e^{-t/\tau} \mathbf{j}$$

What is the induced emf and the direction of the induced current?

$$\mathcal{E} = -\frac{d\Phi_B}{dt} = -\frac{d}{dt}(5B_0 A e^{-t/\tau})$$

$$\mathcal{E} = \frac{5B_0 A}{\tau} e^{-t/\tau}$$

that is, the emf decays exponentially in time. Since the *external* flux *decreases* in time and is to the right, the induced current is the *reverse* of 1.I, or it must compensate by producing its *own* flux to the right.

2 A rectangular loop of wire of width a and length b is placed in a region where the magnetic field is ⊥ to the plane of the loop and varies in time according to the expression $B = B_0 + B_1 \sin(\omega t)$, where B_0 and B_1 are constants.

2.A

What is the magnetic flux through the loop at t = 0?

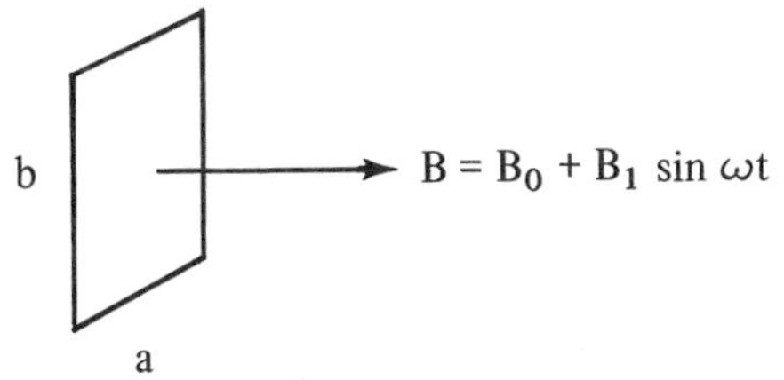

At t = 0, $B = B_0$, and since **B** is ∥ to d**A**, $\theta = 0$ and

$$\Phi_B(0) = AB_0 = abB_0 \qquad (1)$$

2.B

What is the magnetic flux through the loop at any time $t > 0$?

Since $B(t) = B_0 + B_1 \sin(\omega t)$,

$$\Phi_B(t) = AB(t) = ab[B_0 + B_1 \sin(\omega t)] \qquad (2)$$

2.C

Calculate the induced emf in the loop using (2) and Faraday's law.

$$\mathcal{E} = -\frac{d\Phi_B}{dt} = -\frac{d}{dt}[abB_0 + abB_1 \sin(\omega t)]$$

Since abB_0 is a constant, we get

$$\mathcal{E} = -ab\omega B_1 \cos(\omega t) \qquad (3)$$

2.D

Note that the *constant* component of the field, that is, B_0, does *not* affect the emf. $\mathcal{E}$ is generated *only* by a *varying* flux. What is the maximum emf?

From (3), we see that

$$|\mathcal{E}_{max}| = ab\omega B_1$$

This occurs at t = 0, or integral multiples of $\frac{\pi}{\omega}$.

2.E

Make rough sketches of B *vs* t and $\mathcal{E}$ *vs* t for this system. (Note that if the loop has a resistance R, then $I = \frac{\mathcal{E}}{R}$ is the induced current and has the *same* time variation as $\mathcal{E}$.)

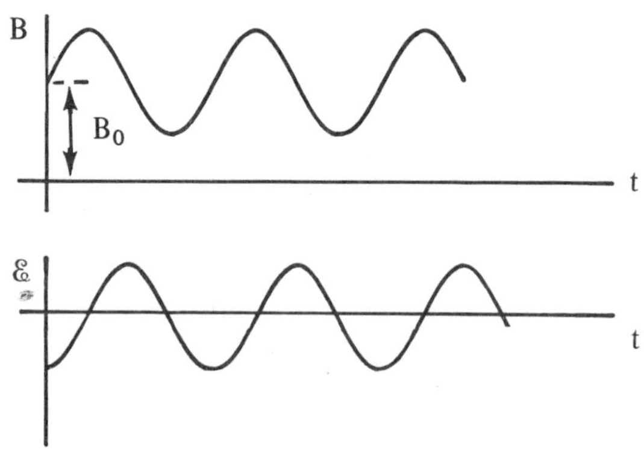

$\mathcal{E}$ is $\frac{\pi}{2}$ out of phase with B.

3 A long solenoid 1 m in length, 20 cm in diameter, has 500 turns of wire in a single layer. The current through the solenoid *increases* at the rate of $0.2 \frac{A}{sec}$. A 5 turn circular coil of radius 5 cm is placed at the center such that it is perpendicular to the field of the solenoid.

3.A

What is the magnetic field due to the solenoid when the current is I?

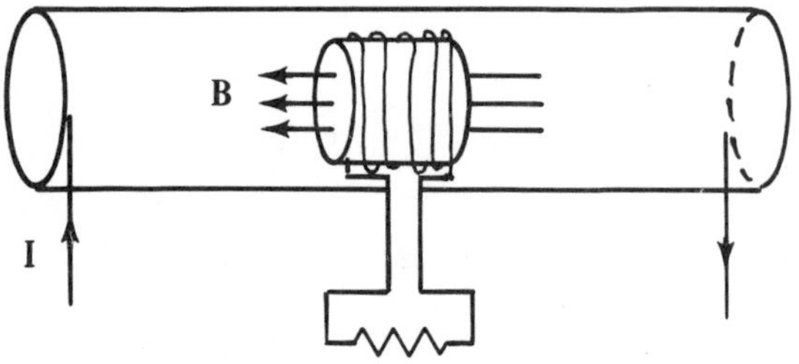

For a *long* solenoid, $B = \mu_0 nI$, the center, where n is the number of turns per unit length.

$$\therefore \quad B = 500\,\mu_0 I \qquad (1)$$

3.B

What is the rate at which B is increasing in time? Use the data stated in the problem.

$$\frac{dB}{dt} = 500\mu_0 \frac{dI}{dt}$$

But $$\frac{dI}{dt} = 0.2 \frac{A}{sec}$$

$$\therefore \quad \frac{dB}{dt} = 100\mu_0 \frac{T}{sec} \quad (3)$$

3.C

Find the magnitude of the induced emf in the 5 turn coil at the center of the solenoid. Recall that

$$\mu_0 = 4\pi \times 10^{-7} \frac{Wb}{A\ m}$$

$$|\mathcal{E}| = N\frac{d\Phi_B}{dt} = N\frac{d}{dt}(\pi r^2 B)$$

$$|\mathcal{E}| = N\pi r^2 \frac{dB}{dt}$$

$$|\mathcal{E}| = 5\pi \times (5 \times 10^{-2})^2 \times 100\mu_0 V$$

$$|\mathcal{E}| = 1.25\pi\mu_0\ V \cong 4.9\mu V$$

3.D

If the coil has a resistance of 0.7Ω, what is the value of the induced current?

$$I = \frac{|\mathcal{E}|}{R} = \frac{4.9\ \mu V}{0.7\ A}$$

$$I = 7\ \mu A$$

3.E

What is the direction of the induced current if the coil is wound as shown in frame 3.A?

Since **B** of the solenoid is increasing to the *left,* the coil must produce a flux to the right.

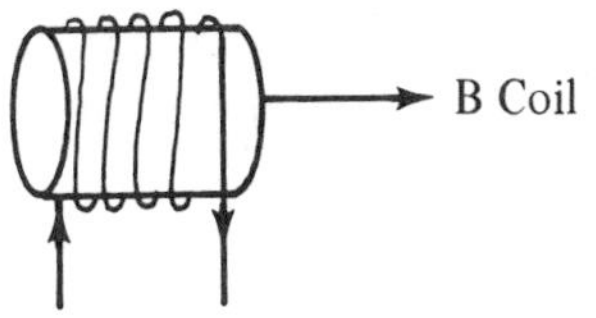

†4 *Motional EMF.* Two parallel conducting rails on a horizontal surface are connected by a resistance R at one end. A conducting crossbar of mass M and length l is placed across the rails and is given an initial velocity v_0 as shown below. A magnetic field B is directed normal out of the paper, and the rails are assumed to be frictionless with no resistance.

4.A

What is the magnetic flux through the circuit?

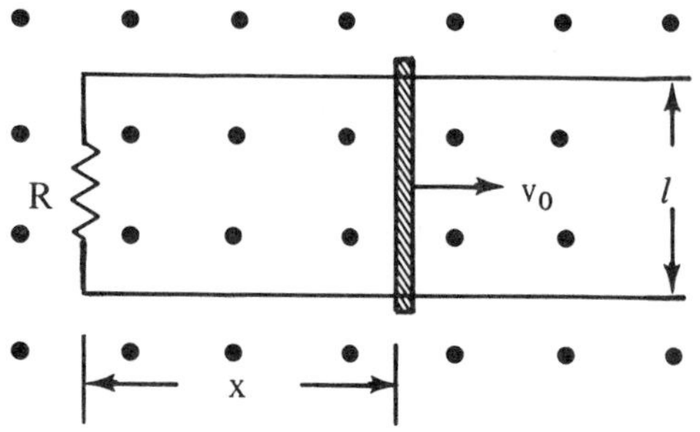

$$\Phi_B = BA$$

or,

$$\Phi_B = Blx \quad (1)$$

4.B

What is the magnitude of the induced emf in the circuit at $t = 0$, when $v = v_0$?

$$|\mathcal{E}| = \frac{d\Phi_B}{dt} = Bl\frac{dx}{dt}$$

$$|\mathcal{E}| = Blv \quad (2)$$

$$|\mathcal{E}|_{t=0} = Blv_0 \quad (3)$$

4.C

What is the direction of the induced current? What is the direction of the magnetic force on the crossbar?

By Lenz's law, the induced current is directed *clockwise*. Therefore, the magnetic force $F = Il \times B$ is to the *left* in the figure, since l is downwards in the direction of I.

4.D

What is the magnitude of the induced current in the wire when the speed of the crossbar is v?

$$I = \frac{|\mathcal{E}|}{R} = \frac{Blv}{R} \quad (4)$$

4.E

Will I and $|\mathcal{E}|$ remain constant in time? Explain.

No. The magnetic force is a *retarding* force, and therefore slows the bar down. Therefore, $v < v_0$, and both I and $|\mathcal{E}|$ will decrease in time.

4.F

Write an equation of motion (that is, Newton's second law) for the crossbar. Note that the magnetic force is *not* constant in time, since I decreases in time.

Since $F = IlB$ is to the *left*, we have

$$\Sigma F_x = M\frac{dv}{dt} = -IlB \qquad (5)$$

Using (4) in (5) gives

$$M\frac{dv}{dt} = -\frac{B^2 l^2}{R} v \qquad (6)$$

4.G

Solve (6) for v. Note that M, B, l and R are *constants.*

Let $\tau = \dfrac{MR}{B^2 l^2}$

We can rewrite (6) as

$$\frac{dv}{v} = -\frac{B^2 l^2}{MR} dt \qquad (7)$$

Integrating (7) with $v = v_0$ at $t = 0$ gives

$$\ln\left(\frac{v}{v_0}\right) = -\frac{t}{\tau}$$

or,

$$v = v_0 e^{-t/\tau} \qquad (8)$$

4.H

How do $|\mathcal{E}|$ and I vary in time?

From (2), (4) and (8), we get

$$|\mathcal{E}| = Blv = Blv_0 e^{-t/\tau} \qquad (9)$$

$$I = \frac{Blv}{R} = \frac{Blv_0}{R} e^{-t/\tau} \qquad (10)$$

4.I

What is the *total* distance the bar will travel before coming to rest? Note that you must use $x = \int v dt$, since $v = v(t)$.

Since the velocity decays exponentially in time according to (8), we get

$$x = \int_0^\infty v_0 e^{-t/\tau} dt = \tau v_0$$

or,

$$x = \frac{MR}{B^2 l^2} v_0 \qquad (11)$$

4.J

The initial kinetic energy of the bar is $\frac{1}{2}Mv_0{}^2$. What happens to this energy when the bar comes to rest? How would you calculate this energy?

It is dissipated in the form of heat in the resistance R. Since $P = \frac{dW}{dt} = I^2R$, the total energy dissipated is $W = \int^{\infty} I^2 R dt$. If you substitute (10) into this integral, the result will be $\frac{1}{2}Mv_0{}^2$. Try it!

5.A

A loop of wire of area A carries a current I as shown. Show the lines of the **B** field associated with the loop.

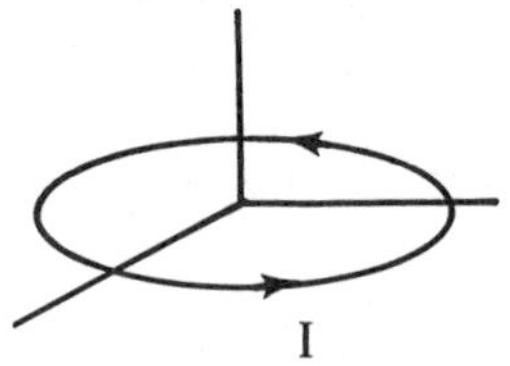

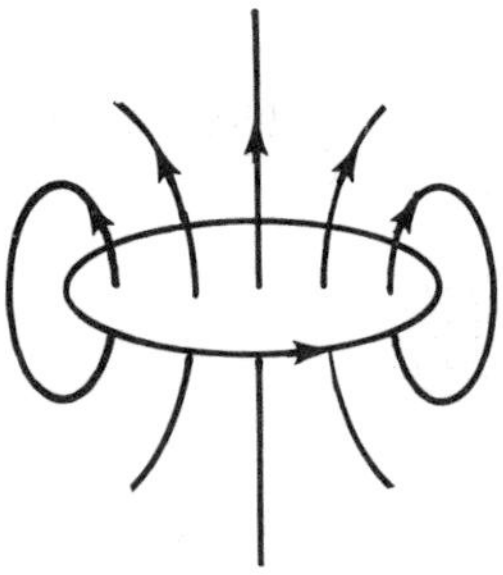

5.B

What is the magnetic flux through the loop due to its "own" **B** field?

$$\Phi_B = \int \mathbf{B}\cdot d\mathbf{A} \qquad (1)$$

5.C

If I is constant, what can you conclude about Φ_B?

Since $B \propto I$, Φ_B is constant if I is constant.

5.D

What is the induced emf in the loop if I is constant?

Zero. $\mathcal{E} = -\frac{d\Phi_B}{dt} = 0$ when Φ_B = constant.

5.E

Suppose I varies in time. Is there an induced emf in this case?

Yes. From Faraday's law,

$$\mathcal{E} = -\frac{d\Phi_B}{dt} \neq 0$$

5.F

What is this induced emf called?

We call this a *self-inducted* emf or *self-inductance.*

5.G

If a coil has N turns, the flux through each is Φ_B and the current is I, define the *inductance* of the coil.

$$L = \frac{N\Phi_B}{I} \qquad (2)$$

5.H

Apply Faraday's law to an inductor and determine the induced emf using (2).

$$\mathcal{E} = -N\frac{d\Phi_B}{dt} = -\frac{d}{dt}(N\Phi_B)$$

$$\mathcal{E} = -\frac{d}{dt}(LI) = -L\frac{dI}{dt} \qquad (3)$$

5.I

What are the units of inductance? Use the defining expression (3) for L.

$$L = \frac{|\mathcal{E}|}{\frac{dI}{dt}}$$

$$[L] = \frac{V}{\frac{A}{sec}} = 1 \text{ henry} = H$$

5.J

How much energy is stored in an inductor, L, carrying a current I?

$$U = \frac{1}{2}LI^2 \qquad (4)$$

5.K

Show that (4) has units of energy. Recall that $1\ V = 1\frac{J}{C}$ and $1\ A = 1\frac{C}{sec}$.

$$[LI^2] = HA^2 = \frac{V\ sec}{A}A^2$$

$$= \frac{J}{C}\ sec\ A = \frac{J}{C}\ sec\ \frac{C}{sec} = J$$

5.L

An air core inductor has an inductance of 8 mH. If the current is 2 A, how much energy is stored in the inductor?

$$U = \frac{1}{2}LI^2 = \frac{1}{2}\left(8 \times 10^{-3}\right)(2)^2$$

$$U = 1.6 \times 10^{-2}\ J$$

5.M

If the inductor in frame 5.L is filled with soft iron, which has a relative permeability of $\kappa_m = 3000$, what is the "new" L and the energy stored in the inductor for I = 2 A?

The field in the inductor is *increased* by a factor 3000; ∴ the "new" inductance is 3000 times larger and U is increased by 3000. So

$$L' = 3000\ L = 24\ H$$

$$U' = 3000\ U = 48\ J$$

6 *The RL Circuit* pictured below contains a switch which can be thrown to either position 1 or 2.

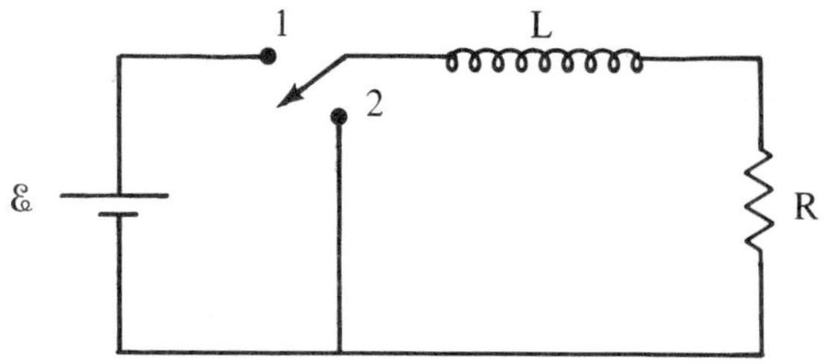

6.A

When the switch is thrown to position 1, describe the behavior of the current in the circuit.

The current will increase from zero and approach the maximum value of $\frac{\mathcal{E}}{R}$.

6.B

How does the *inductor* affect this current change?

The current through L increases from left to right, and there is self-induced emf from right to left in L. That is, L acts as a "battery" whose polarity is *opposite* to ℰ, and this opposes the current.

6.C

Write Kirchhoff's second rule as applied to the circuit when the switch is in position 1.

$$\mathcal{E} - IR - L\frac{dI}{dt} = 0 \qquad (1)$$

where IR is the voltage drop across R and $L\frac{dI}{dt}$ is the self-induced (or back) emf of the inductor.

6.D

The solution to (1) is

$$I = \frac{\mathcal{E}}{R}(1 - e^{-R/L\,t})$$

(see section 6.6) where t = 0 is the time the switch is closed. Is this solution plausible at t = 0 and t = ∞? Explain.

Yes, at t = 0, $I = \frac{\mathcal{E}}{R}(1 - e^{0}) = 0$. That is, the current starts off at 0 when the switch is thrown to position 1. At t = ∞,

$$I = \frac{\mathcal{E}}{R}(1 - e^{-\infty}) = \frac{\mathcal{E}}{R} = I_{max}\,;$$

that is, at this time, the current is not changing and the inductor no longer produces a "back" emf.

6.E

What is the time constant of the LR circuit?

$$\tau = \frac{L}{R} \qquad (2)$$

6.F

Show that τ has units of seconds.

$$[\tau] = \left[\frac{L}{R}\right] = \frac{H}{\Omega} = \frac{\frac{\text{Vsec}}{A}}{\frac{V}{A}} = \text{sec}$$

6.G

Sketch the current as a function of time. Recall that $I_{max} = \frac{\mathcal{E}}{R}$ at $t = \infty$, and

$$I = \frac{\mathcal{E}}{R}(1 - e^{-t/\tau})$$

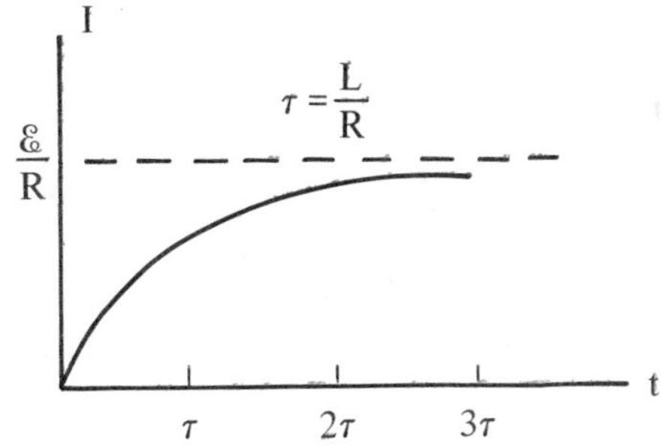

6.H

Suppose the switch is thrown to position 2 after I has reached its *maximum* equilibrium value of $\frac{\mathcal{E}}{R}$. Describe the behavior of the current in the circuit.

The current will decrease from $\frac{\mathcal{E}}{R}$ to zero in an exponential manner. Again, the self-induced emf in L opposes the change in current.

6.I

Write Kirchhoff's second rule for the right loop when the switch is in position 2.

Since there is no battery in the circuit, we get

$$IR + L\frac{dI}{dt} = 0 \qquad (3)$$

6.J

We can write (3) as $\frac{dI}{I} = -\frac{R}{L}dt$. What is the solution to this equation?

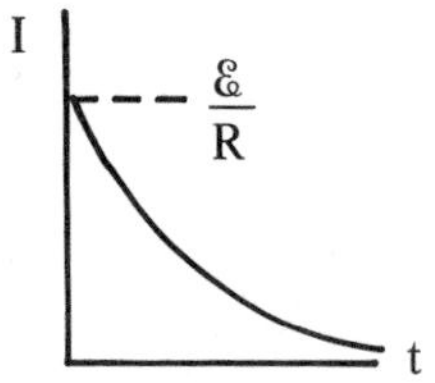

$$\int_{\mathcal{E}/R}^{I} \frac{dI}{I} = -\frac{R}{L}\int_{0}^{t} dt$$

or,

$$I = \frac{\mathcal{E}}{R} e^{-R/L\, t} \qquad (4)$$

where t = 0 corresponds to the time S is thrown to position 2.

6.K

Suppose $R = 3 \times 10^3\ \Omega$, $L = 6$ H and $\mathcal{E} = 27$ V. Determine τ and I_{max}.

$$\tau = \frac{L}{R} = \frac{6\ \text{H}}{3 \times 10^3\ \Omega} = 2\ \text{msec}$$

$$I_{max} = \frac{\mathcal{E}}{R} = \frac{27\ \text{V}}{3 \times 10^3\ \Omega} = 9\ \text{mA}$$

6.L

Write an expression for I(t) corresponding to the switch in position 2 for this circuit.

$$I = \frac{\mathcal{E}}{R} e^{-t/\tau} = 9e^{-t/\tau}\ \text{mA where } \tau = 2\ \text{msec}$$

6.8 SUMMARY

The emf *induced* $\mathcal{E}$ in a coil of N turns in proportional to the rate of change of magnetic flux through the coil. This is known as *Faraday's law* and can be expressed as

$$\mathcal{E} = -N\frac{d\Phi_B}{dt} \tag{6.2}$$

where Φ_B is the magnetic flux linked by the circuit and $\mathcal{E}$ is measured in volts. The negative sign in Faraday's law signifies that the direction of the induced current will be such that the flux produced by the current opposes the original change in flux through the loop.

The self-inductance L of a conductor is defined through the relation

$$L = -\frac{\mathcal{E}}{\frac{dI}{dt}} \tag{6.9}$$

The *energy* stored in an inductor carrying a current I is

$$U = \frac{1}{2}LI^2 \tag{6.12}$$

6.9 PROBLEMS

1. A 20 turn circular coil of radius 5 cm and resistance 0.5Ω is placed in a magnetic field which is perpendicular to the plane of the coil. If the field varies in time, according to the expression $B = 0.20t + 0.05t^2$, find (a) the induced emf in the coil at t = 6 sec and (b) the induced current in the coil at t = 6 sec.

2. A 50 turn rectangular coil of dimensions 10 cm by 20 cm is "dropped" from a region where B = 0 to a new position where B = 0.5 T directed perpendicular to the plane of the coil. If it takes 0.2 sec for this displacement, find the average induced emf in the coil.

3. A common technique used to measure the earth's magnetic field is to "flip" a coil of N turns and resistance R through some angle such that the flux changes from Φ_1 to Φ_2. If the coil is connected to a galvanometer, show that the total charge that flows through the galvanometer is given by $\dfrac{N(\Phi_1 - \Phi_2)}{R}$.

4. In Figure 6–2, suppose an external agent pulls the crossbar to the right with a force **F** such that v remains constant. (a) Show that the power supplied by the external agent is given by $\dfrac{(Blv)^2}{R}$. (b) Verify from energy considerations that this power is dissipated by the resistor in the form of heat.

5. A solenoid of radius 8 cm and length 50 cm is wound with 400 turns of wire in a single layer. The current in the solenoid changes at the rate of $3\ \dfrac{A}{sec}$. Determine the emf induced in a 3 turn insulated coil wrapped around the solenoid at its center.

6. (a) Calculate the inductance of the solenoid described in Problem 5. (b) Determine the emf induced in the solenoid if the current increases linearly at the rate of $5\dfrac{A}{sec}$.

7. A rectangular loop of wire is situated near a long straight wire which carries a current given by $I = I_0 e^{-t/\tau}$, as in Figure 6–7. (a) Find the induced emf in the loop as a function of time. (b) What is the direction of the induced current in the loop for $t > 0$?

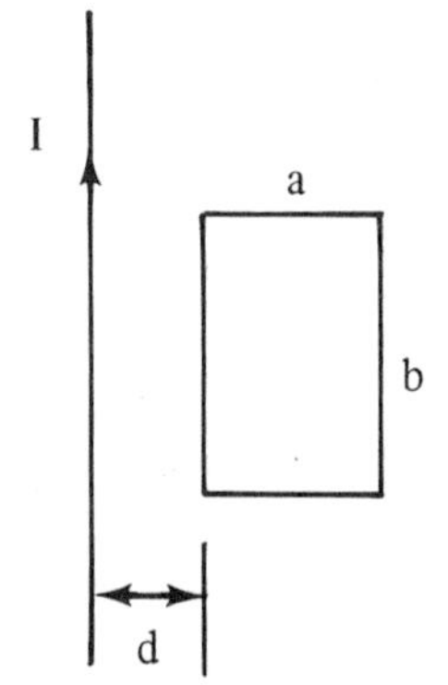

Figure 6–7

8. (a) Energy *density* in a magnetic field is given by $\frac{B^2}{2\mu_0}$ in free space. Use this formula to show that the energy stored in a solenoid carrying a current I is given by $\frac{1}{2}LI^2$. (See section 6.5.) (b) What is the energy stored in a solenoid whose inductance is 30 mH, carrying a current of 200 mA?

9. The square loop of wire shown in Figure 6-8 is connected to a galvanometer and is positioned such that the earth's magnetic field is perpendicular to the loop, with a magnitude of 5×10^{-6} T. The loop has a resistance of 0.3 Ω. If the loop is suddenly forced to collapse to zero area, what is the total charge through the galvanometer?

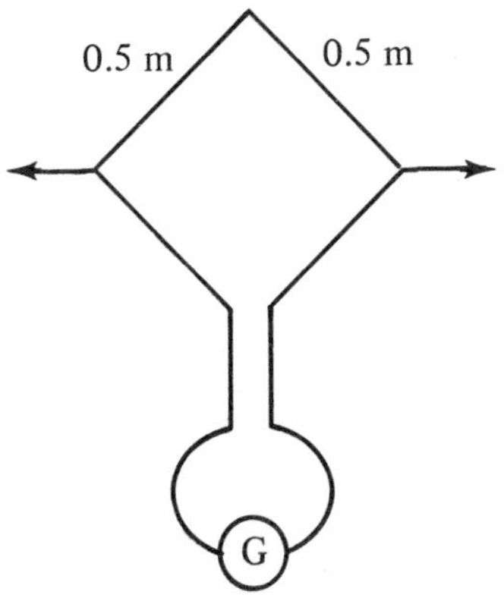

Figure 6-8

10. A series circuit containing an inductor of 12 mH and resistor of 24 Ω is connected to a 6 V battery by a switch. If the switch is closed at t = 0, find (a) the time constant of the circuit, (b) the maximum current in the circuit and (c) the time it takes the current to reach 50 per cent of its final value.

11. An inductor of 8 mH and resistor of 40 Ω are connected to a battery, as in Figure 6-9. The switch is in position 1 for a long time and is then thrown to position 2. Find the time it takes the current in the left loop to decay to 10 per cent of its initial value.

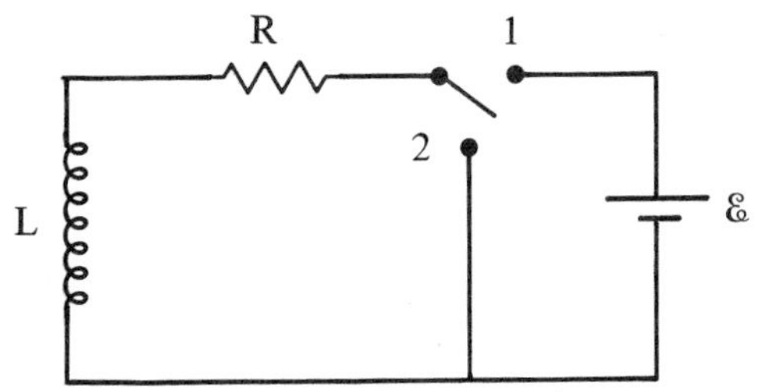

Figure 6-9

12. An LR circuit is connected to a battery of emf $\mathcal{E}$ through a switch. When the switch is closed, find the rate at which energy is being supplied to the inductor.

13. A *long* solenoid has n turns per unit length, radius R, and carries a current which varies at the rate $\frac{dI}{dt}$. Find the *induced* electric field in the regions (a) $r > R$, and (b) $r < R$.

Hint: Use Faraday's law in the form $\mathcal{E} = \oint \mathbf{E} \cdot d\mathbf{s} = -\frac{d\Phi_B}{dt}$, where Φ_B is the magnetic flux through the area enclosed by the path of integration.

7

AC CIRCUITS, OSCILLATIONS AND ELECTROMAGNETIC WAVES

7.1 OSCILLATIONS IN THE LC CIRCUIT

If a *charged* capacitor is connected to an inductor as in Figure 7-1, and the switch is closed, *oscillations* in the current and electric charge on the capacitor will occur. If the resistance of the circuit is neglected, no energy is dissipated in the form of heat and the oscillations will persist. The oscillations will correspond to the continual charge and discharge of the capacitor, where the charge on a plate alternates between $+q_0$ and $-q_0$, q_0 being the maximum charge. When the capacitor is fully charged, the total energy in the circuit is stored in the electric field of the capacitor, or $U_T = \frac{q_0^{\,2}}{2C}$. At this time, $I = 0$, so the energy stored in the inductor is zero. When the capacitor becomes totally discharged, all the energy is stored in the magnetic field of the inductor, so $U_T = \frac{1}{2}LI_0^{\,2}$, where I_0 is the *maximum* current. Therefore, the oscillations correspond to a continual transfer of energy between the capacitor and inductor. At some arbitrary time, when the capacitor has a charge q and the current in the circuit is I, both elements store energy, but the sum is still U_T. That is,

$$U_T = \frac{q^2}{2C} + \frac{1}{2}LI^2 = \text{constant} \tag{7.1}$$

Differentiating this expression with respect to time, and noting that U_T is a constant, gives

$$\frac{q}{C}\frac{dq}{dt} + LI\frac{dI}{dt} = 0 \tag{7.2}$$

But $I = \frac{dq}{dt}$, and $\frac{dI}{dt} = \frac{d^2q}{dt^2}$; therefore, Equation (7.2) can be written as

$$L\frac{d^2q}{dt^2} + \frac{q}{C} = 0 \tag{7.3}$$

We can solve Equation (7.3) for q by noting that it is of the *same* form as the harmonic oscillator equation in mechanics. A *particular* solution is

$$q = q_0 \cos(\omega t) \tag{7.4}$$

where this solution *requires* that $q = q_0$ at $t = 0$. To obtain the value for ω, we substitute Equation (7.4) into Equation (7.3). This gives

$$\omega = \frac{1}{\sqrt{LC}} \tag{7.5}$$

That is, the angular frequency of the oscillation in q depends solely on the values of L and C.

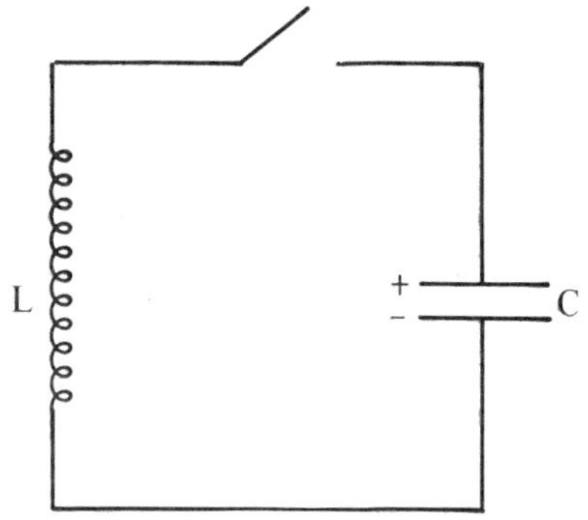

Figure 7-1 If the capacitor is initially charged, and the switch is closed, the charge and current in the LC circuit will oscillate with a characteristic frequency given by $\omega = \frac{1}{\sqrt{LC}}$.

Since q varies in a periodic fashion, the current also varies periodically. Differentiating Equation (7.4) with respect to time gives

$$I = \frac{dq}{dt} = -\omega q_0 \sin(\omega t) = -I_0 \sin(\omega t) \tag{7.6}$$

where $I_0 = \omega q_0$ is the maximum current in the circuit. We see that $q = q_0$ and $I = 0$ at $t = 0$. This corresponds to *all* of the energy stored in the capacitor when it is fully charged. Also, $I = I_0$ when $q = 0$, corresponding to *all* of the energy stored in the inductor when the current is a maximum. We can also show that our solution is consistent with energy conservation, that is, Equation (7.1).

$$U_T = \frac{q^2}{2C} + \frac{1}{2}LI^2 = \frac{q_0^{\ 2}}{2C}\cos^2 \omega t + \frac{LI_0^{\ 2}}{2}\sin^2 \omega t \tag{7.7}$$

That is, the *maximum* energy stored in the capacitor is $\frac{q_0^2}{2C}$, whereas the *maximum* energy stored in the inductor is $\frac{LI_0^2}{2}$. However, at any time t the sum of the energies stored in L and C is *a constant,* given by

$$U_T = \frac{q_0^2}{2C} = \frac{LI_0^2}{2} \tag{7.8}$$

It should be noted that U_T = constant *only* if losses in energy due to radiation of electromagnetic waves are neglected. Ideally, if these losses together with the circuit resistance are neglected, the oscillations in the circuit persist *indefinitely.*

Example 7.1

Suppose the circuit in Figure 7-1 has L = 2.81 mH and C = 9 pF. The capacitor is initially charged with a 20 V battery when the switch is open. At t = 0, the switch is closed. (a) Determine the frequency of the oscillations.

$$f = \frac{\omega}{2\pi} = \frac{1}{2\pi\sqrt{LC}} = \frac{1}{2\pi\sqrt{2.81 \times 9 \times 10^{-15}}} = 10^6 \text{ Hz}$$

(b) Find the maximum charge on the capacitor and the maximum current in the circuit.

$$q_0 = C\mathcal{E}_0 = 9 \times 10^{-12} \text{ F} \times 20 \text{ V} = 1.8 \times 10^{-10} \text{ C}$$

$$I_0 = \omega q_0 = 2\pi f q_0 = 2\pi \times 10^6 \times 1.8 \times 10^{-10} = 1.3 \times 10^{-3} \text{ A}$$

(c) Determine the charge and current as functions of time.

$$q = q_0 \cos(\omega t) = 1.8 \times 10^{-10} \cos(\omega t) \text{ C}$$

$$I = -I_0 \sin(\omega t) = -1.13 \times 10^3 \sin(\omega t) \text{ A}$$

where

$$\omega = 2\pi f = 2\pi \times 10^6 \ \frac{\text{rad}}{\text{sec}}$$

(d) Calculate the maximum energy stored in the circuit and show that Equation (7.8) is satisfied.

$$U_T = \frac{q_0^2}{2C} = \frac{(1.8 \times 10^{-10})^2}{2 \times 9 \times 10^{-12}} = 1.8 \times 10^{-9} \text{ J}$$

or,

$$U_T = \frac{LI_0^2}{2} = \frac{2.81 \times 10^{-3} \times (1.13 \times 10^{-3})^2}{2} = 1.8 \times 10^{-9} \text{ J}$$

7.2 SERIES AC CIRCUITS WITH SINGLE ELEMENTS

An AC circuit is one which is powered by an alternating voltage, designated by the symbol (~). If the emf of the source varies sinusoidally in time, we can write

$$\mathcal{E} = V\sin\omega t \tag{7.9}$$

where V is the maximum emf of the source and $\omega = 2\pi f$, f being the frequency in Hz. Most power circuits operate at a frequency of 60 Hz. Since $\mathcal{E}$ varies sinusoidally in time, it follows that the current behaves similarly. If the circuit contains *only* a resistor and AC source as in Figure 7–2(*a*), then the *instantaneous* current is given by

$$i = \frac{\mathcal{E}}{R} = \frac{V}{R}\sin\omega t = I\sin\omega t \tag{7.10}$$

where I is the *maximum* current in the circuit. We say that i and $\mathcal{E}$ *are in phase,* since they both vary as sin ωt. The rotor diagram below the circuit in Figure 7–2(*a*) shows that i and $\mathcal{E}$ are in phase. The *projections* of I and V along the y axis give the instantaneous values i and $\mathcal{E}$, respectively.

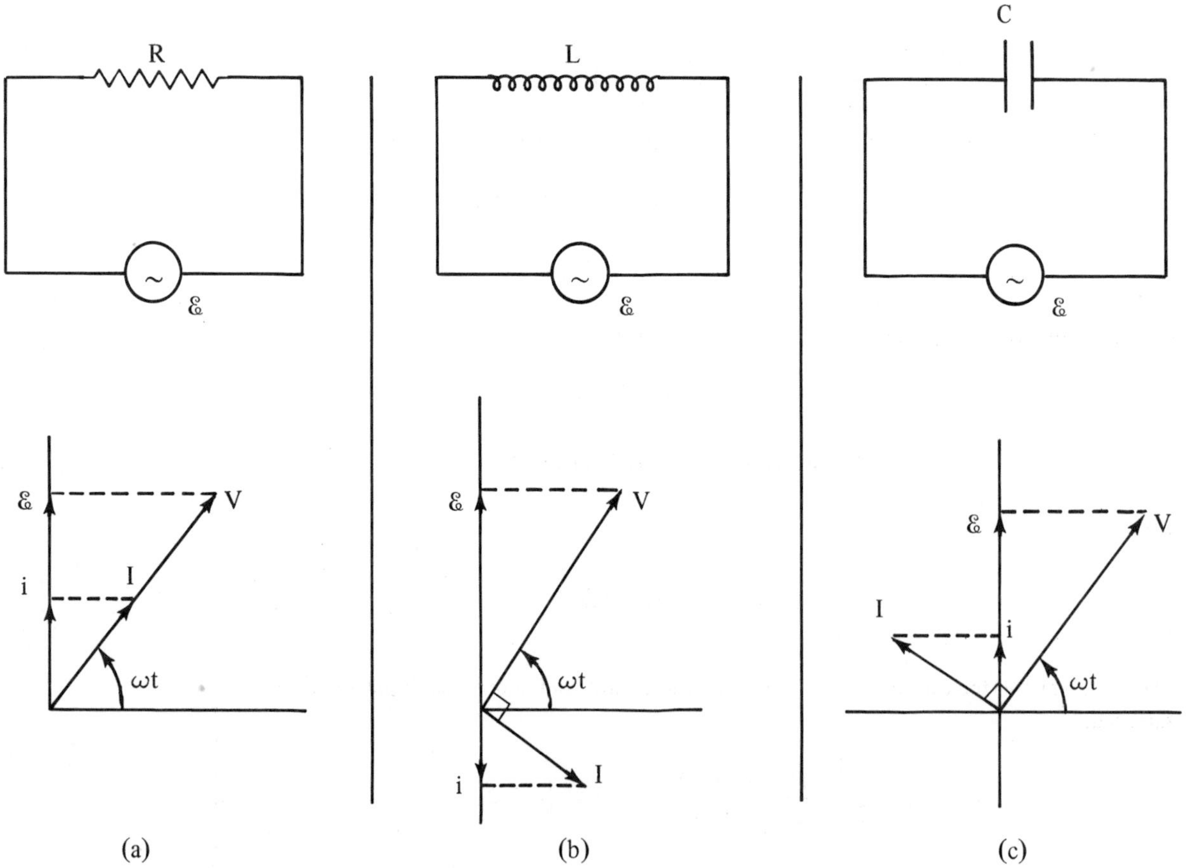

Figure 7–2 Single element AC circuits, where in case (*a*) the current and voltage are in phase, in case (*b*) the current "lags" the voltage by $\frac{\pi}{2}$ and in case (*c*) the current "leads" the voltage by $\frac{\pi}{2}$. The diagrams below each circuit represent the voltages and currents at "rotating vectors."

If the circuit contains *only* an inductor and AC source as in Figure 7-2(*b*), Kirchhoff's second rule applied to the closed loop gives

$$V\sin\omega t = L\frac{dI}{dt} \tag{7.11}$$

or,

$$dt\sin\omega t = \frac{L}{V}dI \tag{7.12}$$

Integrating Equation (7.12) gives

$$i = -\frac{V}{L}\cos\omega t = -I\cos\omega t \tag{7.13}$$

where $I = \frac{V}{\omega L}$ is the maximum current. The negative sign in Equation (7.13) can be interpreted by saying that the current *lags* the voltage by $\frac{\pi}{2}$. This is equivalent to writing Equation (7.13) as $i = I\sin\left(\omega t - \frac{\pi}{2}\right)$, where $\phi = -\frac{\pi}{2}$. The rotor diagram in Figure 7-2(*b*) shows that the current is $\frac{\pi}{2}$ out of phase with $\mathcal{E}$.

If we define the *inductive reactance* X_L as $X_L = \omega L$, we can express the maximum current as

$$I = \frac{V}{\omega L} = \frac{V}{X_L} \tag{7.14}$$

where the unit of X_L is the ohm. We can think of Equation (7.14) as an analogy to Ohm's law for an inductive circuit.

If an AC circuit contains *only* a capacitor and AC source, as in Figure 7-2(*c*), we can write the instantaneous charge on the capacitor as

$$q = C\mathcal{E} = CV\sin\omega t \tag{7.15}$$

Therefore, the instantaneous current is given by

$$i = \frac{dq}{dt} = \omega CV\cos\omega t = I\cos\omega t \tag{7.16}$$

where $I = \omega CV$ is the *maximum* current. Since i can also be written as $i = I\sin\left(\omega t + \frac{\pi}{2}\right)$, the current *leads* the voltage by $\frac{\pi}{2}$, as shown in Figure 7-2(*c*).

If we define the *capacitive reactance* X_C as $X_C = \frac{1}{\omega C}$, then the maximum current is given by

$$I = \omega CV = \frac{V}{X_C} \tag{7.17}$$

The unit of X_C is also the ohm.

Example 7.2

(a) The inductive reactance of a 5 mH inductor at an angular frequency of $3 \times 10^3 \frac{\text{rad}}{\text{sec}}$ is given by

$$X_L = \omega L = 3 \times 10^3 \frac{\text{rad}}{\text{sec}} \times 5 \times 10^{-3} \text{ H} = 15\ \Omega$$

(b) The capacitive reactance of a 2μF capacitor at an angular frequence of $3 \times 10^3 \frac{\text{rad}}{\text{sec}}$ is given by

$$X_C = \frac{1}{\omega C} = \frac{1}{3 \times 10^3 \times 2 \times 10^{-6}} = 167\ \Omega$$

Note that X_C *decreases* with increasing ω, while X_L *increases* with increasing ω. Why?

7.3 THE RLC SERIES CIRCUIT

Consider the AC circuit shown in Figure 7–3(*a*) which is a series connection of a resistor, an inductor and a capacitor. Kirchhoff's second rule applied to the closed loop requires that $\mathcal{E} = v_R + v_L + v_C$. $\mathcal{E}$ is the instantaneous emf of the source, and v_R, v_L and v_C are the instantaneous potential differences across R, L and C. If V, V_R, V_L and V_C are represented by rotating vectors, as in Figure 7–3, then $\mathcal{E}$, v_R, v_L and v_C are the *projection* of these vectors along the y axis.

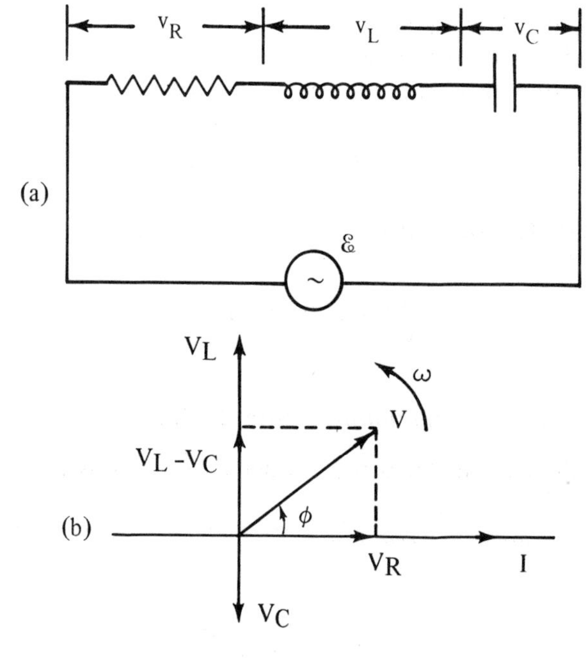

Figure 7–3

Since the voltage amplitudes must add *vectorially,* the total magnitude of the "rotating vector" V is related to V_R, V_L and V_C by the expression

$$V = \sqrt{V_R{}^2 + (V_L - V_C)^2} \tag{7.18}$$

Since $V_R = IR$, $V_L = IX_L$ and $V_C = IX_C$, Equation (7.18) can also be written as

$$V = I\sqrt{R^2 + (X_L - X_C)^2} = I\sqrt{R^2 + X^2} \tag{7.19}$$

where $X = X_L - X_C$ is the *total reactance* of the circuit. The *impedance* of the circuit is defined as $Z = \sqrt{R^2 + X^2}$, so Equation (7.19) can be written as

$$V = IZ \tag{7.20}$$

where,

$$Z = \sqrt{R^2 + X^2} = \sqrt{R^2 + (X_L - X_C)^2} \tag{7.21}$$

Obviously, Z must have the unit of ohm.

It should be noted that the "rotating vector" diagram in Figure 7–3 is drawn for the case where I *lags* V by an angle ϕ, so if $\mathcal{E} = V\sin\omega t$, then $i = I\sin(\omega t - \phi)$. This is valid only if $X_L > X_C$, since in this case

$$\tan\phi = \frac{V_L - V_C}{V_R} = \frac{X_L - X_C}{R} = \frac{X}{R} \tag{7.22}$$

If $X_C > X_L$, then the "rotating vector" diagram becomes one in which I *leads* V by an angle ϕ, corresponding to a negative value for X, hence a negative value for $\tan\phi$.

Example 7.3

Suppose that the inductor and capacitor in Example 7.2 are in a series circuit for which $\omega = 3 \times 10^3 \frac{\text{rad}}{\text{sec}}$, $R = 200\ \Omega$ and $V = 300$ V. Then $X_L = 15\ \Omega$, $X_C = 127\ \Omega$ and the impedance is given by

$$Z = \sqrt{R^2 + (X_L - X_C)^2} = \sqrt{200^2 + (-112)^2} = 230\ \Omega$$

The current *amplitude* is $I = \frac{V}{Z} = \frac{300}{230} = 1.30$ A.

The voltage amplitudes across R, L and C are

$$V_R = IR = 1.30\,(200) = 260 \text{ V}$$

$$V_L = IX_L = 1.30\,(15) = 19.5 \text{ V}$$

$$V_C = IX_C = 1.30\,(127) = 165 \text{ V}$$

Since $X_C > X_L$, I *leads* V by an angle ϕ, where

$$\tan\phi = \frac{X_L - X_C}{R} = -\frac{112}{200} = -0.56$$

or,

$$\phi \cong -27^\circ$$

If the time-varying emf varies as $\sin\omega t$, then the instantaneous emf is given as

$$\mathcal{E} = V\sin\omega t = 300\sin\omega t \text{ V}$$

and the instantaneous current is given as

$$i = I\sin(\omega t - \phi) = 1.30\sin(\omega t - \phi) \text{ A}$$

where $\omega = 3 \times 10^3 \frac{\text{rad}}{\text{sec}}$ and $\phi = -27^\circ = -0.47$ rad. Finally, the *instantaneous* potential differences across R, L and C are given by

$$v_R = iR = 260\sin(\omega t - \phi) \text{ V}$$

$$v_L = L\frac{di}{dt} = IX_L\cos(\omega t - \phi) = 19.5\cos(\omega t - \phi) \text{ V}$$

$$v_C = \frac{q}{C} = \frac{1}{C}\int i dt = -\frac{1.30}{\omega C}\cos(\omega t - \phi)$$

$$v_C = -165\cos(\omega t - \phi) \text{ V}$$

7.4 POWER IN AN AC CIRCUIT

The *instantaneous power* which is delivered to an RLC circuit by the source is given by

$$P = i\mathcal{E} = IV\sin\omega t \sin(\omega t - \phi) \tag{7.23}$$

The *average power* is obtained by integrating Equation (7.23) according to the expression

$$P_{av} = \frac{1}{T}\int_0^T P dt$$

where $T = \frac{1}{f}$ is the time for one cycle. The student should verify that the integration gives

$$P_{av} = \frac{1}{2} IV\cos\phi \tag{7.24}$$

Measurements of voltage and current in AC circuits are usually made with instruments calibrated to read root mean square (rms) values (not to be confused with average values which are zero for sinusoidal variations). The rms values are given by

$$V_{rms} = \sqrt{<v^2>_{av}} = \left[\frac{1}{T}\int_0^T V^2 \sin^2 \omega t dt\right]^{1/2} = \frac{V}{\sqrt{2}} \tag{7.25}$$

$$I_{rms} = \sqrt{<i^2>_{av}} = \left[\frac{1}{T}\int_0^T I^2 \sin^2 (\omega t - \phi)\, dt\right]^{1/2} = \frac{I}{\sqrt{2}} \tag{7.26}$$

Therefore, the average power can also be written as

$$P_{av} = V_{rms} I_{rms} \cos\phi \tag{7.27}$$

However, we note from Figure 7-3 the $V\cos\phi = V_R = IR$; therefore, Equation (7.27) can be written as

$$P_{av} = I_{rms}{}^2 R \tag{7.28}$$

In other words, the power loss in an AC circuit occurs only in the resistor. There is no power loss in the inductor or capacitor. The constant $\cos\phi$ is referred to as the *power factor.* For a *pure resistive* circuit, $\phi = 0$. However, ϕ is *nonzero* for any AC circuit containing an inductor or capacitor.

7.5 RESONANCE IN A SERIES RLC CIRCUIT

We see from Equations (7.20) and (7.21) that the current in a series RLC circuit is given by

$$I = \frac{V}{Z} = \frac{V}{\sqrt{R^2 + (X_L - X_C)^2}}$$

so I depends on the frequency of the AC source. When $X_L = X_C$, we see that $Z = R$ and I has a *maximum* value of $\frac{V}{R}$. The frequency, ω_R, for which $X_L = X_C$ is called the *resonant frequency* of the circuit, which corresponds to the frequency for which

I is a *maximum.* If $R = 0$, the current would *theoretically* be infinite. At resonance, the condition $X_L = X_C$ corresponds to

$$\omega L = \frac{1}{\omega C} \quad or \quad \omega_R = \sqrt{\frac{1}{LC}} \tag{7.29}$$

Note that this is identical to the natural oscillator frequency given by Equation (7.5). In other words, if the frequency of the source *matches* the natural oscillator frequency determined only by L and C, the current will be a maximum.

7.6 MAXWELL'S EQUATIONS

The laws of electromagnetism can be summarized in a set of expressions called *Maxwell's equations.* For simplicity, we will write the equations assuming that no dielectric or magnetic materials are present.

$$\oint \mathbf{E} \cdot d\mathbf{A} = \frac{q}{\epsilon_0} \tag{7.30}$$

$$\oint \mathbf{B} \cdot d\mathbf{A} = 0 \tag{7.31}$$

$$\oint \mathbf{B} \cdot d\mathbf{s} = \mu_0 \left[\epsilon_0 \frac{d\Phi_E}{dt} + I\right] \tag{7.32}$$

$$\oint \mathbf{E} \cdot d\mathbf{s} = -\frac{d\Phi_B}{dt} \tag{7.33}$$

Note that we have already described these laws in previous chapters. Equation (7.30) is Gauss's law in electricity and describes how to relate the **E** field to a charge distribution (which, you should recall, follows from Coulomb's inverse square law). Equation (7.31) represents Gauss's law in magnetism, and essentially means that isolated magnetic monopoles do not exist, for if they did, a term analogous to q in Equation (7.30) would appear on the right side of Equation (7.31). Equation (7.32) is the general form of Ampere's law, and includes the displacement current term $\mu_0 \epsilon_0 \frac{d\Phi_E}{dt}$. This expression relates the magnetic field to a current *or* changing electric field. That is, a current-carrying conductor and/or a changing **E** field set up a magnetic field. Finally, Equation (7.33) is Faraday's law of induction, which describes the relationship between a changing **B** field and an **E** field. Note that Equations (7.30) and (7.31) are symmetric in the sense that both surface integrals are equal to a constant. Likewise, Equations (7.32) and (7.33) are symmetric in that the line integrals of **E** and **B** around a closed path are related to the time derivative of magnetic (or electric) flux. This symmetry is consistent with the Lorentz transformation equations and Einstein's theory of relativity.

7.7 PROPERTIES OF ELECTROMAGNETIC WAVES

When charged particles accelerate or decelerate, electromagnetic waves are radiated from the source which might be a radio or a radar antenna. These waves propagate with a speed c equal to the speed of light. They carry energy and momentum, which, in turn, can be transferred to other objects located in the path of the waves. From Faraday's law of induction, it can be shown that the instantaneous values of E and B are related by the expression

$$\frac{\partial E}{\partial x} = -\frac{\partial B}{\partial t} \tag{7.34}$$

where the derivatives here are partial derivatives. For a *plane wave* traveling in the x direction, we can write the magnitudes of E and B as

$$E = E_0 \cos(kx - \omega t) \tag{7.35}$$

and

$$B = B_0 \cos(kx - \omega t) \tag{7.36}$$

where $k = \frac{2\pi}{\lambda}$, $\omega = 2\pi f$ and E_0 and B_0 are the maximum values of E and B, respectively. Substituting these expressions into Equation (7.34) gives

$$E = cB \tag{7.37}$$

where

$$c = \frac{\omega}{k} = \lambda f \tag{7.38}$$

Likewise, from Ampere's law, Equation (7.32), it can be shown that

$$\frac{\partial B}{\partial x} = -\mu_0 \epsilon_0 \frac{\partial E}{\partial t} \tag{7.39}$$

Substituting Equations (7.35) and (7.36) into Equation (7.39), we find that

$$\frac{E}{B} = \frac{k}{\mu_0 \epsilon_0 \omega} = \frac{1}{\mu_0 \epsilon_0 c} \tag{7.40}$$

But from Equation (7.37), $\frac{E}{B} = c$; therefore, substitution of this into Equation (7.40) gives

$$c = \frac{1}{\sqrt{\mu_0 \epsilon_0}} = 3 \times 10^8 \frac{m}{sec} \tag{7.41}$$

In other words, Maxwell's equations *predict* the speed of light and relate it to the fundamental constants μ_0 and ϵ_0. More recently, the speed of light was determined to be $2.9979250 \times 10^8 \frac{\text{m}}{\text{sec}}$.

Electromagnetic waves can transfer energy to objects placed in their path. The rate at which energy passes through a unit area of a surface is described by *Poynting's vector,* **S**, given by

$$\mathbf{S} = \frac{1}{\mu_0} \mathbf{E} \times \mathbf{B} \tag{7.42}$$

S can also be described as the energy flux. **S** has units of $\frac{\text{watts}}{\text{m}^2}$, and is in the direction of propagation of the wave. For example, in a plane wave, if **E** is in the y direction and **B** is in the z direction, **S** will be in the x direction. For the special case of a plane wave, S is given in magnitude by

$$S = \frac{EB}{\mu_0} \tag{7.43}$$

The *average value* of S over one cycle of the wave can be obtained by substituting Equations (7.35) and (7.36) into Equation (7.43). Since the average value of $\cos^2(kx - \omega t)$ over one cycle is $\frac{1}{2}$, it follows that

$$S_{av} = \frac{E_0 B_0}{2\mu_0} \tag{7.44}$$

Electromagnetic waves transport *linear momentum,* and it follows that they exert *pressure* (radiation pressure) on surfaces. Such radiation pressures (although very small for typical experiments) have been measured in the laboratory by shining light on mirrors suspended on fibers, which in turn rotate through a small angle. If U is the total energy absorbed by a surface in a time t, the momentum delivered to this surface is given by

$$p = \frac{U}{c} \tag{7.45}$$

If the object is a perfect reflector (say, a mirror with a 100 per cent reflecting surface), the net momentum delivered in a time t is $\frac{2U}{c}$. For a perfectly black body, where all the energy is absorbed (none reflected), the net momentum delivered in a time t is $\frac{U}{c}$. In practice, the momentum delivered has a value between $\frac{U}{c}$ and $\frac{2U}{c}$.

Finally, electromagnetic waves can be *polarized,* which means that all the **E** vectors oscillate in a common plane of vibration (likewise for the **B** vectors). For a plane polarized wave, the plane of polarization of **E** is perpendicular to the plane of polarization of **B**. Electromagnetic waves are *transverse* waves (since **E** and **B** oscillate perpendicular to the direction of propagation), so it follows that they can be polarized. Such polarization properties do not exist for longitudinal waves. Radio waves can be polarized by inserting slotted sections of metal between the source (transmitter) and the receiver. Light waves can be polarized with a special polarizing material. In Chapter 9 we will have more to say about the nature of polarized light.

The spectrum of electromagnetic radiation ranges from the long wavelength radio waves (for example, $\lambda = 10^5$ m, $f = \frac{c}{\lambda} = 3 \times 10^3$ Hz) to the short wavelength x-rays (for example, $\lambda = 10^{-12}$ m, $f = \frac{c}{\lambda} = 3 \times 10^{20}$ Hz). Radiations intermediate to these extreme cases include microwaves, infrared radiation, visible light and ultraviolet light.

(See Programmed Exercise 3 for review.)

Example 7.4

A plane electromagnetic wave whose frequency is 6×10^7 Hz travels in free space in the +x direction, as in Figure 7-4. The magnetic field vector **B** has an amplitude of $5 \times 10^{-7} \frac{\text{webers}}{\text{m}^2}$ and, at some instant of time, is directed along the +z axis. (a) Determine the wavelength of this wave. (b) Calculate the magnitude and direction of the **E** vector. (c) Write expressions analogous to Equations (7.35) and (7.36) with appropriate values for k and ω. (d) Determine S_{av} for this wave.

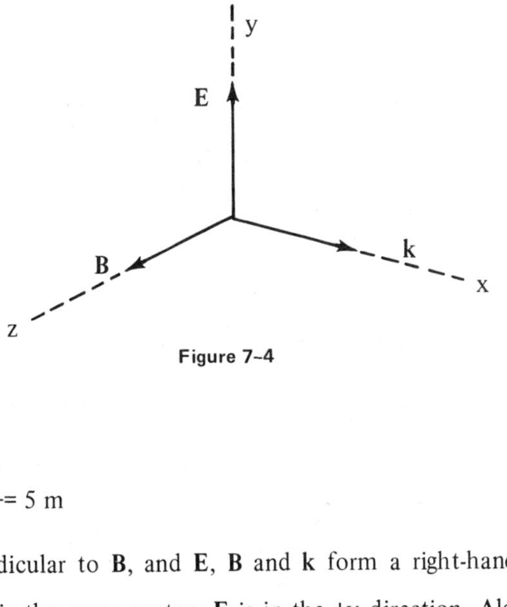

Figure 7-4

Solution

$$\text{(a) } \lambda = \frac{c}{f} = \frac{3 \times 10^8 \frac{\text{m}}{\text{sec}}}{6 \times 10^7 \text{Hz}} = 5 \text{ m}$$

(b) Since **E** is perpendicular to **B**, and **E**, **B** and **k** form a right-handed coordinate as shown, where $k = \frac{2\pi}{\lambda}$ is the wave vector, **E** is in the +y direction. Also, from Equation (7.37), we get

$$E_0 = cB_0 = 3 \times 10^8 \frac{\text{m}}{\text{sec}} \times 5 \times 10^{-7} \frac{\text{webers}}{\text{m}^2}$$

$$E_0 = 1500 \frac{\text{volts}}{\text{m}}$$

(c) For this wave, $\omega = 2\pi f = 12\pi \times 10^7 \frac{\text{rad}}{\text{sec}}$ and $k = \frac{2\pi}{\lambda} = 0.4\pi \text{ m}^{-1}$. Substituting these values, together with the results of (b) into Equations (7.35) and (7.36), gives

$$E = 1500 \cos(0.4\pi x - 12\pi \times 10^7 t) \frac{\text{volts}}{\text{m}}$$

$$B = 5 \times 10^{-7} \cos(0.4\pi x - 12\pi \times 10^7 t) \frac{\text{webers}}{\text{m}^2}$$

where x is in meters and t in seconds.

(d) $S_{av} = \frac{E_0 B_0}{2\mu_0} = \frac{1500 \times 5 \times 10^{-7}}{2 \times 4\pi \times 10^{-7}} \cong 300 \frac{\text{watts}}{\text{m}^2}$

The student should verify that the units of $\frac{E_0 B_0}{\mu_0}$ reduce to $\frac{\text{watts}}{\text{m}^2}$.

Example 7.5

The average solar energy normally incident on the earth's surface per unit time per unit area is given by $1340 \frac{\text{watts}}{\text{m}^2}$. (a) Calculate the total energy that falls on an area whose measurements are 5 m × 5 m in a time of one hour. (b) If all the energy can be converted to electrical energy, how many 60 watt bulbs can be lit with this source? (c) Determine the total momentum delivered to this area in a time of one hour if it represents a perfectly absorbing surface.

Solution

(a) The Poynting's vector has a magnitude given by $1340 \frac{\text{watts}}{\text{m}^2}$; therefore, since S is energy per unit area per unit time, the total energy U is

$$U = SAt = \left(1340 \frac{\text{J}}{\text{sec m}^2}\right) \times (5 \text{ m})^2 \times (3600 \text{ sec})$$

$$U = 1.2 \times 10^8 \text{ J}$$

(b) The total power that the sun delivers to this area is

$$\left(1340 \frac{\text{watts}}{\text{m}^2}\right) \times (5 \text{ m})^2 = 3.35 \times 10^4 \text{ watts}$$

Therefore, the total number of 60 watt bulbs which can be lit is

$$n = \frac{3.35 \times 10^4 \text{ watts}}{60 \frac{\text{watts}}{\text{bulb}}} \cong 560 \text{ bulbs}$$

In reality, the efficiency of the conversion from solar energy to electrical energy is far less than 100 per cent, so the number here is a gross overestimate. However, the power is quite substantial for large areas. Try a calculation for say one square mile (about $2.6 \times 10^6 \text{ m}^2$).

(c) $p = \frac{U}{c} = \frac{1.2 \times 10^8 \text{ J}}{3 \times 10^8 \frac{\text{m}}{\text{sec}}} = 0.4 \text{ kg} \frac{\text{m}}{\text{sec}}$

7.8 PROGRAMMED EXERCISES

1 The charge on the capacitor of the LC circuit shown below is q_0 *before* the switch is closed.

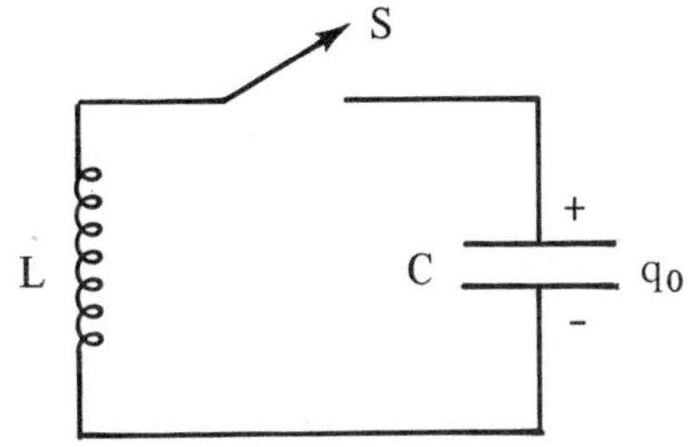

1.A

If the switch is closed at $t = 0$, describe qualitatively what happens to the charge on the capacitor and the current in the circuit.

The capacitor will discharge through the inductor, setting up a current (ccw) in the circuit. When $q = 0$, the current is a maximum. Charge will continue to flow until the capacitor is charged with opposite polarity, $-q_0$. The current will then reverse (cw) and the process continues.

1.B

Describe the *oscillations* that occur for I and q with time for $t > 0$. What do these oscillations mean in terms of the energy in the circuit?

Both q and I vary sinusoidally in time, where q is a maximum when $I = 0$ (initial condition) and I is a maximum when $q = 0$. At $t = 0$, $q = q_0$ and *all* of the energy is stored in C, or $U_T = \frac{q_0^2}{2C}$. Later, when $q = 0$, $I = I_0$ and *all* of the energy is stored in L, or $U_T = \frac{1}{2} L I_0^2$.

1.C

Under what conditions is the *total* energy U_T constant in time?

U_T is constant if we neglect the resistance in the circuit and energy losses due to radiation of EM waves.

1.D

Write an expression for the total energy stored in the circuit for an *arbitrary* charge q and current I.

$$U_T = \frac{q^2}{2C} + \frac{LI^2}{2} \qquad (1)$$

where

$$U_T = \frac{q_0^2}{2C} = \frac{LI_0^2}{2} \qquad (2)$$

1.E

If we differentiate (1) with respect to time, and note that $I = \frac{dq}{dt}$, we obtain the expression

$$L\frac{d^2q}{dt} + \frac{q}{C} = 0$$

Show that this result also follows from the *loop theorem.*

Kirchhoff's loop theorem gives

$$-L\frac{dI}{dt} - \frac{q}{C} = 0$$

But $I = \frac{dq}{dt}$, so this reduces to

$$L\frac{d^2q}{dt^2} + \frac{q}{C} = 0 \qquad (3)$$

This again shows that the loop theorem is an energy conservation concept.

1.F

If we rewrite (3) as

$$\frac{d^2q}{dt^2} = -\frac{1}{LC}q$$

we note that it is the *simple harmonic oscillator* equation. Show that a *particular* solution of (3) is

$$q = q_0 \cos\omega t$$

where $\omega = \sqrt{\frac{1}{LC}}$.

$$q = q_0 \cos \omega t$$

$$\frac{dq}{dt} = -q_0 \omega \sin\omega t$$

$$\frac{d^2q}{dt^2} = -q_0 \omega^2 \cos\omega t$$

$$\therefore \quad -q_0 \omega^2 \cos\omega t = -\frac{1}{LC} q_0 \cos\omega t$$

which *requires* that

$$\omega = \sqrt{\frac{1}{LC}} \qquad (4)$$

1.G

What is ω?

ω is the oscillator frequency, which depends *only* on L and C. $\omega = 2\pi f$, where f is the usual frequency in Hz and ω is in $\frac{\text{rad}}{\text{sec}}$.

1.H

If you compare the electrical oscillator equation with the mechanical oscillator equation, what are the *analogous* parameters?

$$\frac{d^2q}{dt^2} = -\frac{1}{LC}q; \quad \frac{d^2x}{dt^2} = -\frac{k}{m}x$$

$$q \sim x$$

$$L \sim m$$

$$C \sim \frac{1}{k}$$

Since $I = \frac{dq}{dt}$ and $v = \frac{dx}{dt}$, we also see that $I \sim v$.

1.I

We use the solution $q = q_0 \cos\omega t$, since this meets the condition that $q = q_0$ at $t = 0$. What is the current as a function of time?

$$I = \frac{dq}{dt} = \frac{d}{dt}(q_0 \cos\omega t)$$

$$I = -q_0 \omega \sin\omega t \qquad (5)$$

or,

$$I = -I_0 \sin\omega t$$

1.J

Note that the maximum value of q is q_0, and the maximum value of I is $q_0\omega$. Plot q and I *vs* t.

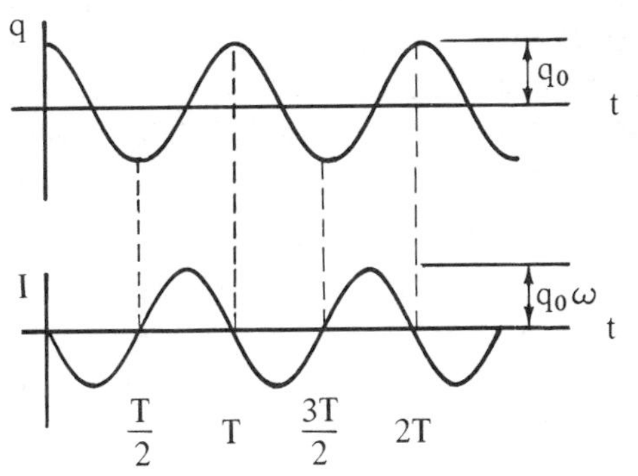

1.K

Write expressions for the energies stored in C and L as functions of time, assuming $q = q_0 \cos\omega t$.

$$U_C = \frac{q^2}{2C} = \frac{q_0^2}{2C}\cos^2\omega t \qquad (6)$$

$$U_L = \frac{LI^2}{2} = \frac{LI_0^2}{2}\sin^2\omega t \qquad (7)$$

1.L

Plot U_C and U_L *vs* t. Note that both U_C and U_L are always *positive* and

$$U_T = U_C + U_L = \text{constant.}$$

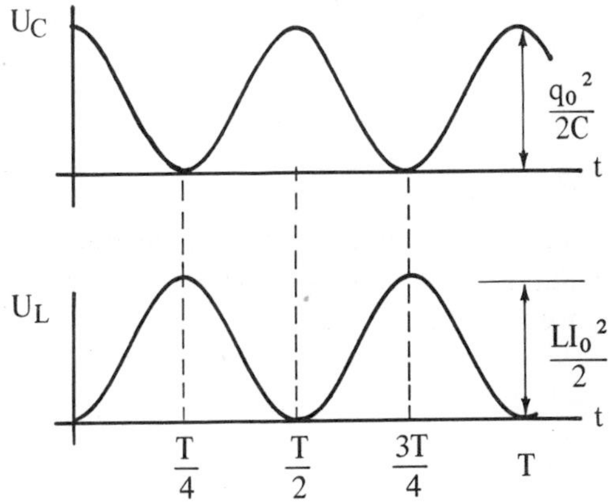

1.M

What are the maximum values of U_C and U_L?

From (6) and (7) we see that

$$U_C\ (\max) = \frac{q_0^{\,2}}{2C} \tag{8}$$

$$U_L\ (\max) = \frac{LI_0^{\,2}}{2} \tag{9}$$

Note that *both* of these maximum values equal U_T.

1.N

Suppose a 5μF capacitor is given a charge of 30μC. If it is then connected to an inductor of 5 H as in frame 1, and the switch is closed at t = 0, what is the oscillator frequency?

$$\omega = \sqrt{\frac{1}{LC}} = \sqrt{\frac{1}{5 \times 5 \times 10^{-6}}}$$

$$\omega = \frac{10^3}{5} = 200\ \frac{\text{rad}}{\text{sec}}$$

or,

$$f = \frac{\omega}{2\pi} = \frac{100}{\pi}\ \text{Hz}$$

1.O

Determine the charge on the capacitor as a function of time.

At t = 0, $q = q_0 = 30\mu C$, so

$q(t) = q_0 \cos\omega t$

$$q(t) = 30\cos(200t)\ \mu C \tag{10}$$

1.P

What is the *maximum* current in the oscillator circuit and the current as a function of time?

$I_0 = q_0\omega = 30 \times 10^{-6} \times 200$

$I_0 = 6 \times 10^{-3}\ A = 6\ mA$

$I(t) = -I_0 \sin\omega t$

$$I(t) = -6\sin(200t)\ mA \tag{11}$$

1.Q

Find the *maximum* energy stored in the capacitor, which of course is the total energy stored in the circuit.

$$U_T = \frac{q_0^{\,2}}{2C} = \frac{(30 \times 10^{-6})^2}{2 \times 5 \times 10^{-6}}$$

$$U_T = 9 \times 10^{-5}\ J$$

The student should also show that this is *also* equal to the maximum energy in the inductor, given by (9).

2.A

If an AC voltage is applied across a resistor as shown, how does the current vary in time?

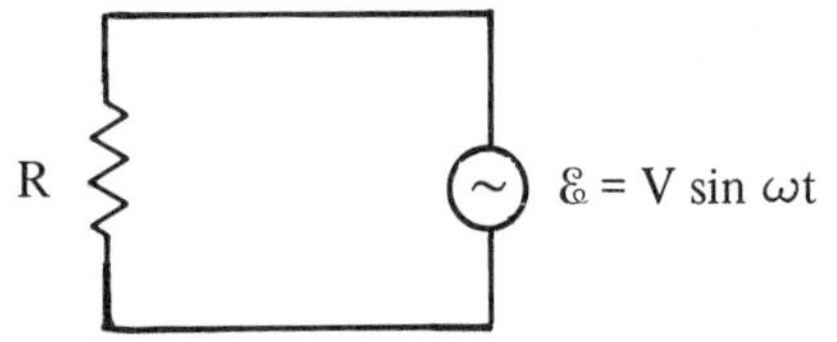

$$i = \frac{\mathcal{E}}{R} = \frac{V}{R}\sin\omega t \qquad (1)$$

where $I = \frac{V}{R}$ is the maximum current in the circuit and V is the maximum voltage.

2.B

Draw a rotating vector diagram for V and I corresponding to the circuit in frame 2.A.

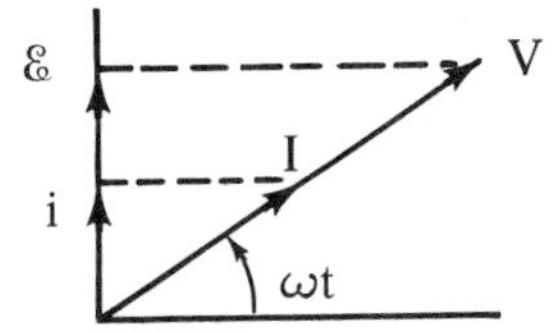

V and I are *in phase.* That is, i is a maximum when $\mathcal{E}$ is a maximum.

2.C

Now consider the circuit below, containing an AC source and an inductor. What is the instantaneous potential difference across L?

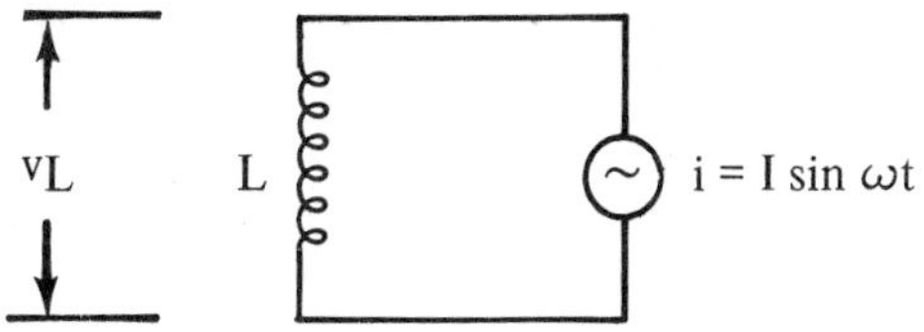

By definition, since $i = I\sin\omega t$,

$$v_L = L\frac{di}{dt} = LI\frac{d}{dt}(\sin\omega t)$$

$$v_L = IL\cos\omega t \qquad (2)$$

Note that i and $\mathcal{E}$ are the projections of I and V along the y axis, as in frame 2.B.

2.D

Is v_L in phase with i? Explain.

No. From (2) we see that v_L is $\frac{\pi}{2}$ out of phase with i. That is, v_L is zero if $i = I$, and $v_L = I\omega L$ when $i = 0$.

2.E

What is ωL called and what are its units?

ωL is defined to be the *inductive reactance* X_L and has the unit of ohm.

$$X_L = \omega L \qquad (3)$$

where $\omega = 2\pi f$.

2.F

What is the maximum value of v_L in terms of X_L?

$$v_L\ (\max) = V_L = I\omega L$$

or,

$$V_L = IX_L \qquad (4)$$

2.G

Draw a rotating vector diagram for V_L and I corresponding to the circuit in frame 2.C.

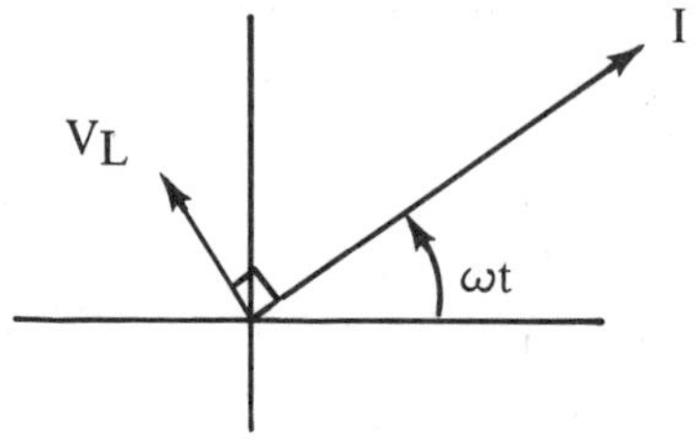

V_L leads I by $\frac{\pi}{2}$, *or* by $\frac{1}{4}$ of a cycle.

2.H

Now, consider a circuit containing an AC source and a capacitor. What is the instantaneous *charge* on the capacitor?

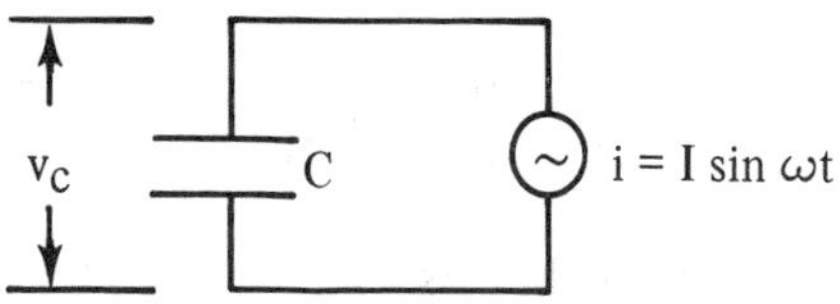

By definition, $i = \frac{dq}{dt}$, so integration gives

$$q = \int_0^t i dt = \int_0^t I \sin\omega t dt$$

$$q = -\frac{I}{\omega}\cos\omega t \qquad (5)$$

where $q_0 = \frac{I}{\omega}$ is the *maximum* charge on C.

2.I

What is the *instantaneous* voltage across the capacitor? What is the maximum voltage across C?

$$v_C = \frac{q}{C} = -\frac{I}{\omega C}\cos\omega t \qquad (6)$$

where $V_C = \frac{I}{\omega C}$ is *maximum* voltage across C.

2.J

Describe the phase relation between v_C and i.

Here again, v_C is $\frac{\pi}{2}$ out of phase with i. But the *negative* sign in (6) means that v_C *lags* I by $\frac{1}{4}$ of a cycle.

2.K

What is the constant $\frac{1}{\omega C}$ called and what are its units?

$\frac{1}{\omega C}$ is defined to be the *capacitive reactance* X_C and has the unit of ohm.

$$X_C = \frac{1}{\omega C} \tag{7}$$

2.L

Draw a rotating vector diagram for V_C and I corresponding to the circuit in frame 2.H.

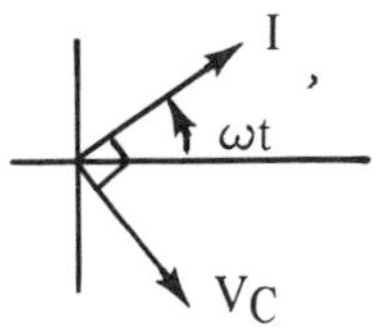

V_C *lags* I by $\frac{\pi}{2}$.

2.M

Suppose $i = I\sin\omega t$ in the RLC circuit shown. What is the instantaneous potential difference v?

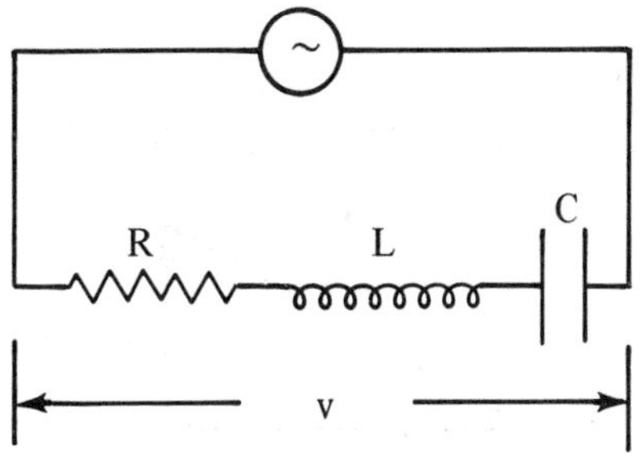

$$v = v_R + v_L + v_C$$

Using (1), (2) and (6) gives

$$v = I[R\sin\omega t + (X_L - X_C)\cos\omega t] \tag{8}$$

where $X_L = \omega L$ and $X_C = \frac{1}{\omega C}$.

2.N

Equation (8) can be put into a simpler form by noting the trigonometric relation

$$A\sin\theta + B\cos\theta = \sqrt{A^2 + B^2}\,\sin(\theta + \phi)$$

where

$$\tan\phi = \frac{B}{A}$$

Use these relations and write (8) in this form.

If we let $\theta = \omega t$, $A = R$, $B = X_L - X_C$, (8) becomes

$$v = I[R^2 + (X_L - X_C)^2]^{1/2}\sin(\omega t + \phi) \tag{9}$$

where

$$\tan\phi = \frac{X_L - X_C}{R} \tag{10}$$

2.O

Describe the significance of the angle ϕ.

ϕ is the phase angle between V and I. That is, the current and voltage are generally *out of phase* in an RLC circuit.

2.P

Write an expression for the *maximum* value of v in terms of I, R, X_L and X_C.

$$V = I[R^2 + (X_L - X_C)^2]^{1/2} \qquad (11)$$

or,

$$V = [V_R{}^2 + (V_L - V_C)^2]^{1/2}$$

where

$$V_R = IR, \; V_L = IX_L \text{ and } V_C = IX_C$$

2.Q

Sketch a vector diagram of V in terms of V_R, V_L and V_C for the RLC circuit. Assume that $V_L > V_C$.

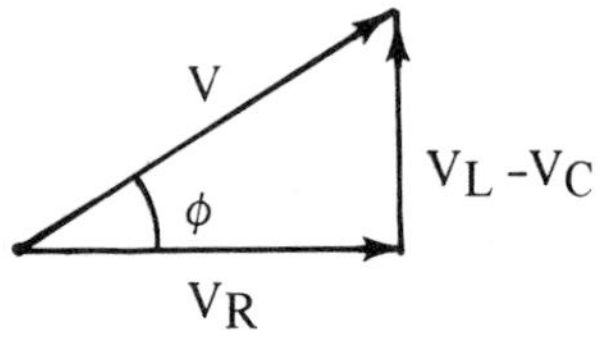

2.R

Define the *impedance* Z of the series RLC circuit, and represent Z by a vector diagram.

$$Z = [R^2 + (X_L - X_C)^2]^{1/2} \qquad (12)$$

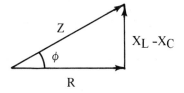

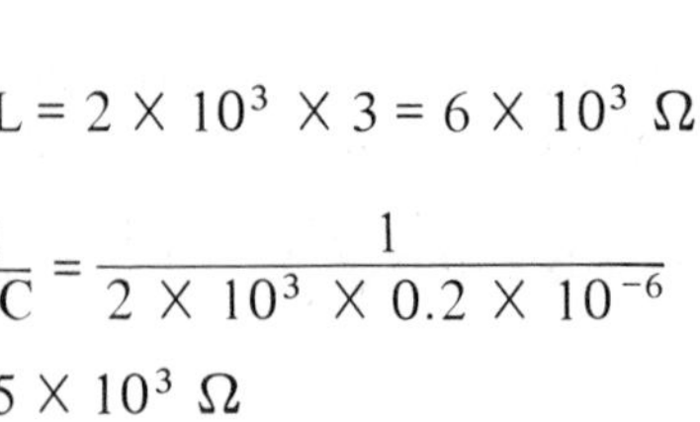

2.S

Consider a series RLC circuit with $R = 2000\ \Omega$, $L = 3$ H, $C = 0.2\ \mu F$, $\omega = 2 \times 10^3 \dfrac{\text{rad}}{\text{sec}}$ and $V = 80$ volts. Determine X_L, X_C and Z.

$$X_L = \omega L = 2 \times 10^3 \times 3 = 6 \times 10^3\ \Omega$$

$$X_C = \frac{1}{\omega C} = \frac{1}{2 \times 10^3 \times 0.2 \times 10^{-6}}$$

$$= 2.5 \times 10^3\ \Omega$$

$$Z = \sqrt{R^2 + (X_L - X_C)^2}$$

$$Z = \sqrt{(2 \times 10^3)^2 + (3.5 \times 10^3)^2}$$

$$\cong 4 \times 10^3\ \Omega$$

2.T

Determine the phase angle ϕ between V and I for the series circuit described in frame 2.S.

$$\tan\phi = \frac{X_L - X_C}{R} = \frac{(6 - 2.5) \times 10^3}{2 \times 10^3}$$

$$\tan\phi = 1.75$$

$$\phi \cong 60.3°$$

2.U

Calculate the voltage amplitudes across R, L and C.

$$I = \frac{V}{Z} = \frac{80}{4 \times 10^3} = 2 \times 10^{-2}\ \text{A};\therefore$$

$$V_R = IR = 2 \times 10^{-2} \times 2 \times 10^3 = 40\ \text{V}$$

$$V_L = IX_L = 2 \times 10^{-2} \times 6 \times 10^3 = 120\ \text{V}$$

$$V_C = IX_C = 2 \times 10^{-2} \times 2.5 \times 10^3 = 50\ \text{V}$$

2.V

Show that (11) is satisfied for this numerical exercise.

$$V = [V_R{}^2 + (V_L - V_C)^2]^{1/2}$$

$$V = [40^2 + (70)^2]^{1/2} \cong 80\ \text{V}$$

2.W

What is the condition for the *resonance frequency* of the RLC circuit?

The resonance frequency of the circuit is that for which I is a *maximum*. From (11) we see that this occurs for

$$X_L = X_C, \quad \text{or} \quad \omega_R = \sqrt{\frac{1}{LC}}$$

2.X

Determine the resonance frequency of the circuit described in frame 2.S.

$$\omega_R = \sqrt{\frac{1}{LC}} = \sqrt{\frac{1}{3 \times 0.2 \times 10^{-6}}}\ \frac{\text{rad}}{\text{sec}}$$

$$\omega_R = 1.3 \times 10^3\ \frac{\text{rad}}{\text{sec}}$$

2.Y

What is the *average power* delivered by the source?

$$P_{av} = \frac{1}{2} IV\cos\phi$$

$$P_{av} = \frac{1}{2}(2 \times 10^{-2}) \times (80)\cos(60.3)$$

$$P_{av} \cong 40\ \text{W}$$

3.A

An electromagnetic wave can be described by oscillating electric and magnetic vectors. What is the speed of electromagnetic waves in free space, and what are the orientations of **E** and **B** relative to each other and to the direction of propagation?

The speed equals the speed of light, c. **E** and **B** are perpendicular to each other and each is perpendicular to the direction of propagation.

3.B

Describe the orientation of **E** and **B** for a plane polarized wave traveling in the +x direction.

E oscillates in the xy plane.

B oscillates in the xz plane.

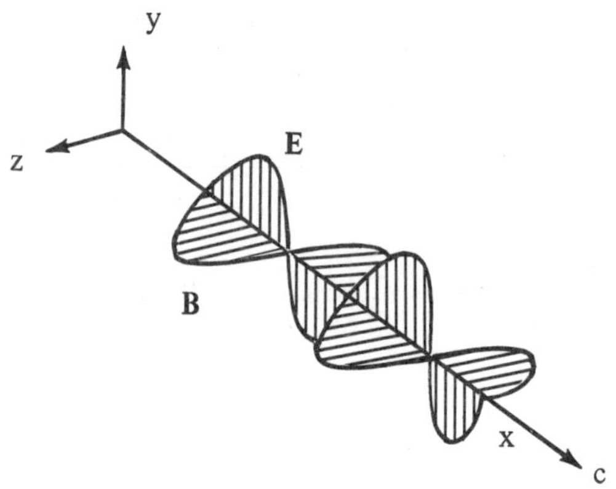

E X **B** is in the direction of **c**.

3.C

How are electromagnetic waves created?

They are produced whenever electrical charges accelerate or decelerate. Therefore, there is no electromagnetic field associated with charges in equilibrium.

3.D

Give mathematical expressions for the **E** and **B** fields associated with a plane electromagnetic wave traveling in the x direction as shown in frame 3.B. Define the terms in your equations.

$$\mathbf{E} = E_0 \cos(kx - \omega t)\mathbf{j} \qquad (1)$$

$$\mathbf{B} = B_0 \cos(kx - \omega t)\mathbf{k} \qquad (2)$$

where E_0 and B_0 are the maximum amplitudes of the waves, $k = \frac{2\pi}{\lambda}$ and $\omega = 2\pi f$. λ is the wavelength of the wave and f its frequency in Hertz (sec^{-1}).

3.E

Maxwell's equations provide relations between E and B for electromagnetic waves. In the case of the plane wave traveling in free space, what is the relation? What is c equal to in terms of μ_0 and ϵ_0 defined earlier?

$$E = cB \tag{3}$$

where

$$c = \frac{1}{\sqrt{\mu_0 \epsilon_0}} \tag{4}$$

The numerical value of c can be calculated from (4).

$$c \cong 3 \times 10^8 \frac{m}{sec},$$

3.F

How is c related to ω and k for an electromagnetic wave traveling in free space?

$$c = \frac{\omega}{k} \tag{5}$$

Since $\omega = 2\pi f$ and $k = \frac{2\pi}{\lambda}$, we can also write

$$c = \lambda f \tag{6}$$

3.G

Are electromagnetic waves longitudinal or transverse? Explain.

They are transverse waves, since the disturbance (oscillating fields) is perpendicular to the direction of propagation.

3.H

What are some of the important properties of electromagnetic waves?

They transport energy, carry momentum and can be polarized (because they are transverse waves). The energy and momentum can be transferred to objects placed in the path of the waves.

3.I

What quantity describes the rate at which electromagnetic energy passes through a unit area, where the area is perpendicular to the direction of propagation?

Poynting's vector, S, given by

$$S = \frac{E \times B}{\mu_0} \tag{7}$$

S has units of $\frac{\text{watts}}{m^2}$. Since 1 watt = $1 \frac{J}{sec}$, Equation (7) reduces to units of $\frac{J}{sec}$ m^2; that is, energy per unit area per unit time.

3.J

What is the direction of **S**? For example, what is the direction of **S** for the wave described in frame 3.B?

S is in the direction defined by $\mathbf{E} \times \mathbf{B}$. But $\mathbf{E} \times \mathbf{B}$ is in the direction of propagation, so this also defines the direction of **S**. For the plane wave shown in frame 3.B, **S** is in the +x direction.

3.K

What is the magnitude of S for a plane wave in terms of E, B and μ_0?

Since E is $\perp$ to B, it follows that $|\mathbf{E} \times \mathbf{B}| = EB$, so from (7) we get

$$S = \frac{EB}{\mu_0} \qquad (8)$$

3.L

E and B given in (7) are instantaneous values of the fields. What is S when averaged over one cycle of the wave?

From (1) and (2), we see that $EB = E_0 B_0 \cos^2(kx - \omega t)$. Therefore, the time average of S over *one cycle* is proportional to

$$< \cos^2(kx - \omega t) >_T = \frac{1}{2}$$

and it follows that

$$S_{av} = \frac{E_0 B_0}{2\mu_0} \qquad (9)$$

(The student should verify that

$$<\cos^2(kx - \omega t)>_T = \frac{1}{T}\int_0^T \cos^2(kx - \omega t)dt = \frac{1}{2}$$

where T is the period or time for one oscillation of the wave.)

3.M

If a flat surface has an area A, and a plane electromagnetic wave is incident normally on the surface, what is the total energy delivered to the surface in a time t? Assume that the energy flux S_{av} is known.

From the definition of S,

$$[S] = \frac{\text{Energy}}{\text{Area} \times \text{time}}$$

$$\therefore \text{Energy} = U = S_{av} At \qquad (10)$$

3.N

If the surface is perfectly absorbing (that is, a black body), what is the momentum delivered to the surface in the time t?

$$p = \frac{U}{c} \qquad (11)$$

$$\therefore \quad p = \frac{S_{av}At}{c} \qquad (12)$$

(The student should verify that the units of p are kg $\frac{m}{sec}$ in the SI system.)

3.O

If the surface is perfectly reflecting, what is the momentum delivered to the surface in a time t?

$$p = \frac{2U}{c} = \frac{2S_{av}At}{c} \qquad (13)$$

That is, momentum is delivered by the incident wave and the reflected wave.

3.P

What is the average force exerted on (a) the perfectly absorbing surface and (b) the perfectly reflecting surface in the time t?

Newton's second law states that $F = \frac{dp}{dt}$. Since we are dealing with average values, we have

$$F_{av} = \frac{p}{t}$$

Using (12) and (13), we get

$$\text{(a) } F_{av} = \frac{S_{av}A}{c} \quad \text{(absorbing)} \qquad (14)$$

$$\text{(b) } F_{av} = \frac{2S_{av}A}{c} \quad \text{(reflecting)} \qquad (15)$$

3.Q

Calculate the average pressure (radiation pressure) on the two surfaces.

By definition, pressure = $\frac{\text{Force}}{\text{Area}}$,

$$\therefore \quad P = \frac{F_{av}}{A} = \frac{S_{av}}{c} \quad \text{(absorbing)} \qquad (16)$$

$$P = \frac{2S_{av}}{c} \quad \text{(reflecting)} \qquad (17)$$

3.R

A plane wave has an energy flux of $1000 \frac{\text{watts}}{\text{m}^2}$ (roughly the flux of sunlight at the earth). If the wave is incident normally on a mirror of area 0.3 m^2, what is the total energy delivered in a time of one minute?

Using (10), we have

$$U = S_{av} At$$

$$U = 1000 \frac{\text{watts}}{\text{m}^2} \times 0.3 \text{ m}^2 \times 60 \text{ sec}$$

But 1 watt = $1 \frac{\text{J}}{\text{sec}}$, so

$$U = 1.8 \times 10^4 \text{ J}$$

3.S

Calculate the average force on the mirror and the radiation pressure.

Using (15) and (17) gives,

$$F_{av} = \frac{2 \times 1000 \frac{\text{watts}}{\text{m}^2} \times 0.3 \text{ m}^2}{3 \times 10^{10} \frac{\text{m}}{\text{sec}}}$$

$$F_{av} = 2 \times 10^{-8} \text{ N}$$

$$P = \frac{2S_{av}}{c} = \frac{2 \times 1000 \frac{\text{watts}}{\text{m}^2}}{3 \times 10^{10} \frac{\text{m}}{\text{sec}}}$$

$$P = 6.7 \times 10^{-8} \frac{\text{N}}{\text{m}^2}$$

7.9 SUMMARY

The natural angular frequency of oscillation of an LC circuit is given by

$$\omega = \frac{1}{\sqrt{LC}} \tag{7.5}$$

The maximum current in an AC circuit containing only an inductor and voltage source is given by

$$I = \frac{V}{\omega L} = \frac{V}{X_L} \tag{7.14}$$

The maximum current in an AC circuit containing only a capacitor and voltage source is given by

$$I = \omega CV = \frac{V}{X_C} \tag{7.17}$$

X_L and X_C are called the inductive reactance and capacitance reactance, respectively, and have the unit ohm.

The voltage amplitude in an AC circuit is given by

$$V = IZ \tag{7.20}$$

where Z is the impedance of the circuit. For a series R, L, C circuit, Z is given by

$$Z = \sqrt{R^2 + (X_L - X_C)^2} \tag{7.21}$$

The phase between the voltage and current in a series RLC circuit is given by

$$\tan\phi = \frac{X_L - X_C}{R}$$

The average power delivered to an RLC circuit is given by

$$P_{av} = \frac{1}{2} IV\cos\phi$$

The electric and magnetic field vectors of a plane electromagnetic wave traveling in free space are related by the expression

$$E = cB \tag{7.37}$$

where $c = 3 \times 10^8 \frac{\text{m}}{\text{sec}}$ is the speed of light in free space. The wavelength, λ, and frequency, f, of the electromagnetic wave are related by the equation

$$c = \lambda f \tag{7.38}$$

The energy flux through a surface, or the rate at which energy passes through a unit area, is given by Poynting's vector, S:

$$S = \frac{E \times B}{\mu_0} \tag{7.42}$$

The momentum delivered to a perfectly absorbing surface which absorbs a total energy U is given by

$$p = \frac{U}{c} \tag{7.45}$$

while the momentum delivered to a perfectly reflecting surface is *twice* this amount.

7.10 PROBLEMS

1. An 8.0 μF capacitor is charged with a 24 V battery. The battery is then disconnected and the charged capacitor is connected across a 20 mH inductor. Neglecting the resistance of the LC circuit, calculate (a) the frequency of the oscillating current and (b) the maximum current in the coil.

2. What size inductor would you connect to a 4 μF capacitor to make it oscillate at a frequency of 10^3 Hz?

3. Verify by direct substitution that Equations (7.4) and (7.5) represent a solution to Equation (7.3).

4. A 4 μF capacitor is charged to a potential difference of 50 V, as in Figure 7-5. It is then discharged through a 10 mH inductor. Neglecting the resistance in the circuit, determine (a) the period of oscillation, (b) the maximum charge on the capacitor, (c) the maximum current in the circuit, (d) the total energy stored in the circuit and (e) the time it takes for the energy to first be *equally* shared between the capacitor and inductor.

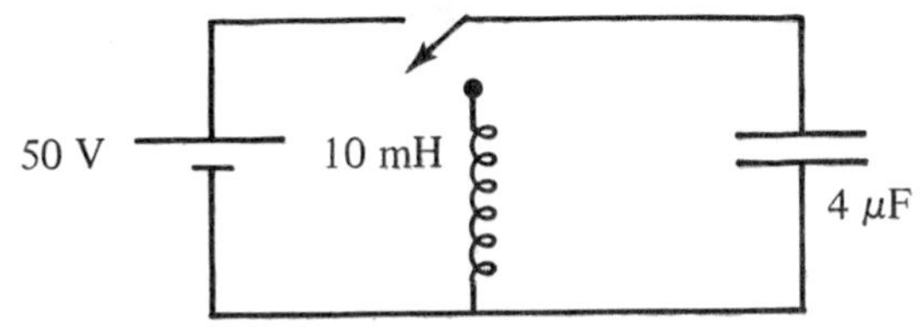

Figure 7-5

5. (a) Calculate the reactances of a 3 μF capacitor at frequencies of $\frac{100}{\pi}$ Hz and $\frac{5000}{\pi}$ Hz. (b) Calculate the reactances of a 3 H inductor at frequencies of $\frac{100}{\pi}$ Hz and $\frac{5000}{\pi}$ Hz.

6. A 1000 ohm resistor is connected in series to a 0.6 H inductor and a 2.5 μF capacitor. The frequency of the source is $\frac{1000}{\pi}$ Hz. (a) Calculate the impedance of the circuit. (b) Determine the phase angle between the current and line voltage and determine whether I leads or lags V. (c) What is the average power delivered to the circuit if the rms value of V is 80 V.

7. Suppose that the voltage source in the circuit described in Problem 6 varies as $\mathcal{E} = 80 \sin \omega t$. (a) Determine the voltage amplitude across the resistor, inductor and capacitor. (b) Calculate the instantaneous potential differences across the resistor, inductor and capacitor.

8. A series RLC circuit consists of an 8 Ω resistor, a 5 μF capacitor and a 50 mH inductor. A variable frequency source of 40 V (rms) is applied across the combination. Find the rms current and phase angle when the frequency of the source is (a) equal to and (b) one-half of the resonance frequency.

9. Using the procedure outlined in section 7.4, verify that the rms voltage and rms current are given by Equations (7.25) and (7.26), respectively.

10. A series RLC circuit consists of a 50 Ω resistor, a 4 μF capacitor and a 0.1 mH inductor. (a) For what frequency will the average power delivered to the circuit be a maximum? (b) If the rms voltage is 30 V, determine the average power delivered at this frequency.

11. Show that for a plane wave, Equation (7.37) is a satisfactory solution of Equation (7.34), and that E = cB. Likewise, verify Equation (7.40).

12. Perform a dimensional analysis on the Poynting's vector, and show that it has units of $\frac{\text{watts}}{\text{m}^2}$ in the SI system. Recall that in this system, E has units of $\frac{\text{volts}}{\text{m}}$, B has units of $\frac{\text{webers}}{\text{m}^2}$, and μ_0 has units of $\frac{\text{webers}}{\text{amp m}}$.

13. Calculate the total power radiated by the sun, using the fact that the average solar energy normally incident on the earth's surface per unit area per unit time is $1340 \frac{\text{watts}}{\text{m}^2}$ and that the mean earth-sun separation is 1.49×10^{11} m. *Hint:* Treat the sun as a point source, and note that the power passing through a sphere of radius r is $4\pi r^2 S_{av}$.

14. A microwave transmitter emits monochromatic electromagnetic waves. The maximum electric field at a distance 1 km from the transmitter is $6.0 \frac{\text{V}}{\text{m}}$. Assuming the transmitter is a point source, and neglecting waves reflected from the earth, calculate (a) the maximum magnetic field at this distance and (b) the total power emitted by the transmitter.

15. A thin tungsten filament of length 1 m radiates 60 watts of energy in the form of electromagnetic waves. A perfectly absorbing surface in the form of a hollow cylinder of radius 5 cm and length 1 m is placed concentric with the filament. Calculate the radiation pressure acting on the cylinder. (Assume that the radiation is emitted in the radial direction, and neglect end effects.)

16. A plane electromagnetic wave has an energy flux of $300 \frac{\text{watts}}{\text{m}^2}$. A flat rectangular surface of dimensions 20 cm × 40 cm is placed perpendicular to the direction of the plane wave. If the surface absorbs half of the energy, and reflects half (that is, it is a 50 per cent reflecting surface), calculate (a) the total energy delivered to the surface in a time of one minute and (b) the momentum delivered in this time.

17. A community plans to build a facility to use solar radiation and convert it into electrical power. They require 1 megawatt of power (10^6 watts), and the system to be installed has an efficiency of 30 per cent (30 per cent of the solar energy incident on a surface is converted to electrical energy). What is the effective area of a perfectly absorbing surface which would be used in such an installation, assuming a constant energy flux of $1340 \frac{\text{watts}}{\text{m}^2}$?

18. An astronaut in a spacecraft moving with constant velocity wishes to increase the speed of his vehicle using a laser beam attached to the spaceship. The laser beam emits 100 joules of electromagnetic energy per pulse, and the laser is pulsed at the rate of $0.2\,\frac{\text{pulse}}{\text{sec.}}$. If the mass of the spaceship plus its contents is 5000 kg, how long must the beam be on in order to increase the speed of the vehicle by $1\,\frac{\text{m}}{\text{sec}}$ in the direction of its initial motion? In what direction should the beam be pointed to achieve this?

8

GEOMETRICAL OPTICS

8.1 INTRODUCTION

Although light waves are electromagnetic waves, we can deal with the subject of geometrical optics by assuming that light travels in straight lines as *rays*. That is, for the time being we will neglect the "bending effects" of light when it meets certain obstacles or apertures. These latter effects will be dealt with in the following chapter on physical optics.

In an earlier chapter on waves, we found that the speed of propagation of matter waves through a medium depends on the physical properties of that medium. Light waves differ in that they can propagate through a vacuum. However, the velocity of light depends on the permittivity and permeability of the substance through which it propagates. This dependence of velocity on the nature of the medium gives rise to the phenomena of *reflection* and *refraction*. These are basic phenomena in the formulation of geometrical optics.

8.2 REFLECTION AND REFRACTION

Consider a light ray traveling through medium 1 incident on medium 2, such as glass, as in Figure 8-1. The *incident* light ray makes an angle θ_1 with the normal, while the *reflected* light ray makes an angle θ_1' with the normal. That part of the light that enters medium 2 is called the *refracted* ray, which makes an angle θ_2 with the normal.

The following laws based on experiments pertain to the reflection and refraction of light:

1. The reflected and refracted rays lie in the plane formed by the normal and the incident ray.

2. The angle of incidence equals the angle of reflection:

$$\theta_1 = \theta_1' \tag{8.1}$$

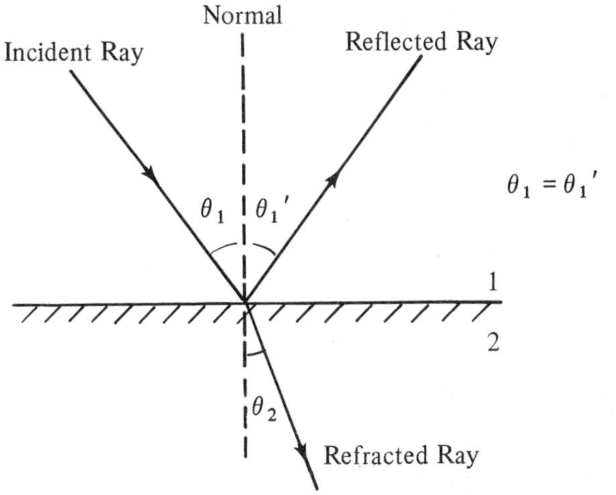

Figure 8–1 The reflection and refraction of light waves at an interface. The directions of the rays are perpendicular to the wave fronts.

3. The ratio of $\frac{\sin\,\theta_1}{\sin\,\theta_2}$ is a constant called the index of refraction n_{12} of medium 2 relative to medium 1.

$$\frac{\sin\theta_1}{\sin\theta_2} = n_{12} \tag{8.2}$$

This law of refraction is sometimes referred to as Snell's law. The student should note that Equation (8.2) is valid for *monochromatic light,* that is, light of one wavelength. The index of refraction *depends* on the wavelength of light under consideration. Usually, the index is measured with yellow light at a wavelength of 5890 Å (light from a sodium flame). (1Å = 1 Angstrom = 10^{-10} m.) A compilation of some indices of refraction is given in Table 8–1. Note that $n \cong 1$ for air, and is greater than one for all other substances. These values are measured with respect to a vacuum. Therefore, when light passes from a vacuum to one of the substances listed in Table 8–1, the rays are always bent *towards* the normal. This follows from Equation (8.2), where $n_{12} > 1$; therefore, $\theta_1 > \theta_2$.

TABLE 8–1 INDICES OF REFRACTION FOR VARIOUS SUBSTANCES MEASURED AT λ = 5890 Å

Medium	Index of Refraction	Medium	Index of Refraction
Air	1.0003	Crown Glass	1.50
Ice	1.31	Sodium Chloride	1.53
Water	1.33	Flint Glass	1.58
Fused Quartz	1.46	Zircon	1.92
Benzene	1.50	Diamond	2.42

The time it takes a light wave to travel one wavelength is given by $\frac{\lambda_1}{v_1}$, while the time it takes the light wave to travel one wavelength in medium 2 is $\frac{\lambda_2}{v_2}$. But these times must be equal, since the frequency of the wave is the same in both media. Therefore,

$$\frac{\lambda_1}{v_1} = \frac{\lambda_2}{v_2} \tag{8.3}$$

From geometrical considerations, it is found that

$$\frac{\sin\theta_1}{\sin\theta_2} = \frac{\lambda_1}{\lambda_2} = \frac{v_1}{v_2} \tag{8.4}$$

We now define the index of refraction of medium 2 relative to medium 1 as

$$n_{12} = \frac{v_1}{v_2} \tag{8.5}$$

If medium 1 is a vacuum, $v_1 = c = 2.99793 \times 10^8 \frac{m}{sec}$, and the absolute *index of refraction* of a medium relative to a *vacuum* becomes

$$n = \frac{c}{v} \tag{8.6}$$

Thus, a more general form of Snell's law can be written, using $n_1 = \frac{c}{v_1}$ and $n_2 = \frac{c}{v_2}$ in Equation (8.4). This gives

$$n_1 \sin\theta_1 = n_2 \sin\theta_2 \tag{8.7}$$

Example 8.1

Yellow light ($\lambda = 5890$ Å) travels from a vacuum, through water and then through crown glass, as in Figure 8-2. The angle of incidence is 30°. (a) Determine the speed of light in water and in the glass. (b) Calculate the angle that the final emerging ray from the glass makes with the normal.

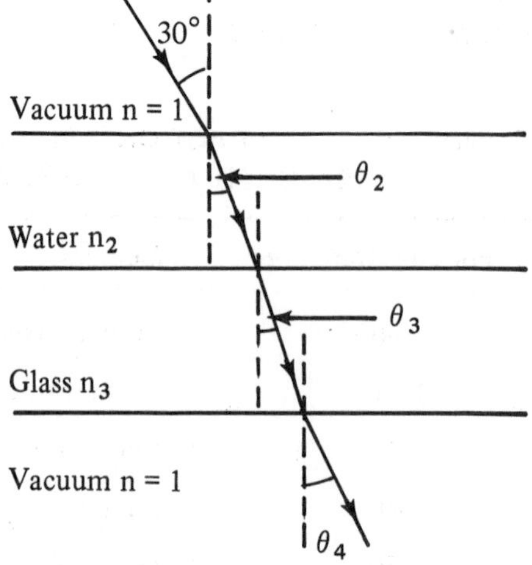

Figure 8-2

Solution

(a) Using Equation (8.6) and the values $n_2 = 1.33$ for water and $n_3 = 1.52$ for crown glass gives

$$v_2 = \frac{c}{n_2} \cong \frac{3 \times 10^8 \frac{m}{sec}}{1.33} = 2.26 \times 10^8 \frac{m}{sec}$$

$$v_3 = \frac{c}{n_3} = \frac{3 \times 10^8 \frac{m}{sec}}{1.52} = 1.97 \times 10^8 \frac{m}{sec}$$

(b) Applying Equation (8.7) to the first two media (vacuum and water) gives

$$\sin\theta_1 = n_2 \sin\theta_2 \qquad (1)$$

Likewise, for the next two interfaces (water-glass, glass-vacuum) we get

$$n_2 \sin\theta_2 = n_3 \sin\theta_3 \qquad (2)$$

$$n_3 \sin\theta_3 = \sin\theta_4 \qquad (3)$$

Comparing (1), (2) and (3) we see that $\sin\theta_1 = \sin\theta_4$, or $\theta_1 = \theta_4 = 30^\circ$. That is, the emerging light makes the *same* angle with the normal as the incident light, since they both correspond to propagation in a vacuum. The student should also verify that $\theta_2 = 22^\circ$ and $\theta_3 = 19^\circ$.

If light rays traveling in an optically dense medium strike an interface which has a medium of lower optical density (such as water to air), the rays can be *totally reflected* at or greater than some critical angle. This is illustrated in Figure 8–3. This *critical angle* θ_c is found by substituting $\theta_2 = \frac{\pi}{2}$ in Equation (8.7), corresponding to total reflection (that is, no refracted rays in medium 1). It follows from Equation (8.7) that

$$\sin\theta_c = \frac{n_2}{n_1} \qquad \text{(where } n_1 > n_2\text{)} \qquad (8.8)$$

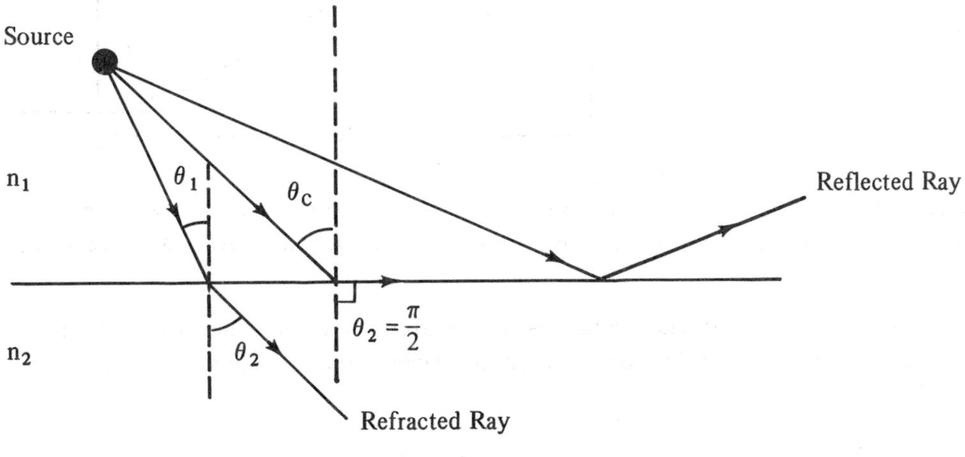

Figure 8–3 Total internal reflection of light occurs for angles of incidence $\theta_1 \geqslant \theta_c$, where $n_1 > n_2$.

Example 8.2

Calculate the critical angle for diamond, assuming $n_2 = 1.00$ for air.

$$\sin\theta_c = \frac{n_2}{n_1} = \frac{1}{2.42} = 0.413$$

$$\theta_c \cong 24°$$

The student should show that the critical angle for crown glass is 41°.

8.3 SPHERICAL AND PLANE MIRRORS

In this section we will treat reflections from spherical and plane surfaces. Spherical mirrors can be either concave or convex (Figs. 8–4 and 8–5), while plane mirrors have a radius of ∞. We will assume that the object is to the left of the concave mirror, as in Figure 8–4, at a distance p from the surface (p is called the object distance). An image is formed at a distance q from the surface, while the mirror has a radius of curvature R.

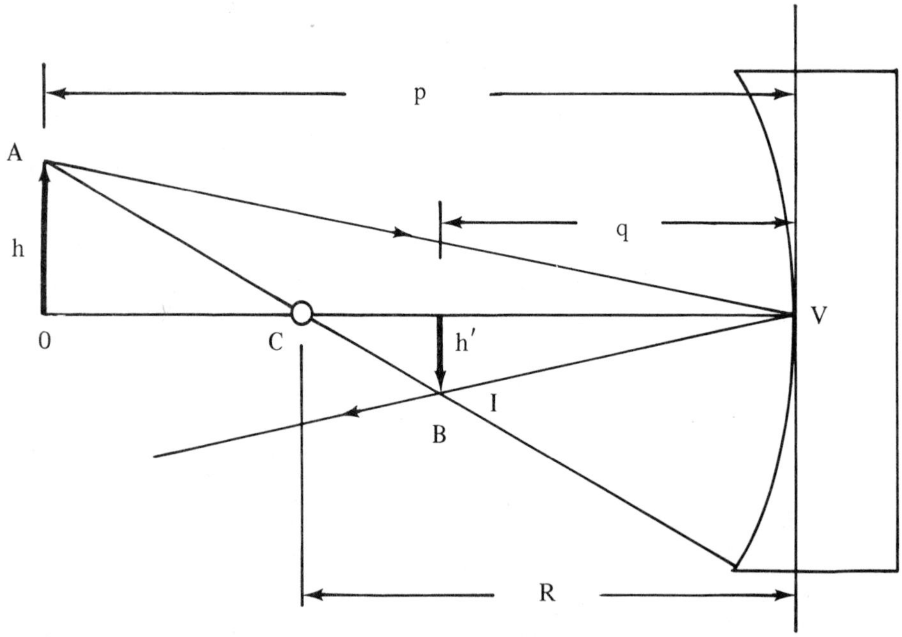

Figure 8–4 Schematic diagram of a concave mirror. Note that p, q and R are all positive in this example. If the object 0 lies within the focal distance $\frac{R}{2}$, the image I is virtual and erect, and q is negative.

The object and image distances are related to each other through the *mirror equation* which can be derived from geometrical considerations. The result is

$$\frac{1}{p} + \frac{1}{q} = \frac{2}{R} \tag{8.9}$$

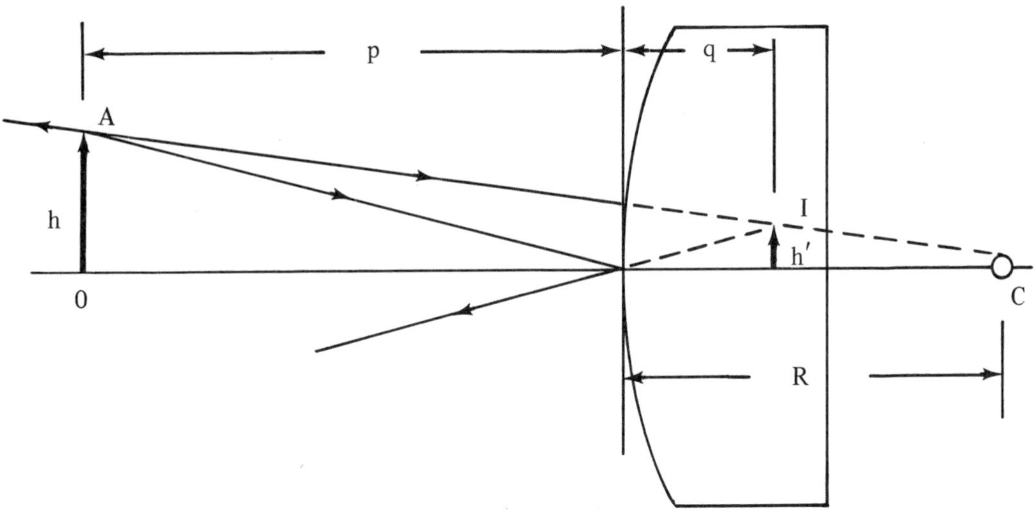

Figure 8-5 Schematic diagram for a convex mirror. Note that both q and R are taken to be negative in the mirror equation. The image is always virtual and erect.

If the object is far away from a concave mirror, as in Figure 8–6(*a*), $p \to \infty$ and $q = \frac{R}{2}$. This point is called the *focal point* F, and the distance of the point from the mirror is called the focal length f. Therefore, since $f = \frac{R}{2}$, Equation (8.9) can be written as

$$\frac{1}{p} + \frac{1}{q} = \frac{1}{f} \tag{8.10}$$

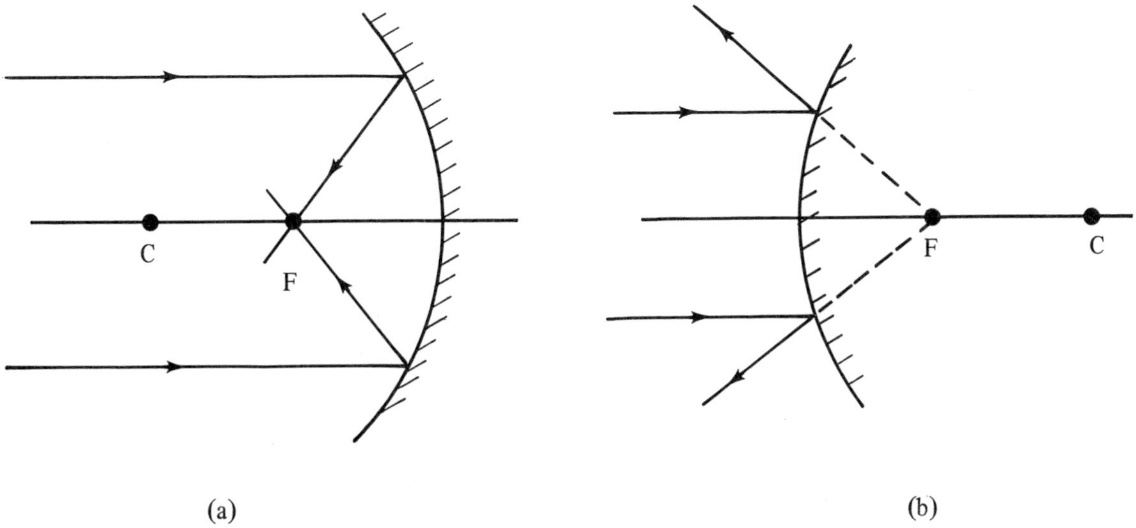

Figure 8-6 (*a*) The real focus of a concave mirror. (*b*) The "virtual" focus of a convex mirror. Note that the object distance $p = \infty$, since the rays are parallel.

If the height of the object is taken to be h, and the height of the image is h′, then the *lateral magnification* of the mirror is given by

$$M = \frac{h'}{h} = \frac{q}{p} \tag{8.11}$$

Let us adopt the following sign conventions for mirrors:

1. R and f are positive for a concave mirror and negative for a convex mirror.

2. The object distance p is positive if the object is in front of the mirror (that is, if the object is real).

3. The image distance q is positive when the image is real, that is, in front of the mirror (see, for example, Fig. 8–4); q is negative if the image is virtual, that is, behind the mirror (see, for example, Fig. 8–5).

The student should note that *concave mirrors* will always form real and inverted images of objects located outside the focal point, that is, for $p > f$. However, if $p < f$, the image is virtual, erect and enlarged. On the other hand, *convex mirrors* always produce virtual, erect and smaller images.

Example 8.3

An object is placed 3 m in front of a *concave* mirror whose radius of curvature is 40 cm, as in Figure 8-4. Determine (a) the position of the image and (b) the magnification of the mirror.

Solution

Using Equation (8.9), we get

(a) $$\frac{1}{p} + \frac{1}{q} = \frac{1}{3} + \frac{1}{q} = \frac{2}{0.4}$$

$$q = \frac{3}{16} \text{ m} \cong 0.187 \text{ m}$$

(b) $$M = \frac{q}{p} = \frac{\frac{3}{16}}{3} = \frac{1}{16}$$

Therefore, the image is $\frac{1}{16}$ as large as the object.

Example 8.4

A convex mirror as shown in Figure 8-5 has a magnification of $\frac{1}{5}$ for an object distance of 50 cm. (a) Determine the position of the virtual image formed by a real object placed 50 cm from the mirror. (b) Find the focal length of the mirror.

Solution

(a) Since the image must be virtual for a convex mirror, we take q as negative. Therefore, the magnification formula gives

$$M = \frac{q}{p} = -\frac{1}{5}$$

$$q = -\frac{1}{5}p = -10 \text{ cm}$$

That is, the virtual image is 10 cm behind the mirror, as in Figure 8-5.

(b)

$$\frac{1}{f} = \frac{1}{p} + \frac{1}{q} = \frac{1}{50} - \frac{1}{10}$$

$$f = -\frac{50}{4} = -12.5 \text{ cm}$$

The negative sign for f is consistent with the fact that the mirror is *convex*.

8.4 RAY DIAGRAMS FOR MIRRORS

Images formed by mirrors can be determined by using *ray diagrams*. All that is required is a knowledge of the locations of the center of curvature and the focal point. Two rays are then constructed to locate the image. A few examples as shown in Figure 8-7. Any two of the following three rays will locate the image:

1. Draw a ray from the top of the object, A (tip of arrow), parallel to the x axis. This ray reflects through the focal point, as do all parallel rays.
2. Draw a ray from the center of curvature through A. This ray is perpendicular to the mirror and reflects on itself.
3. Draw a ray from 0 through the focal point. This ray reflects parallel to the x axis.

The intersection of the three rays discussed corresponds to the top of the image; hence, it describes the position and size of the image. Note that only two of the three rays are required.

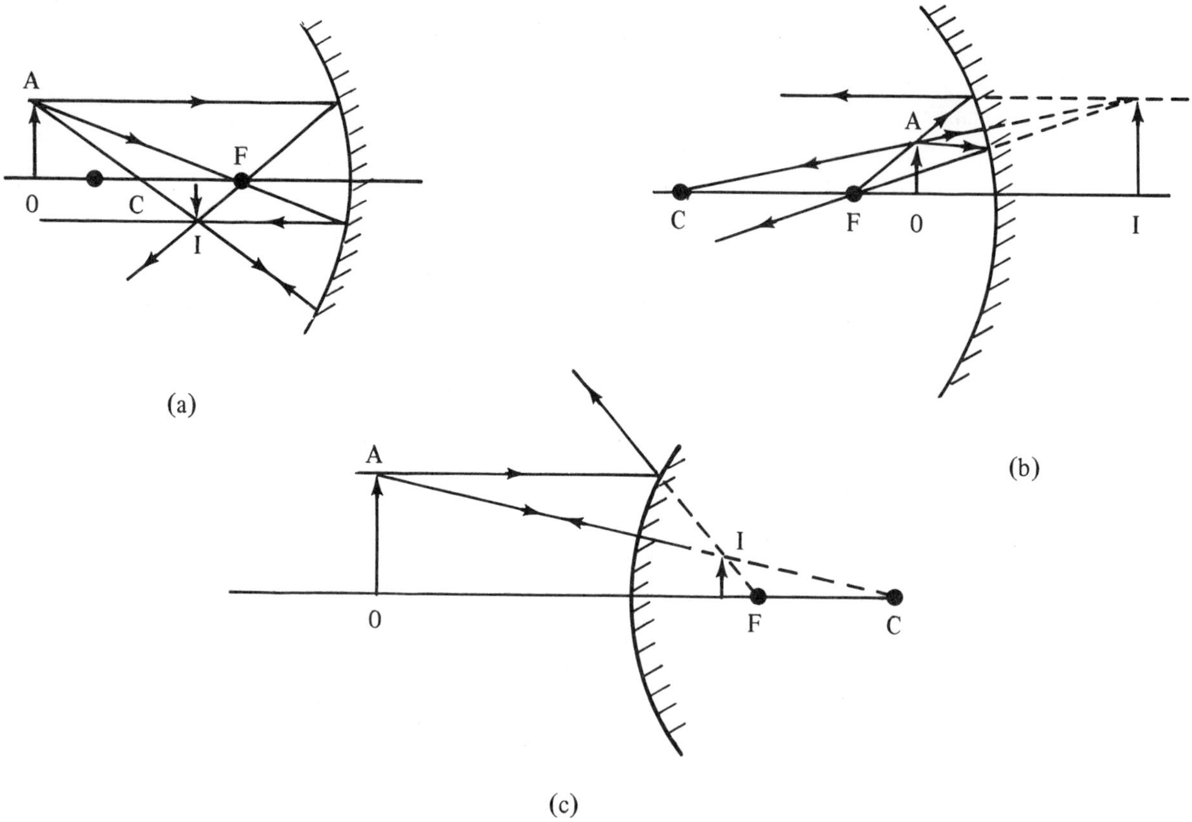

Figure 8-7 Some examples of ray diagrams for concave and convex mirrors. Note that in cases (*b*) and (*c*) the image is virtual and erect.

8.5 IMAGES FORMED BY REFRACTION

Consider an object in a medium whose index of refraction is n_1, and a refracting convex spherical surface of index of refraction n_2 placed in front of it as in Figure 8-8. We assume the medium to the right of the vertex is continuous. The expression that relates p and q is given by

$$\frac{n_1}{p} + \frac{n_2}{q} = \frac{n_2 - n_1}{R} \tag{8.12}$$

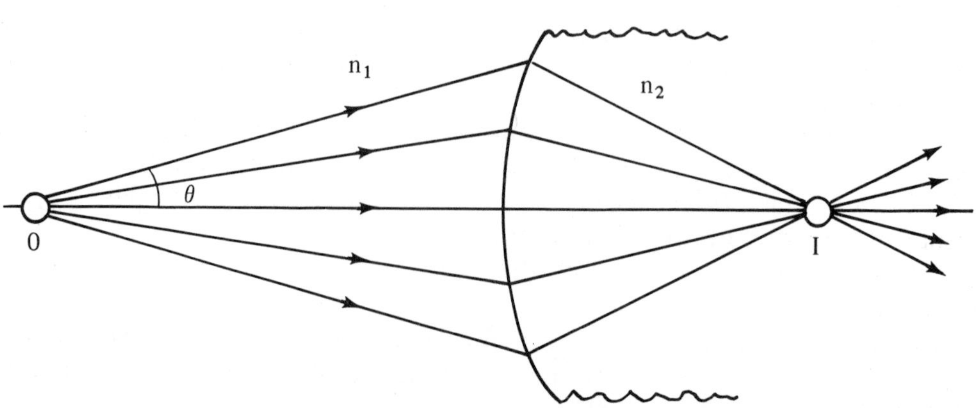

Figure 8-8 Formation of an image by a spherical refracting surface.

Note that Equation (8.12) is valid *only* if the size of the object is small compared to R. This is equivalent to the assumption that θ in Figure 8–8 is small; hence, the image distance is the same for all paraxial rays (paraxial rays are rays nearly parallel to the axis). In addition, Equation (8.12) holds for $n_2 > n_1$ and for $n_1 > n_2$.

Note that if the surface is a *plane surface,* then $R = \infty$ and Equation (8.12) reduces to

$$\frac{n_1}{p} = -\frac{n_2}{q} \qquad \text{Plane Refracting Surface} \qquad (8.13)$$

The student should note that Equation (8.13) is equivalent to Snell's law [that is, Equation (8.7)] for small angles of incidence and refraction, where $\sin\theta \cong \tan\theta$.

Also, for spherical refracting surfaces, real images are formed on the *opposite* side of the surface from which the light comes, whereas for mirrors, real images are formed on the same side of the surface. Therefore, the same sign convention can be used for spherical refracting surfaces as for mirrors, recognizing the change in roles of the real and virtual images. For example, in Figure 8–8, p, q and R are all positive.

In general, q is positive if the (real) image falls on the side where a real image is formed. q is negative if the (virtual) image falls on the side where a virtual image is formed. R is positive if C lies on the side where a real image is formed. R is negative if C lies on the side where a virtual image is formed.

The *lateral magnification* of a refracting surface is given by

$$M = \frac{h'}{h} = -\frac{n_1 q}{n_2 p}$$

Example 8.5

A small object is located in a medium whose index of refraction is 1.50, 20 cm from the surface whose radius is 30 cm, as in Figure 8–9. (a) Find the image of the object as viewed from the right. Assume paraxial rays in the calculation.

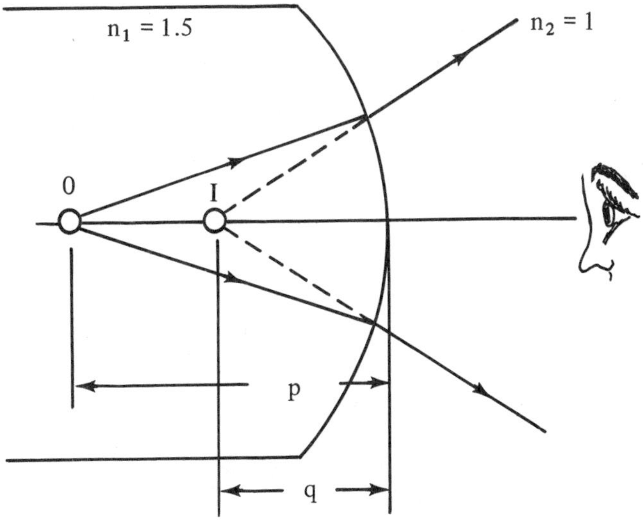

Figure 8–9

(a) Since the virtual image is located in the same medium as the center of curvature, R is *negative.* Applying Equations (8.12) and noting that $n_1 = 1.50$, $n_2 = 1$, $p = 20$ cm gives (we choose n_1 to be the side from which the light comes)

$$\frac{n_1}{p} + \frac{n_2}{q} = \frac{n_2 - n_1}{R}$$

$$\frac{1.50}{20} + \frac{1}{q} = \frac{1 - 1.50}{-30}$$

$$q = -17 \text{ cm}$$

The negative sign for q indicates that the image is in the same medium as the object.

(b) If the object height is 2 cm, what is the height of the image?

$$M = -\frac{n_1 q}{n_2 p} = -\frac{(1.50)(-17)}{(1)(20)} = \frac{h'}{h}$$

$$h' = 0.75\,(1.7)\,h = 2.56 \text{ cm}$$

Example 8.6

A small fish is at the bottom of a pond whose depth is d (Fig. 8-10). What is the apparent depth of the fish as viewed from above the pond?

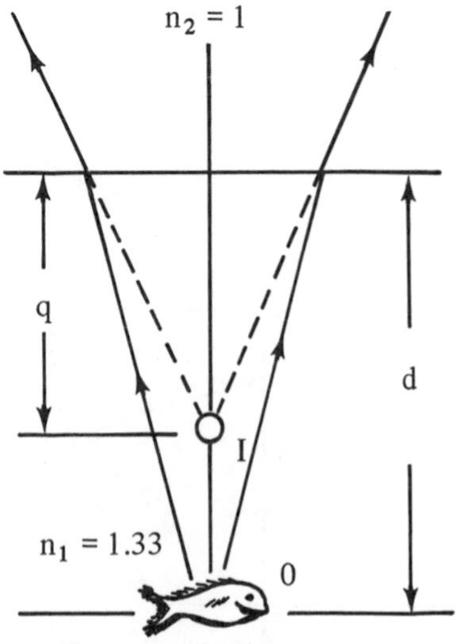

Figure 8-10

Solution

In this example, the surface is a plane, so $R = \infty$ and we can apply Equation (8.13). Since $p = d$, and $n_1 = 1.33$ for water, we get

$$\frac{n_1}{p} + \frac{n_2}{q} = \frac{1.33}{d} + \frac{1}{q} = 0$$

$$q = -0.75d$$

Again, the image is virtual, as in Example 8.5. If d = 10 m, q = -7.5 m.

8.6 THIN LENSES

A thin lens has two spherical surfaces whose radii of curvature are R_1 and R_2, as in Figure 8–11. The object is in a medium whose index of refraction is n_1, while the lens has an index of refraction n_2. If the image of the surface of radius R_1 is treated as the object of the surface whose radius is R_2, Equation (8.12) can be applied twice (see Programmed Exercise 3). This gives the expression

$$\frac{1}{p} + \frac{1}{q} = \left(\frac{n_2}{n_1} - 1\right)\left(\frac{1}{R_1} - \frac{1}{R_2}\right) \tag{8.14}$$

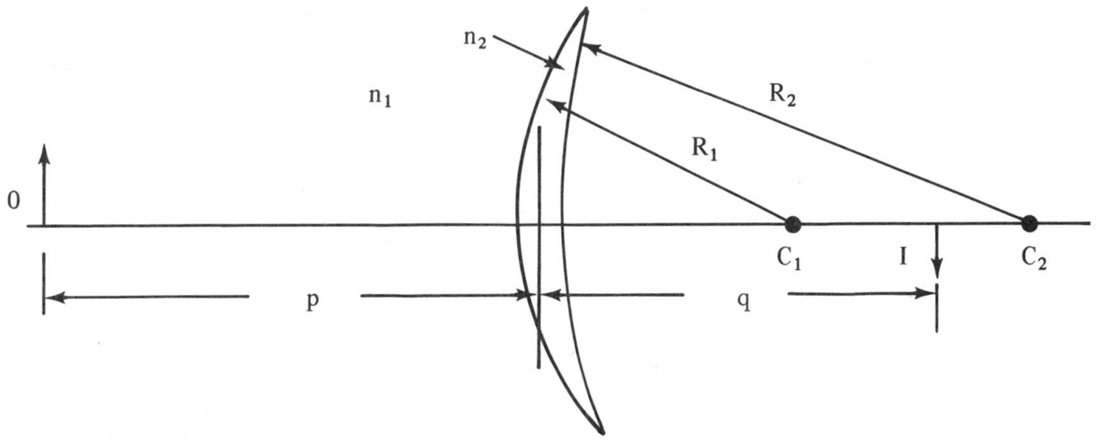

Figure 8–11 Representation of a thin lens (convex-concave) with two radii of curvature R_1 and R_2. *Note:* p, q and R_2 are positive in this case.

If medium 1 is air, $n_1 \cong 1$. Taking $n_2 = n$, Equation (8.14) can be written as

$$\frac{1}{p} + \frac{1}{q} = (n - 1)\left(\frac{1}{R_1} - \frac{1}{R_2}\right) \tag{8.15}$$

The *focal length* of the lens is given by

$$\frac{1}{f} = \frac{1}{p} + \frac{1}{q} = (n - 1)\left(\frac{1}{R_1} - \frac{1}{R_2}\right) \tag{8.16}$$

Note that f is a property of the lens, and Equation (8.16) enables us to calculate f from the radii of curvature and index of refraction of the lens material. Equation (8.16) is known as the *lensmaker's equation.* A thin lens has *two* focal points corresponding to parallel rays emerging from the left or right. This is illustrated in Figure 8–12 for a converging and diverging lens.

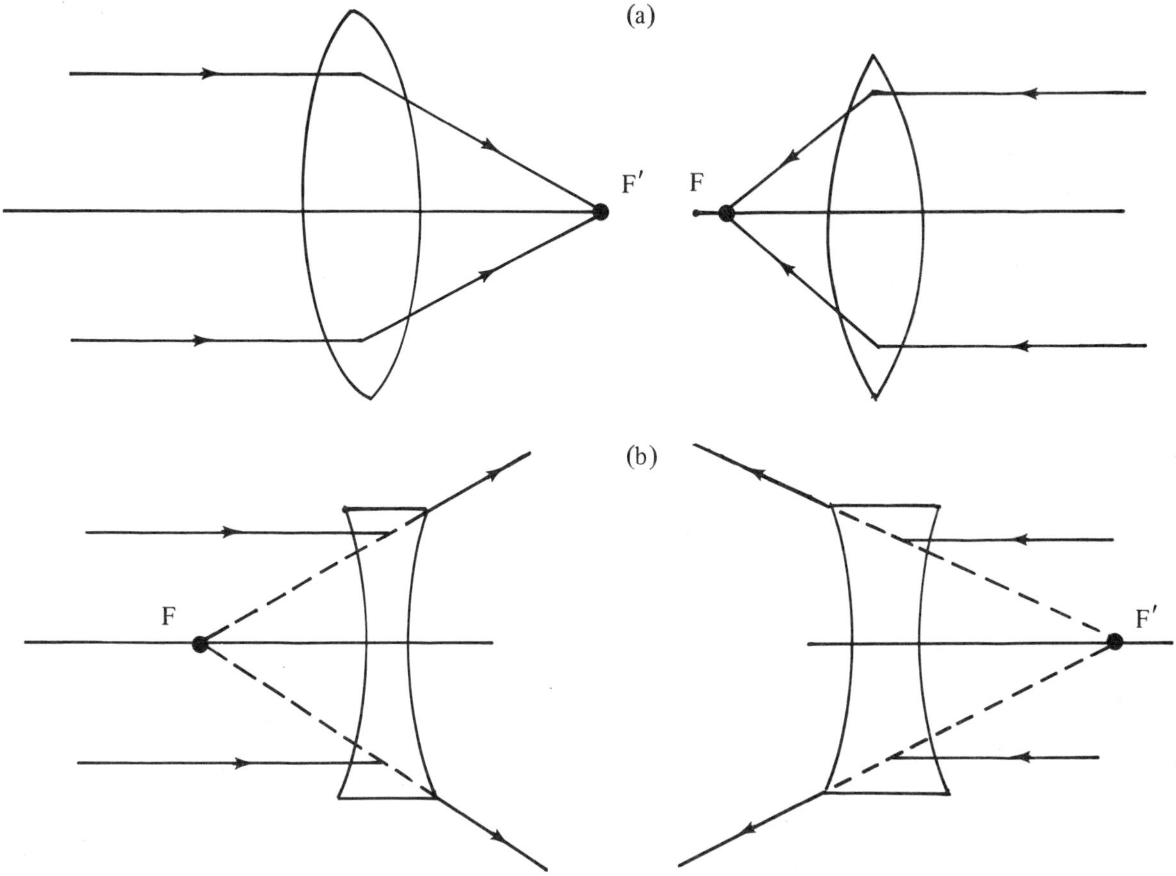

Figure 8–12 Focal points for (*a*) a converging lens (real focus) and (*b*) a diverging lens (virtual focus).

The sign conventions that are used in applying this equation are the same as those discussed in the previous section. For example, in Figure 8–11, if all the quantities p, q, f, R_1 and R_2 were to the left of the lens, that is, on the opposite side of the refracted light, R_2 would be negative. Other examples of sign choices for various lenses are illustrated in Figure 8–13.

The *lateral magnification* of a thin lens is the ratio of the image height h′ to the object height h, and is given by the product of the magnifications of each surface. The resulting magnification is given by

$$M = -\frac{q}{p} = \frac{h'}{h} \tag{8.17}$$

When M is positive, the image is erect. When M is negative, the image is inverted.

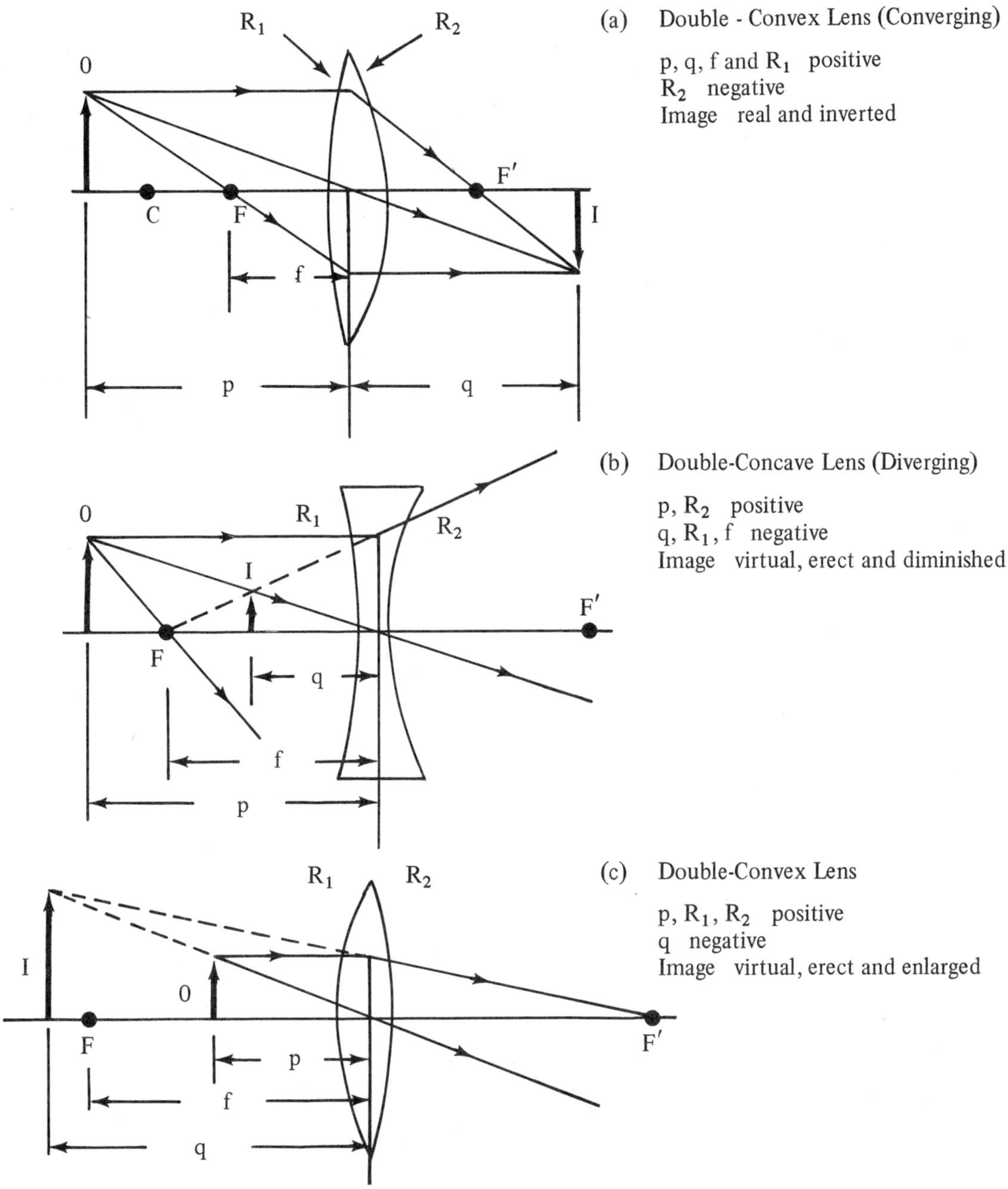

Figure 8–13 Sign conventions for various thin lenses.

Graphical methods (ray diagrams) are also very useful when working with thin lenses. The two rays described below and illustrated in Figure 8–13 will locate the image.

1. A ray drawn from the top of 0 parallel to the axis. This ray passes through one of the focal points after refraction by the lens.

2. A ray from 0 through the center of the lens. This ray doesn't deviate appreciably from a straight line for a thin lens.

Another ray which is useful when the object lies outside of F is the one drawn from the top of 0 through F, as shown in Figure 8–13(*a*). This ray is parallel to the axis on the right.

Example 8.7

A thin lens has surfaces of radii 30 cm and -50 cm, as in Figure 8-13(*a*). (a) Determine the focal length of the lens if it is made of glass of index of refraction 1.52. (b) Calculate the position of the image of an object placed 80 cm from the lens. (c) What is the magnification of the lens for an object distance of 80 cm?

Solution

(a) Since $R_1 = +30$ cm and $R_2 = -50$ cm, we have

$$\frac{1}{f} = (n - 1)\left(\frac{1}{R_1} - \frac{1}{R_2}\right) = (1.52 - 1)\left(\frac{1}{30} + \frac{1}{50}\right) = 0.028$$

$$f = +36 \text{ cm}$$

(b)

$$\frac{1}{p} + \frac{1}{q} = \frac{1}{f}$$

$$\frac{1}{80} + \frac{1}{q} = \frac{1}{36}$$

$$\therefore \quad q = +67 \text{ cm}$$

(c)

$$M = -\frac{q}{p} = -\frac{67}{80} = -0.84$$

The image is diminished and *inverted.*

Example 8.8

A thin lens has a focal length of -20 cm (diverging). (a) Determine the position of an object placed 40 cm in front of the lens, as in Figure 8-12(*b*). (b) What is the height of the virtual image if the height of the object is 2 cm?

Solution

(a)

$$\frac{1}{p} + \frac{1}{q} = \frac{1}{f}$$

$$\frac{1}{40} + \frac{1}{q} = -\frac{1}{20}$$

$$q = -13.3 \text{ cm}$$

(b)

$$M = -\frac{q}{p} = \frac{h'}{h}$$

$$-\frac{(-13.3)}{40} = \frac{h'}{2}$$

$$h' = 0.65 \text{ cm}$$

Example 8.9

The lens described in Example 8.8 is made of fused quartz of index of refraction 1.46. Determine the focal length of the lens if it is immersed in water, which has an index of refraction of 1.33.

Solution

Assuming the index of refraction of air is 1, we can use Equation (8.14) twice. In air, the lens formula gives

$$\frac{1}{f} = -\frac{1}{20} = (n - 1)\left(\frac{1}{R_1} - \frac{1}{R_2}\right) \qquad (1)$$

In water, we get

$$\frac{1}{f'} = (n' - 1)\left(\frac{1}{R_1} - \frac{1}{R_2}\right) \qquad (2)$$

But $n = 1.46$ and $n' = \frac{1.46}{1.33} = 1.10$. Dividing (1) by (2) gives

$$\frac{f'}{f} = \frac{n - 1}{n' - 1} = \frac{1.46 - 1}{1.10 - 1} = 4.6$$

$$\therefore \quad f' = 4.6\, f = 4.6\,(-20) = -92 \text{ cm}$$

8.7 COMBINATION OF THIN LENSES

If two thin lenses are used to form an image, the system can be treated in the following manner. First, the image of the first lens is calculated as if the second lens were not present. Second, the image of the first lens is treated as the object of the second lens. The image of the second lens is the final image position of the system. If the image of the first lens lies behind the second lens, it is treated as a virtual object for the second lens (that is, p negative). The same procedure is used for a system of three or more lenses. Systems of two or more lenses are common in such optical instruments as mircoscopes, telescopes and cameras.

Example 8.10

Two converging lenses of focal lengths +10 cm and +20 cm are separated by 20 cm, as in Figure 8–14. An object is placed 15 cm in front of the first lens. (a) Find the position of the final image. (b) Calculate the magnification of the system.

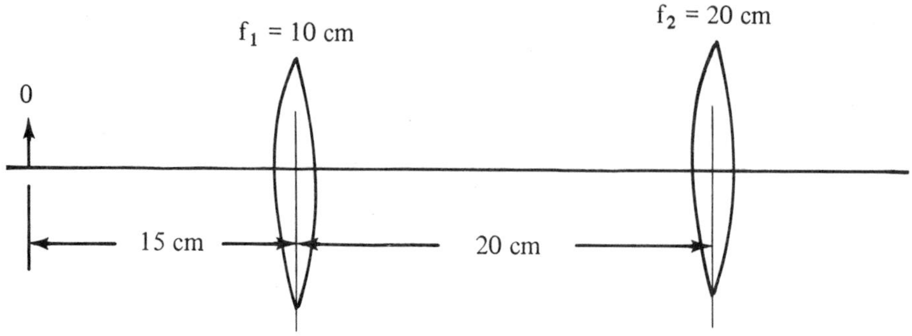

Figure 8–14

Solution

For the first lens we have

$$\frac{1}{p_1} + \frac{1}{q_1} = \frac{1}{15} + \frac{1}{q_1} = \frac{1}{10}$$

$$q_1 = 30 \text{ cm}$$

where q_1 is measured from the first lens.

Since the image of the first lens lies to the right of the second lens, and we take this as the object of the second lens, $p_2 = -10$ cm, where distances are now measured from the second lens of focal length 20 cm. Applying the lens equation to the second lens gives

$$\frac{1}{p_2} + \frac{1}{q_2} = -\frac{1}{10} + \frac{1}{q_2} = \frac{1}{20}$$

$$q_2 = \frac{20}{3} \text{ cm}$$

That is, the final image lies $\frac{20}{3}$ cm to the right of the second lens.

(b) The magnification of the first lens is

$$M_1 = -\frac{q_1}{p_1} = -\frac{30}{15} = -2$$

The magnification of the second lens is

$$M_2 = -\frac{q_2}{p_2} = -\frac{\frac{20}{3}}{-10} = \frac{2}{3}$$

Hence, the total magnification is the product of the individual magnifications.

$$M = M_1 \times M_2 = -2 \times \frac{2}{3} = -\frac{4}{3}$$

The final image is inverted and larger than the object.

When two thin lenses are in *contact,* then we can apply the lens equation to both lenses in the following manner:

$$\frac{1}{p_1} + \frac{1}{q_1} = \frac{1}{f_1}$$

$$\frac{1}{p_2} + \frac{1}{q_2} = \frac{1}{f_2}$$

But the image of the first lens is the object of the second lens, so $p_2 = -q_1$. Using this and the relations above gives the equivalent focal length f of the combination of two lenses:

$$\frac{1}{f} = \frac{1}{f_1} + \frac{1}{f_2} \tag{8.18}$$

Example 8.11

A converging lens of focal length +20 cm is in contact with a diverging lens of focal length -30 cm. (a) Calculate the focal length of the combination. (b) Determine the final image position of an object placed 15 cm to the left of the lenses.

Solution

(a)

$$\frac{1}{f} = \frac{1}{f_1} + \frac{1}{f_2} = \frac{1}{20} - \frac{1}{30}$$

$$f = 60 \text{ cm}$$

(b)

$$\frac{1}{p} + \frac{1}{q} = \frac{1}{f}$$

$$\frac{1}{15} + \frac{1}{q} = \frac{1}{60}$$

$$q = -20 \text{ cm}$$

The final image is virtual and 20 cm to the left of the lenses.

8.8 PROGRAMMED EXERCISES

1.A

Light of one wavelength travels from one medium to another. The *incident* ray makes an angle θ_1 with the normal. What is the angle of the *reflected* ray with respect to the normal?

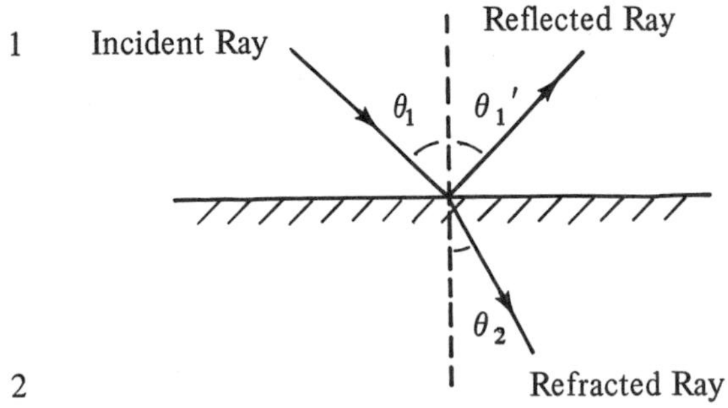

The angle of incidence *equals* the angle of reflection, so

$$\theta_1' = \theta_1 \qquad (1)$$

This is the law of reflection.

1.B

The ray that enters the second medium is called the ________ ray.

refracted

1.C

The refracted ray makes an angle θ_2 with the normal. Is θ_2 equal to θ_1?

No. The speed of light depends on the properties of that medium.

1.D

What can be said about the ratio

$$\frac{\sin\theta_1}{\sin\theta_2}?$$

The ratio is a *constant* for a given wavelength of light. This constant is called the relative index of refraction of medium 2 relative to medium 1. That is,

$$\frac{\sin\theta_1}{\sin\theta_2} = n_{12} \qquad (2)$$

1.E

What is the absolute index of refraction of a medium in terms of the speed of light in that medium?

$n = \frac{c}{v}$, where c is the speed of light in vacuum; $c \cong 3.0 \times 10^8 \frac{m}{sec}$ and v is the speed of light in the medium.

1.F

Defining $n_1 = \frac{c}{v_1}$ and $n_2 = \frac{c}{v_2}$ to be the indices of refraction of the two media, rewrite (2) in a more convenient form. What is this law called?

$$n_1 \sin\theta_1 = n_2 \sin\theta_2 \qquad (3)$$

This is *Snell's law* or the law of refraction. *Note:* It applies to light at one wavelength.

1.G

Is θ_2 less or greater than θ_1? Explain.

The answer to this can be either less or greater, depending on the ratio $\frac{n_1}{n_2}$. If

$$\frac{n_1}{n_2} > 1, \theta_2 > \theta_1$$

If

$$\frac{n_1}{n_2} < 1, \theta_2 < \theta_1$$

1.H

Calculate the speed of light in a medium whose index of refraction is 2.0. What is the wavelength of light in this medium if the wavelength in vacuum is 5000 Å? Note that the frequency is independent of the medium and 1Å = 10^{-10}m.

$$v = \frac{c}{n} = \frac{3.0 \times 10^8}{2} \frac{m}{sec}$$

$$v = 1.5 \times 10^8 \frac{m}{sec}$$

In vacuum we have

$$f = \frac{c}{\lambda} = \frac{3.0 \times 10^8 \frac{m}{sec}}{5000 \times 10^{-10} \text{ m}}$$

$$f = 6 \times 10^{14} \text{ Hz}$$

so

$$\lambda_m = \frac{v}{f} = \frac{1.5 \times 10^8 \frac{m}{sec}}{6 \times 10^{14} \text{ Hz}} = 2500 \text{ Å}$$

1.I

Light is incident of a medium of thickness d as shown, where $n_2 > n_1$. Show that $\theta_1 = \theta_3$.

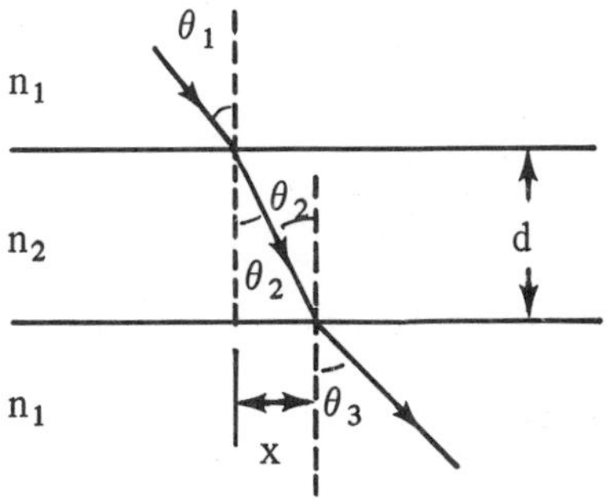

Applying (3) to the two interfaces gives

$$n_1 \sin\theta_1 = n_2 \sin\theta_2$$

$$n_2 \sin\theta_2 = n_1 \sin\theta_3$$

Comparing the two expressions, we see that $\sin\theta_1 = \sin\theta_3$, or $\theta_1 = \theta_3$.

1.J

What is the lateral displacement x of the beam that emerges from medium 2?

From the figure in frame 1.I, we see that

$$\tan\theta_2 = \frac{x}{d}$$

or $x = d\tan\theta_2$. θ_2 can be obtained from Snell's law and a knowledge of n_1, n_2 and θ_1.

1.K

Light passes from medium 1 to medium 2 where $n_1 > n_2$. Sketch the incident reflected and refracted rays.

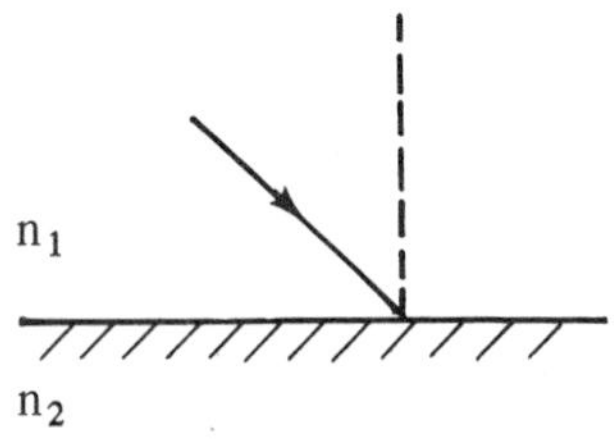

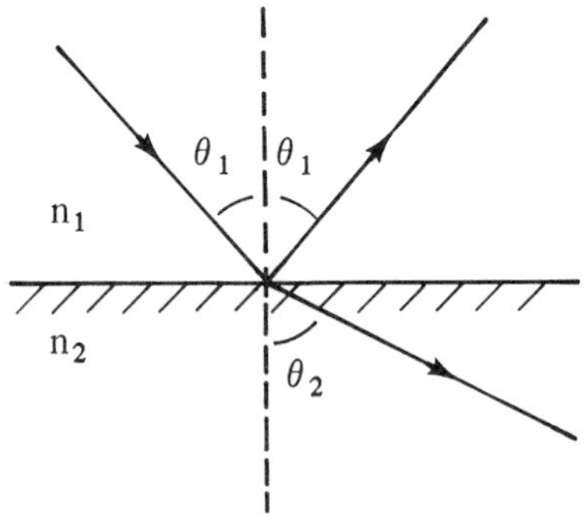

Note: $\theta_2 > \theta_1$ since $n_1 > n_2$.

1.L

For what critical angle of incidence θ_c will *total* reflection at the interface occur when $n_1 > n_2$? (Note that total reflection occurs for angles larger than θ_c.)

At this critical angle, $\theta_2 = \frac{\pi}{2}$ (that is, no light enters medium 2). Substituting this angle in Snell's law gives

$$n_1 \sin\theta_c = n_2 \sin\left(\frac{\pi}{2}\right)$$

or

$$\theta_c = \arcsin\left(\frac{n_2}{n_1}\right)$$

1.M

Obtain a value for θ_c for a water-air interface. Assume $n_1 = 1.33$ for water and $n_2 = 1.00$ for air.

$$\theta_c = \arcsin\left(\frac{1}{1.33}\right)$$

$$\theta_c = \arcsin(0.75)$$

$$\theta_c = 49°$$

2 A *spherical mirror* has a radius R. An object of height h is placed at a distance p from the mirror. An image of height h′ is formed a distance q from the mirror.

2.A

p is called ________________, q is the ________________ and C is the ________________. The mirror shown below is a ________________ mirror.

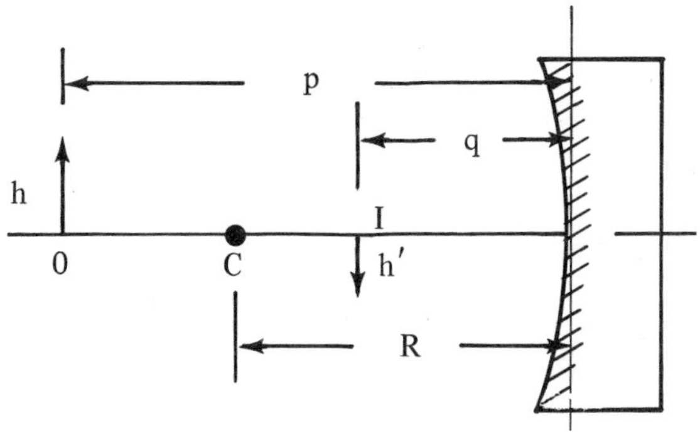

object distance

image distance

center of curvature

concave

2.B

Write the expression that relates p and q to R. This is the so-called *mirror equation.*

$$\frac{1}{p}+\frac{1}{q}=\frac{2}{R} \qquad (1)$$

2.C

How do you find the focal point F of the mirror? Describe this procedure in a ray diagram and determine the distance f of the point from the mirror.

The focal point is the point where parallel rays intersect the x axis. Parallel rays correspond to an object distance p = ∞, so from (1),

$$f = q = \frac{R}{2}$$

f is called the *focal length.*

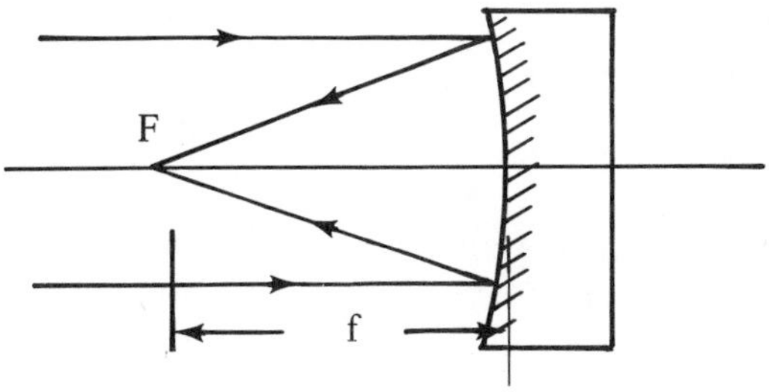

2.D

Rewrite (1) in terms of the focal length f.

$$\frac{1}{p} = \frac{1}{q} = \frac{1}{f} \quad (2)$$

2.E

The spherical mirror shown below is a ____________ mirror.

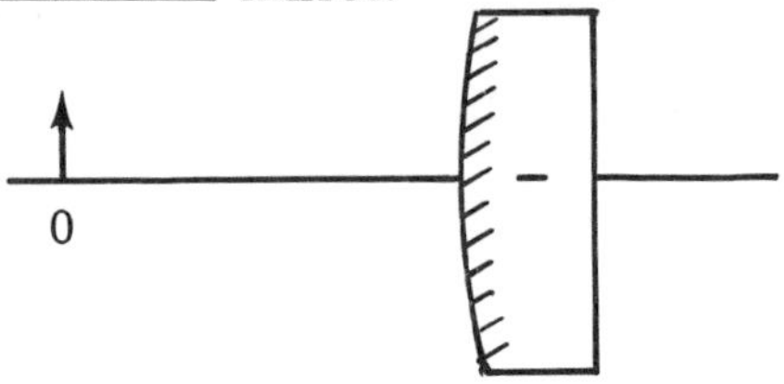

Convex

2.F

What are the signs of R and f for concave and convex mirrors?

concave ⟶ positive R and f

convex ⟶ negative R and f

2.G

If the object is in front of the mirror (that is, a real object), the sign of p is ____________.

positive

This corresponds to a real object.

2.H

If the image lies in front of the mirror, the sign of q is ____________. If the image lies behind the mirror, the sign of q is ____________.

positive (real image)

negative (virtual image)

2.I

If the height of the object is h and the height of the image is h′, the *magnification* of the mirror M is given by ____________. An alternate expression for M in terms of p and q is ____________.

$$M = \frac{h'}{h}$$

or

$$M = \frac{q}{p} \quad (4)$$

2.J

Is an image always larger than the object? That is, is $M > 1$ always? Explain.

No. An image may be smaller, equal to or larger than the object depending on the location of the object relative to the mirror and the focal length of the mirror.

2.K

An object placed 2 m from a *concave* mirror produces a real image 1.5 m in front of the mirror. Find the radius of curvature and focal length of the mirror.

Both p and q are positive, so (1) gives

$$\frac{1}{p}+\frac{1}{q}=\frac{1}{2}+\frac{1}{1.5}=\frac{2}{R}$$

or

$$R=\frac{12}{7}\text{ m}$$

$$f=\frac{R}{2}=\frac{6}{7}\text{ m}$$

2.L

Find the magnification of the mirror described in frame 2.K.

$$M=\frac{q}{p}=\frac{1.5}{2}=0.75$$

That is, the image size is $\frac{3}{4}$ times the size of the object.

2.M

Can the image of the concave mirror described in frame 2.K be virtual for some other value of p? Explain.

Yes. From (2), we see that q is negative (virtual image) when $p < f$, that is, when the object distance is *less* than the focal length. The student should verify this with a ray diagram [See Fig. 8–7(*b*)].

3.A

A small object 0 is in a medium whose index of refraction is n_1. The object is at a distance p from a *convex* spherical surface of radius R. The medium beyond the surface has an index of refraction n_2. What expression gives the position of the image? Where is this measured from?

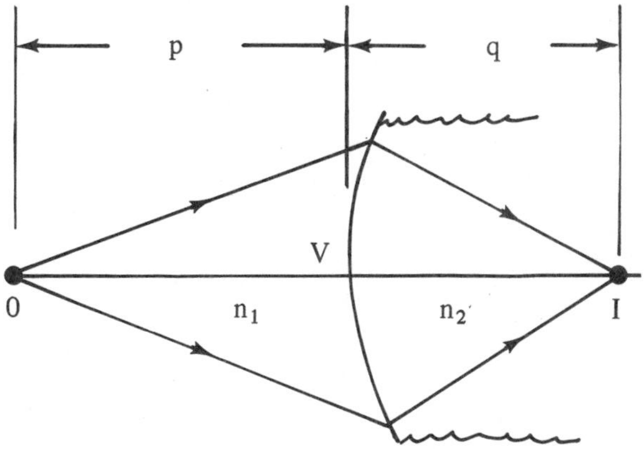

$$\frac{n_1}{p}+\frac{n_2}{q}=\frac{n_2-n_1}{R} \qquad (1)$$

where q is the image distance measured from V. [Note that (1) applies for $n_2 > n_1$ and for $n_2 < n_1$.]

3.B

For the position of the image shown in frame 3.A, what are the signs of p, q and R in (1)?

They are all *positive*. Note that a real image (+q) lies to the right of V in this case.

3.C

Can the image of 0 ever be virtual for a convex surface? Explain.

From (1), we see that q can be negative when the following condition is met:

$$\left|\frac{n_2 - n_1}{R}\right| < \left|\frac{n_1}{p}\right|$$

3.D

For the *concave* refracting surface shown below, with air at the left ($n_1 = 1$) and glass at the right ($n_2 = n$), what are the signs of p, q and R?

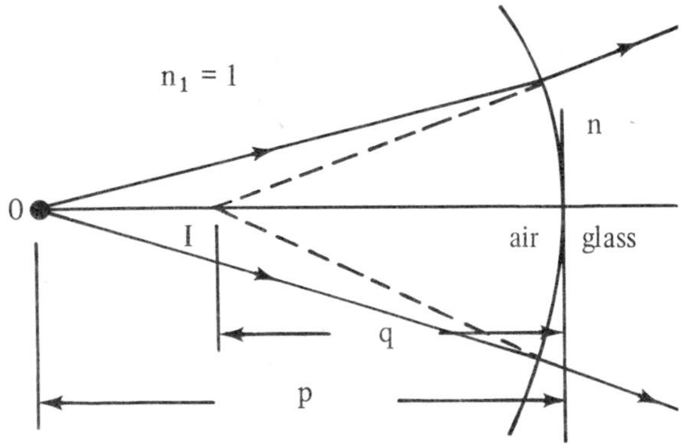

p is positive (real object).

q is negative (virtual image).

R is negative for a concave surface.

Note that the image is *always* virtual for a concave surface if the object lies in the less dense medium.

3.E

Now apply (1) to the thin *double-convex* lens shown below. First, write the expression which gives the "image" of the first surface of radius R_1.

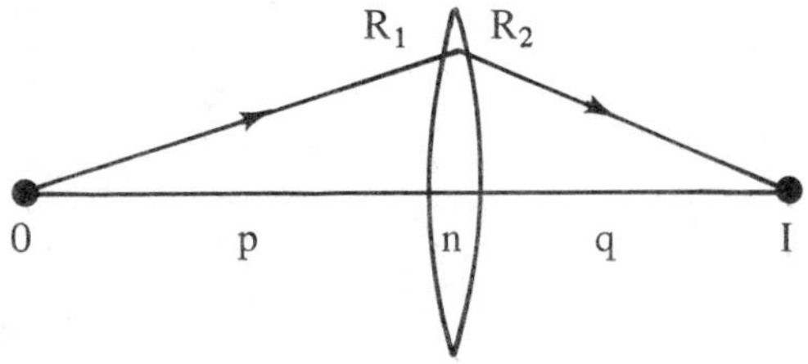

Since $n_1 = 1$ and $n_2 = n_1$, we have

$$\frac{1}{p_1} + \frac{n}{q_1} = \frac{n-1}{R_1} \qquad (2)$$

where q_1 would be the image distance for the left surface if the lens were very thick.

3.F

Now use the image distance q_1 as the object distance for the second surface of radius R_2 to obtain the *thin lens formula.* Note that $p_2 = -q_1$, since the image of the left surface is a virtual object for the surface at the right.

$$\frac{n}{p_2} + \frac{1}{q_2} = \frac{1-n}{R_2} \tag{3}$$

But from (2),

$$\frac{n}{p_2} = -\frac{n}{q_1} = \frac{1}{p_1} - \frac{1-n}{R_1} \tag{4}$$

Substituting (4) into (3), and dropping the subscripts on p_1 and q_2, gives

$$\frac{1}{p} + \frac{1}{q} = (n-1)\left(\frac{1}{R_1} - \frac{1}{R_2}\right) \tag{5}$$

where q is the final image distance.

3.G

What are the sign conventions for R_1 and R_2?

R_1 is positive if the center of curvature for this surface is on the same side as the refracted light (as in frame 3.E). Therefore, R_2 in frame 3.E is *negative,* since the center of curvature for this surface is opposite to that of the refracted light.

3.H

According to the sign conventions for R_1 and R_2 described in 3.G, what are the signs for R_1 and R_2 in the *double-concave* lens shown below?

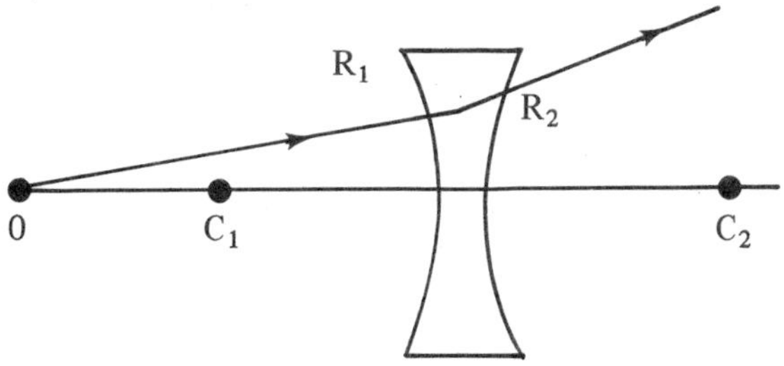

R_1 is *negative,* since C_1 is on the opposite side of the refracted light.

R_2 is *positive,* since C_2 is on the same side as the refracted light.

[The student should note that other sign conventions are sometimes used when (5) is written as

$$\frac{1}{p} + \frac{1}{q} = (n-1)\left(\frac{1}{R_1} + \frac{1}{R_2}\right)$$

4 *Thin Lenses.* In this exercise, we review the properties of thin lenses, the sign conventions and the use of ray diagrams in locating images.

4.A

A *double-convex* lens is shown below. Write an expression for the focal length f of the lens in terms of R_1, R_2 and the index of refraction of the lens material n.

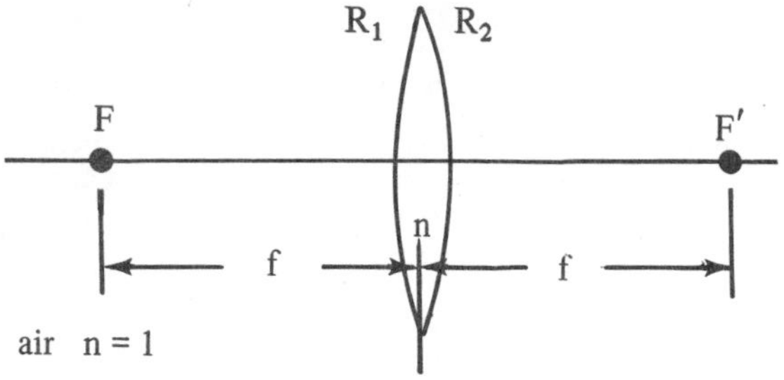

$$\frac{1}{f} = (n - 1)\left(\frac{1}{R_1} - \frac{1}{R_2}\right) \quad (1)$$

This is the so-called *lensmaker's formula.*

4.B

Show in a ray diagram the physical significance of the focal point F′. (F and F′ are also called the principal focus points.)

F′ is the point on the principal axis where parallel rays from the left (corresponding to p = ∞) intersect at a common point.

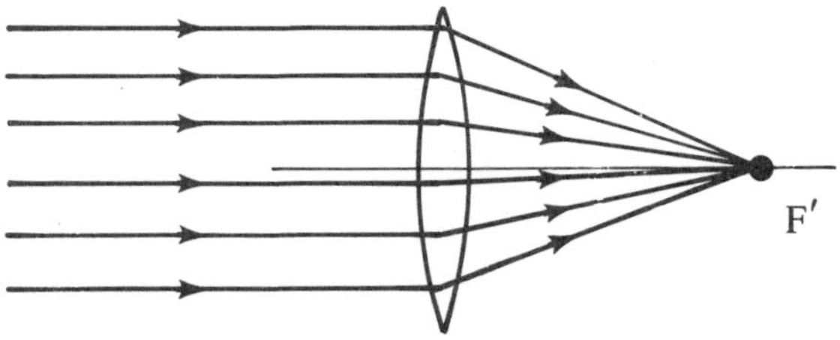

4.C

What is the significance of the focal point F?

F is the point on the axis where parallel rays from the right intersect at a common point.

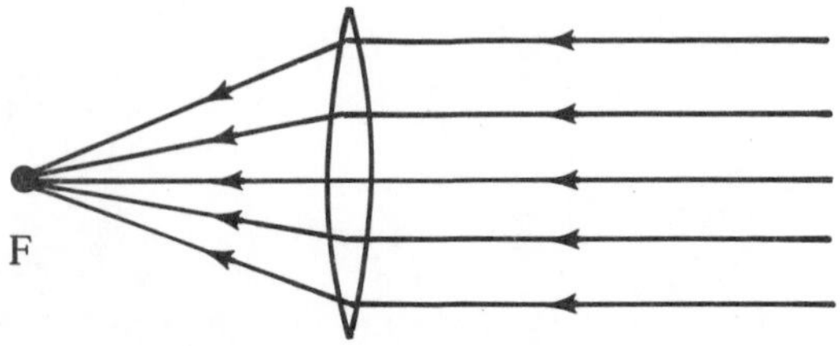

4.D

An object is at a distance p from the lens as shown below. What is the image distance q in terms of p and f?

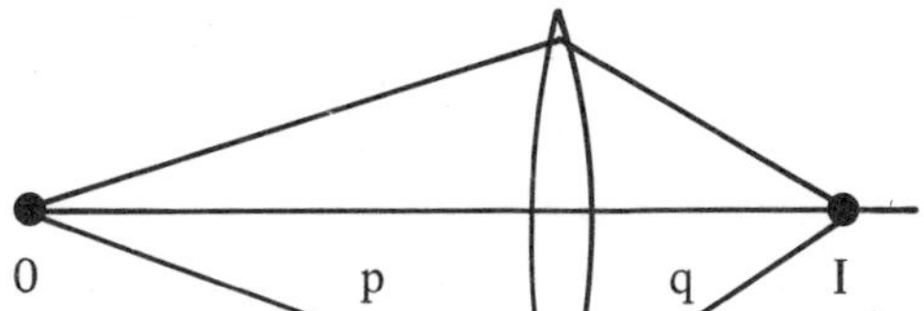

The thin lens formula is given by

$$\frac{1}{p} + \frac{1}{q} = \frac{1}{f} \quad (2)$$

$$\therefore \quad \frac{1}{q} = \frac{1}{f} - \frac{1}{p} = \frac{p - f}{pf}$$

$$q = \frac{pf}{p - f} \quad (3)$$

Likewise, the student should show that

$$p = \frac{qf}{q - f} \quad (4)$$

4.E

Show that f is always *positive* for a double-convex lens as shown in frame 4.A.

First, note that R_1 is *positive* and R_2 is *negative* according to the sign conventions described in Programmed Exercise 3. Therefore, from (1) we see that for $n > 1$, f is always positive for a double-convex lens.

4.F

Show that f is always *negative* for a *double-concave* lens.

In this case, R_1 is negative and R_2 is positive, so from (1) we see that f is *negative.*

4.G

What are the sign conventions for p and q for thin lenses?

p is positive if the object is real (an object is real if it is on the left of the lens and the light rays go to the right). q is positive if the image is real (on the same side as the refracted light). p and q are negative for virtual objects and virtual images, respectively.

4.H

Can a double-concave lens ever have a real image? Explain by a ray diagram.

No. Such a lens is *diverging,* so the image is always *virtual, erect* and diminished in size. ∴ q is always negative.

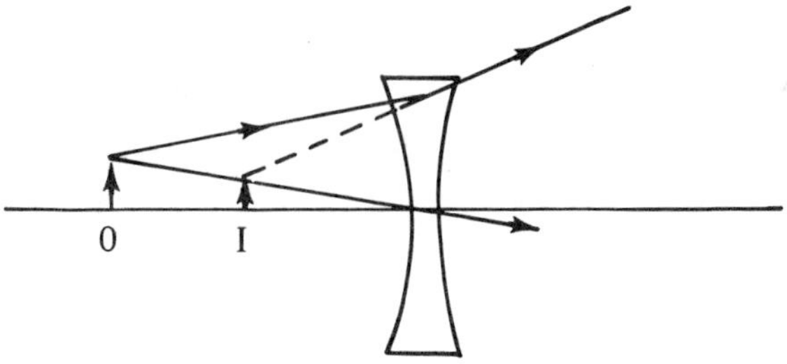

4.I

The magnification of a thin lens is given by M = __________ .

$$M = -\frac{q}{p} \qquad (5)$$

4.J

For a real object and real image, the sign of M is __________ . For a real object and virtual image the sign of M is __________ .

negative

positive

Note that for M positive, the image is *erect.* For M negative, the image is *inverted.*

4.K

If the height of an object is h, and the magnification of a thin lens is M, the height of the image is given by h′ = __________ .

$$h' = Mh \qquad (6)$$

4.L

Draw the three rays that can be used to locate the image of a converging lens when $p > f$ as shown below.

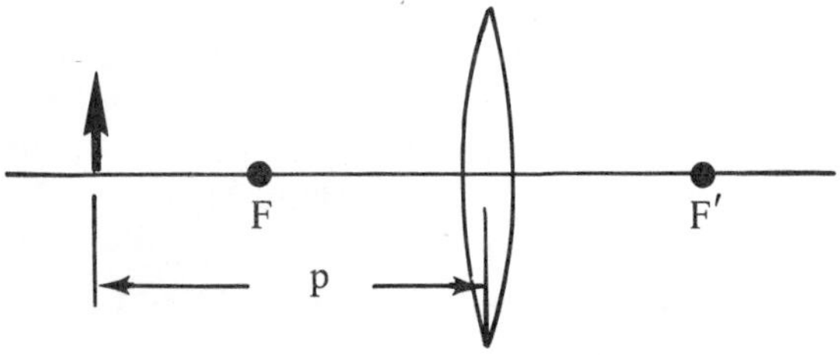

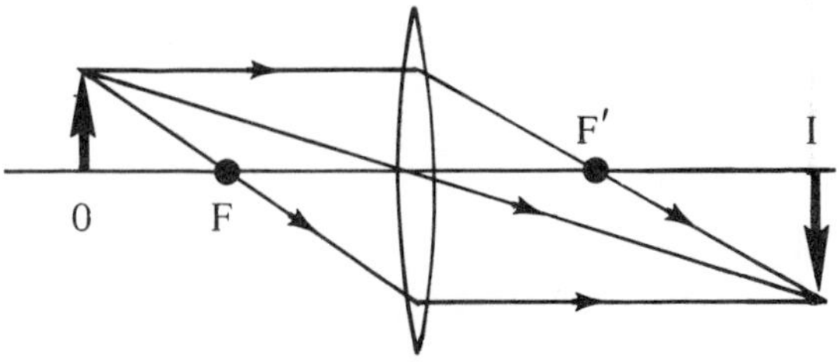

Note: Only two rays are required to locate the image. The image is real and inverted.

4.M

Locate the image of the diverging lens shown below using a ray diagram.

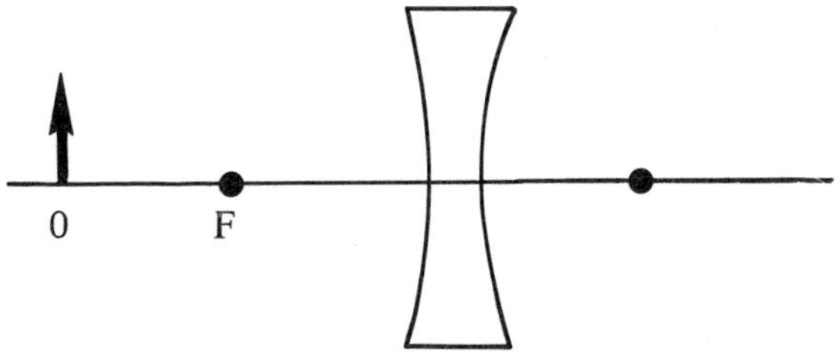

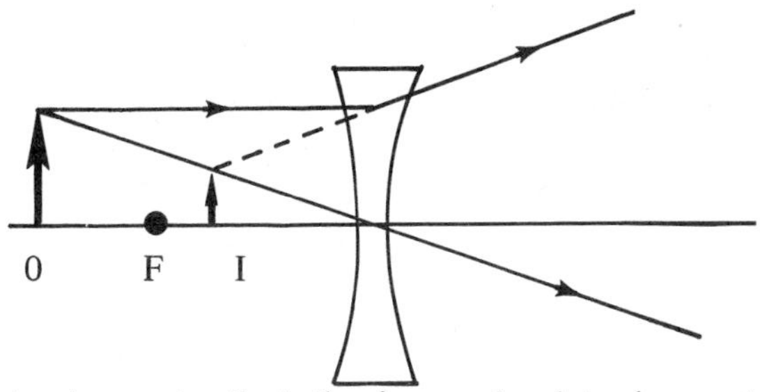

Again, note that the image is virtual, erect and diminished.

4.N

Let us go back to the *converging* lens. From (2), we see that for $p < f$, q is *negative*, that is, the image is virtual. Draw a ray diagram illustrating this point.

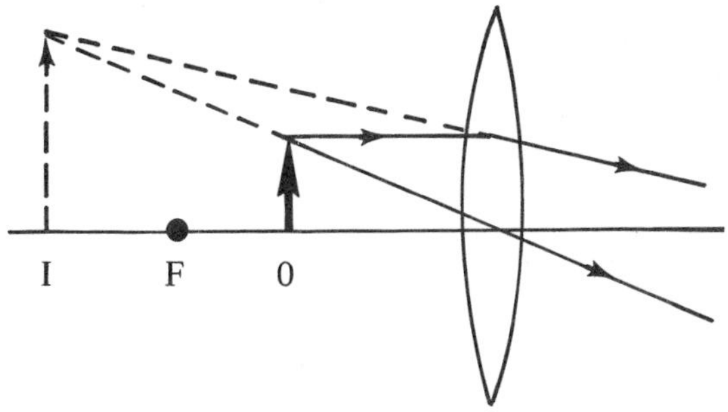

4.O

Since $\frac{1}{p}+\frac{1}{q}=\frac{1}{f}$, we see that *if* the object is at F, $p = f$ and $q = \infty$, that is, the image is at ∞. Draw a ray diagram to show this.

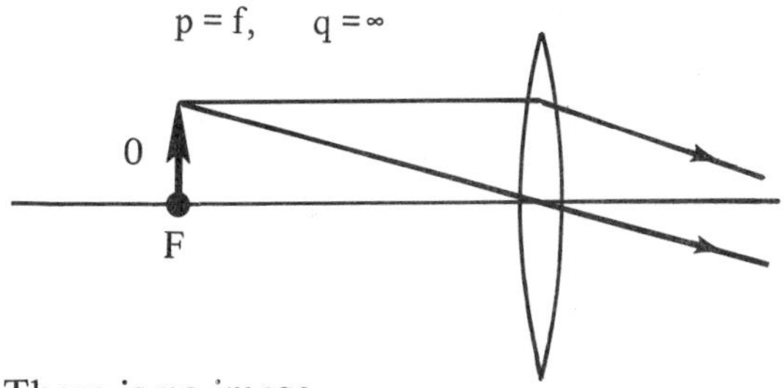

There is *no* image.

4.P

Describe the properties of the image in a ray diagram if the object is at a distance 2f from the lens.

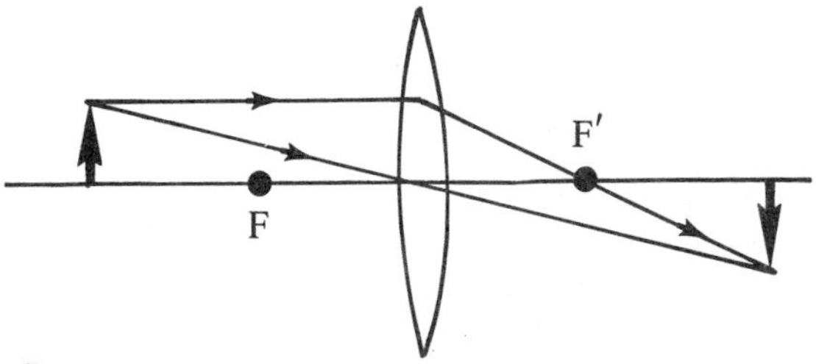

Since $p = 2f$, $q = 2f$. The image is *real*, *inverted* and the *same size* as the object.

4.Q

A combination of two thin lenses are in contact and have focal lengths of f_1 and f_2. Find the focal length of the system.

$$\frac{1}{f} = \frac{1}{f_1} + \frac{1}{f_2} = \frac{f_1 + f_2}{f_1 f_2}$$

$$\therefore \quad f = \frac{f_1 f_2}{f_1 + f_2} \qquad (7)$$

4.R

If the magnification of each lens is M_1 and M_2 in a two lens system, the *total* magnification is ________.

$$M = M_1 M_2 \qquad (8)$$

8.9 SUMMARY

The *index of refraction* of a medium is given by

$$n = \frac{c}{v} \tag{8.6}$$

where $c \cong 3 \times 10^8 \frac{m}{sec}$, the speed of light in vacuum, and v is the speed of light in that medium.

When light travels from one medium to a second, the light is refracted according to *Snell's law,* given by

$$n_1 \sin\theta_1 = n_2 \sin\theta_2 \tag{8.7}$$

where θ_1 is the angle of incidence and θ_2 is the angle of refraction, both measured with respect to the normal to the interface.

Total internal reflection occurs at an interface at or greater than the critical angle of incidence θ_c, given by

$$\sin\theta_c = \frac{n_2}{n_1} \qquad (n_1 > n_2) \tag{8.8}$$

The object and image distances for a spherical mirror of radius R and focal length f are related through the expression

$$\frac{1}{p} + \frac{1}{q} = \frac{2}{R} = \frac{1}{f}$$

The lateral magnification of a mirror is given by

$$M = \frac{q}{p} = \frac{h'}{h} \tag{8.11}$$

where h is the height of the object and h′ is the height of the image.

When light traveling in a medium of index of refraction n_1 meets a spherical interface of radius R beyond which is a medium of index of refraction n_2, the light is refracted. For paraxial rays, the object and image distances are related by

$$\frac{n_1}{p} + \frac{n_2}{q} = \frac{n_2 - n_1}{R} \tag{8.12}$$

The object and image distances of a thin lens are related by the expression

$$\frac{1}{p} + \frac{1}{q} = (n - 1)\left(\frac{1}{R_1} - \frac{1}{R_2}\right) = \frac{1}{f} \tag{8.15}$$

where n is the index of refraction of the lens and R_1 and R_2 are the radii of curvature of the spherical lens surfaces.

The lateral magnification of a thin lens is

$$M = -\frac{q}{p} = \frac{h'}{h} \tag{8.17}$$

When two thin lenses of focal lengths f_1 and f_2 are in contact, the focal length of the combination can be calculated from the expression

$$\frac{1}{f} = \frac{1}{f_1} + \frac{1}{f_2} \tag{8.18}$$

8.10 PROBLEMS

1. (a) Find the speed of light in diamond (n = 2.42). (b) Determine the wavelength of yellow sodium light in diamond, if the wavelength in air is 5890 Å.

2. Light travels in a vacuum and meets an interface at an angle of 20° with the normal, as in Figure 8-1. If the refracted light makes an angle of 16° with the normal, what is the index of refraction of the medium?

3. Light is incident from air onto a flat slab of ice (n = 1.31) at an angle of incidence equal to 30°. What is the angle of refraction?

4. Calculate the critical angles for (a) ice, (b) fused quartz and (c) zircon. See Table 8-1 for values of n, and assume that the opposite side of the interface is air.

5. A ray of light in air is incident on a flat piece of transparent material at an angle of incidence equal to 50°. If the refracted and reflected rays make an angle of 95° with each other, determine the index of refraction of the transparent material.

6. A concave mirror has a radius of curvature of 60 cm. Find the image position and magnification of an object placed in front of the mirror at distances of (a) 90 cm and (b) 20 cm. (c) Draw ray diagrams to obtain the image in each case.

7. Repeat Problem 6, assuming the mirror is convex with a radius of 60 cm.

8. The real image height of a concave mirror is observed to be four times larger than the object height when the object is 30 cm in front of the mirror. What is the radius of the mirror?

9. A concave mirror has a radius of 40 cm. (a) Calculate the image distance q for an arbitrary real object distance p. (b) Obtain values of q for object distances of 5 cm, 10 cm, 20 cm, 40 cm and 60 cm. (c) Make a plot of q *vs* p using the results of part (b).

10. An object placed 10 cm from a concave spherical mirror produces a real image at a distance 8 cm from the mirror. If the object is moved to a new position 20 cm from the mirror, what is the position of the image? Is the final image real or virtual?

11. Prove that the image of a real object placed in front of a spherical concave mirror is virtual and erect when $p < f$. Use a ray diagram to demonstrate this fact.

12. A spherical convex mirror has a radius of 40 cm. Determine the position and magnification of the mirror and the position of the virtual image for object distances of (a) 30 cm and (b) 60 cm. (c) Are the images erect or inverted?

13. A solid transparent sphere has an index of refraction n. What is the *minimum* value of n such that a distant object will form an image at the point Q (Fig. 8-15)? *Hint:* Assume $p = \infty$.

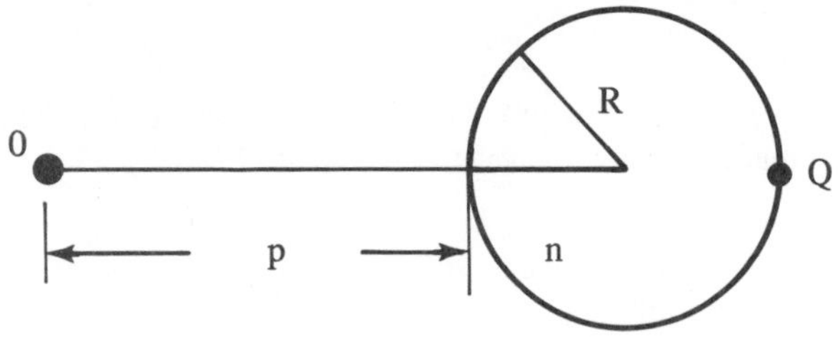

Figure 8–15

14. A "crystal" ball made of solid glass (n = 1.50) has a tiny air bubble located 5 cm from the center, as in Figure 8–16. If the ball has a radius of 15 cm, what is the apparent depth of the bubble when viewed as shown?

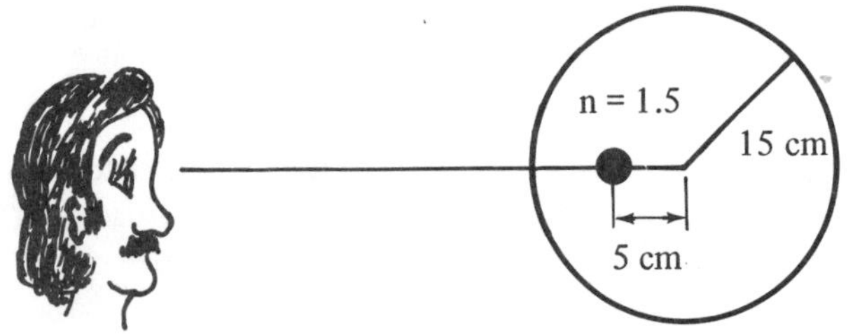

Figure 8–16

15. Repeat Problem 14 if the "crystal" ball is viewed under water.

16. A long glass rod (n = 1.50) has one end in the form of a hemisphere of radius 6 cm as in Figure 8–17 (that is, the diameter of the rod is 12 cm). Find the position of the image of an object placed at distances of (a) 10 m, (b) 18 cm and (c) 3 cm.

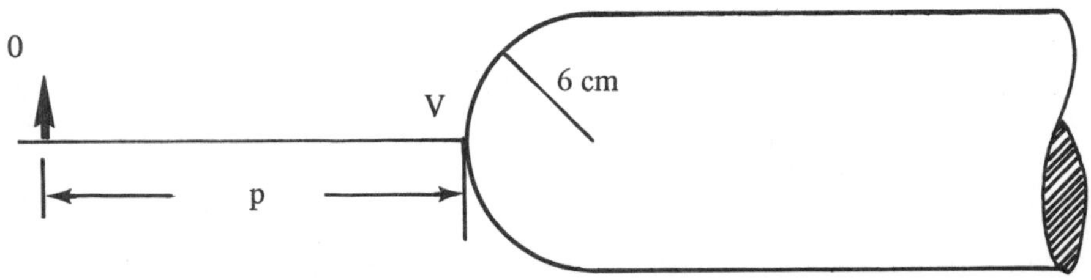

Figure 8–17

17. Repeat Problem 16(a) if the rod and object are immersed in water.

18. Calculate the lateral magnification of the object described in Problem 16 for object distances of (a) 18 cm and (b) 3 cm.

19. A colored marble is dropped in a large tank filled with benzene (n = 1.50). (a) What is the depth of the tank if the apparent depth of the marble when viewed from directly above the tank is 35 cm? (b) If the marble has a diameter of 1.5 cm, what is its apparent diameter when viewed directly above the marble, outside the tank?

20. Prove that, in general, the lateral magnification of an object for a *plane* refracting surface is *one.*

21. A converging lens has a focal length of 20 cm. Find the position of the image for a real object at distances of (a) 50 cm. (b) 30 cm and (c) 10 cm. (d) Determine the magnification of the lens for these object distances and whether the image is erect or inverted. (e) Draw ray diagrams to locate the images for these object distances.

22. Repeat Problem 21 if the lens is diverging, with a focal length of 20 cm.

23. Find the object distances of a thin, converging lens of focal length f if: (a) the image is real and the image distance is four times the focal length; (b) the image is virtual and is three times the focal length. (c) Calculate the magnification of the lens for cases (a) and (b).

24. A thin converging lens has a focal length f. Find the object distance if the image is (a) real and twice as large as the object, (b) virtual and half the size of the object.

25. A converging lens of focal length 20 cm is separated by 50 cm from a converging lens of focal length 5 cm. (a) Find the final position of the image of an object placed 40 cm in front of the first converging lens. (b) If the height of the object is 2 cm, what is the height of the final image? Is it real or virtual?

26. (a) If the two lenses described in Problem 25 are in contact, calculate the focal length of the combination. (b) Determine the image position of an object placed 5 cm from the lenses.

27. An object located 32 cm in front of a lens forms an image on a screen 8 cm behind the lens. (a) Find the focal length of the lens. (b) Determine the magnification of the lens. (c) Is the lens converging or diverging?

28. A converging lens has a focal length of 40 cm. Calculate the size of the real image of an object 4 cm in height for the following object distances: (a) 50 cm, (b) 60 cm, (c) 80 cm, (d) 100 cm, (e) 200 cm and (f) ∞.

29. An object 1 cm in height is placed 4 cm to the left of a converging lens of focal length 8 cm. A diverging lens of focal length -10 cm is located 6 cm to the right of the converging lens. Find the position and size of the final image. Is the image inverted or erect?

30. An object is placed 12 cm to the left of a diverging lens of focal length -6 cm. A converging lens of focal length 18 cm is placed a distance d to the right of the diverging lens. Find the distance d such that the final image is at infinity. Draw a ray diagram for this case.

9

WAVE OPTICS

This chapter is concerned with phenomena which demonstrate the wave nature of light, a field which is sometimes referred to as wave optics. These phenomena include interference, diffraction and polarization. Light waves are electromagnetic and polarization experiments show that they are transverse in nature. The diffraction of waves corresponds to the ability of waves to "bend around corners" when meeting obstacles. For example, we are able to hear sounds generated in another room, although there may be many obstacles between our ear and the source. Interference of waves corresponds to maxima and minima in wave intensity due to the superposition of two or more wave amplitudes at a given point.

9.1 INTERFERENCE—DOUBLE SLIT

The phenomenon of interference of light waves is nicely demonstrated by Young's experiment. A monochromatic (single wavelength) light wave is incident on a pair of slits separated by a distance d, as in Figure 9–1. Each slit then acts as a source of light, sending out radiation in all directions. This follows from *Huygens' principle,* which states that each point on a wave front may be regarded as a new source of secondary waves. The light from the two slits is then viewed on a screen at B, placed a distance L from the slits, where $L \gg a$. What is observed on the screen is a series of bright and dark regions called fringes that can be understood in the following manner. The light intensity at a given point on the screen is the result of the *superposition* of two (or more) wave amplitudes. The two waves can add constructively to give a bright fringe, or destructively to give a dark fringe, depending on the phase difference between the waves.

More specifically, consider the point P on the screen as shown in Figure 9–1, a distance y from the center. The light intensity at P will be the result of light coming from both slits. But the light from the upper slit travels farther than light from the lower slit, by a distance $a\sin\theta$. This distance is called the *path difference* δ. If $\delta = 0, \lambda, 2\lambda, 3\lambda, \ldots$, *constructive interference* results. That is, a bright fringe appears at P corresponding to the two sources being *in phase.* Therefore, the condition for bright fringes is given by

$$\delta = a\sin\theta = n\lambda \qquad (n = 0, 1, 2, 3, \ldots) \tag{9.1}$$

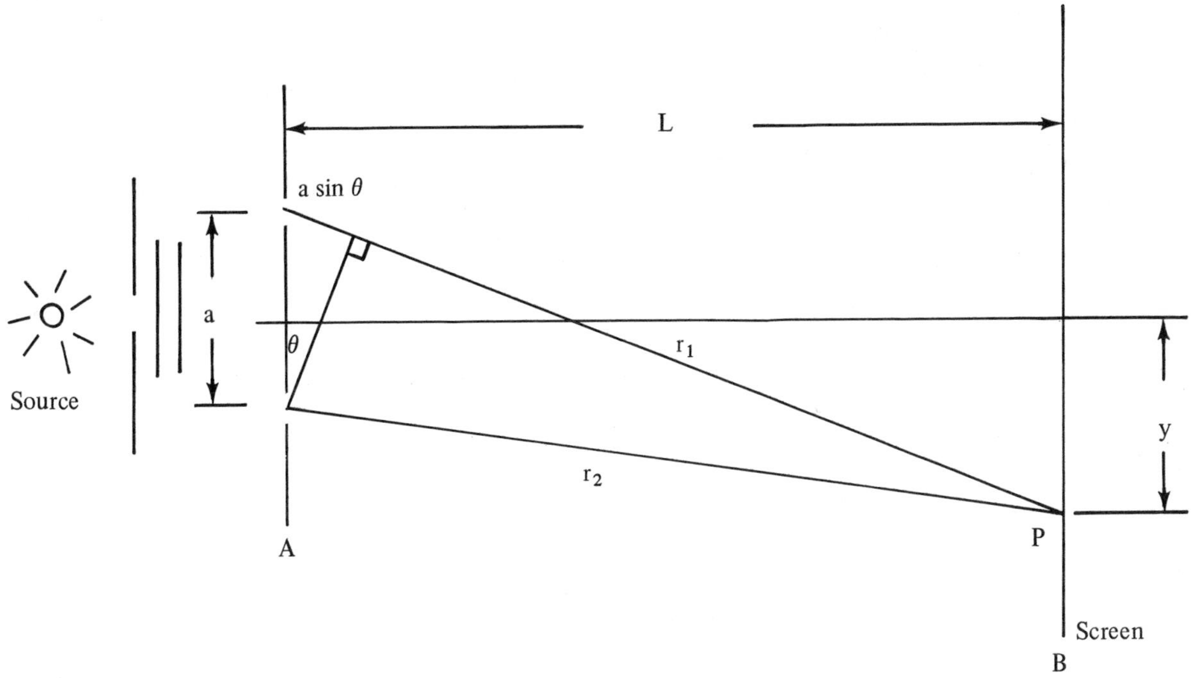

Figure 9–1 Schematic representation of Young's double slit experiment. The light incident on screen A is a plane wave.

The condition for *dark* fringes will be satisfied when the path differences δ is $\frac{\lambda}{2}$, $\frac{3\lambda}{2}, \frac{5\lambda}{2}, \ldots$. That is,

$$\delta = a\sin\theta = \left(n + \frac{1}{2}\right)\lambda \qquad (n = 0, 1, 2, 3, \ldots) \tag{9.2}$$

Dark Fringes

From similar triangles in Figure 9–1, we see that for small θ (that is, $y \ll L$),

$$\frac{a\sin\theta}{a} \cong \frac{y}{L} \tag{9.3}$$

Therefore, it follows from Equation (9.3), Equation (9.1), and Equation (9.2) that the positions of the bright and dark fringes are given by

$$y_n = \frac{\lambda L}{a} n \qquad (n = 0, 1, 2, 3, \ldots) \tag{9.4}$$

Bright Fringes

$$y_n = \frac{\lambda L}{a} \left(n + \frac{1}{2}\right) \tag{9.5}$$

Dark Fringes

Example 9.1

This example illustrates the use of Young's experiment in measuring the wavelength of a light source. A screen is separated from a pair of slits by a distance of 2 m. The slit separation is 0.05 mm. The third bright fringe of the double slit pattern is measured to be 6.5 cm from the center line. (a) Determine the wavelength of the light. (b) Calculate the position of the third dark fringe.

Solution

(a) For the third bright fringe, n = 3 in Equation (9.4), $y_3 = 6.5$ cm, L = 2 m and $a = 5 \times 10^{-5}$ m; therefore,

$$\lambda = \frac{ay_3}{Ln} = \frac{5 \times 10^{-5} \times 6.5 \times 10^{-2}}{2 \times 3} = 5.4 \times 10^{-7} \text{ m}$$

Since $1\text{Å} = 10^{-10}$ m, $\lambda = 5400$ Å.

(b) The third dark fringe corresponds to n = 2 in Equation (9.5), so

$$y_2 = \frac{5\lambda L}{2a} = \frac{5 \times 5.4 \times 10^{-7} \times 2}{2 \times 5 \times 10^{-5}} = 5.4 \times 10^{-2} \text{ m} = 5.4 \text{ cm}$$

To calculate the *intensity distribution* in the interference fringes, we write the expression for the total electric field intensity at P as the *vector superposition* of two waves from the two slits. If these two waves have amplitudes E_1 and E_2, the total E field at P can be calculated by referring to the phasor diagram in Figure 9–2. The waves from the two slits have the same angular frequency ω, and can be written as

$$E_1 = E_0 \sin\omega t$$

$$E_2 = E_0 \sin(\omega t + \phi)$$

In Figure 9–2, t is taken to be zero; however, the resultant wave amplitude E_p is independent of the choice of ωt.

If the phase difference between E_1 and E_2 is ϕ, then we can write $\frac{\phi}{2\pi}$ is equal to the path difference divided by λ. Therefore, we have

$$\phi = \frac{2\pi}{\lambda}\delta = \frac{2\pi}{\lambda} a\sin\theta \qquad (9.6)$$

From the law of cosines, and Figure 9–2, we see that

$$E_P{}^2 = E_1{}^2 + E_2{}^2 + 2E_1E_2\cos\phi$$

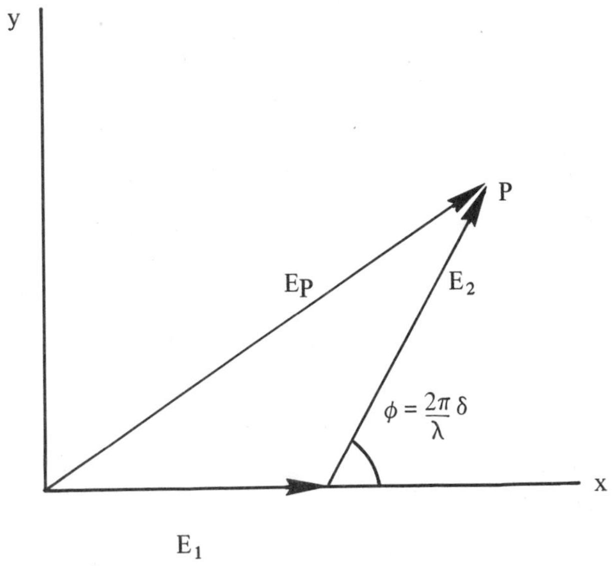

Figure 9–2 Phasor diagram for adding two electric field vectors of amplitudes E_1 and E_2 whose phase difference is ϕ.

But since the sources are coherent from the two slits, $E_1 = E_2 = E$; therefore,

$$E_P{}^2 = 2E^2\ (1 + \cos\phi) = 4E^2 \cos^2 \frac{\phi}{2} \tag{9.7}$$

Since $I_P \sim E_P{}^2$, and taking the *maximum* intensity to be I_0, combining Equations (9.6) and (9.7) gives*

$$I_P = I_0 \cos^2 \left(\frac{\pi a \sin\theta}{\lambda}\right) \tag{9.8}$$

Alternately, since $y = L\sin\theta$, Equation (9.8) can be written as

$$I_P = I_0 \cos^2 \left(\frac{\pi a}{\lambda L}\ y\right) \tag{9.9}$$

Note that the fringes of maximum intensity occur when $\left(\frac{\pi a}{\lambda L}\right)$ y is an integral multiple of 2π, corresponding to $y = \left(\frac{\lambda L}{a}\right)$ n. This is consistent with Equation (9.4) for the positions of the bright fringes. The variation of intensities with y is plotted in Figure 9–3.

---

*Superposition of any two sinusoidal coherent waves, whose phase difference is ϕ (constant in time), gives for the total intensity at P: $I_P = \frac{1}{2}\,[E_1{}^2 + E_2{}^2 + 2E_1 E_2 \cos\phi]$. The factor of $\frac{1}{2}$ comes from the time average of a sinusoidal function.

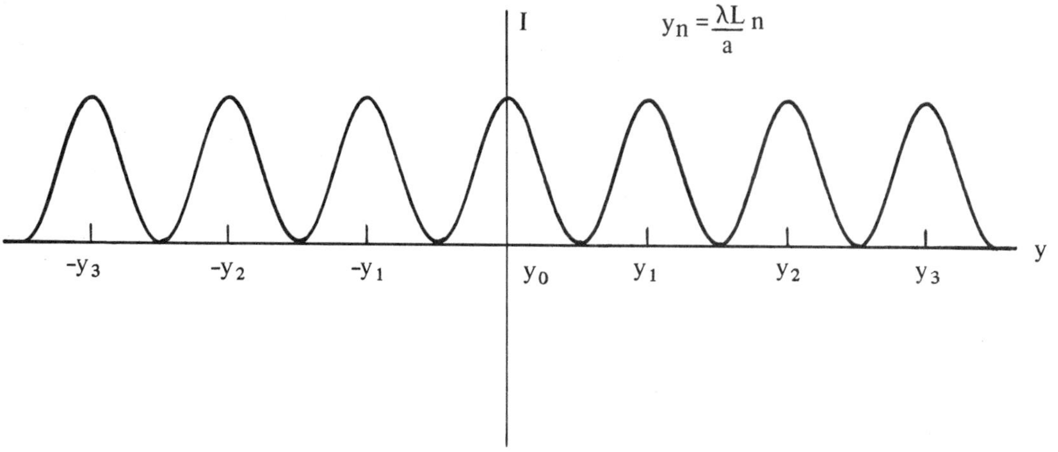

Figure 9-3 Intensity variation of a double slit as a function of y, the distance from the center line as defined in Figure 9-1. The positions of maximum intensity follow from Equation (9.9), and the curve is valid only if $L \gg a$ and $L \gg y$.

In summary, the student should note that the interference pattern produced by two narrow slits a distance d apart consists of equally spaced lines of equal intensity. This result is valid only if $L \gg a$ and $L \gg y$.

9.2 INTERFERENCE IN THIN FILMS

Interference bands are commonly observed in thin films of oil on water or in soap bubbles. These bands arise from interference fringes created by the superposition of light reflected from the upper and lower surfaces.

Consider a film of thickness d and index of refraction n, as in Figure 9–4, where $n > 1$. We will assume that the incident light rays are nearly normal to the surface. To determine whether the reflected rays interfere constructively or destructively, we must note the following facts:

1. A wave which goes from a medium of lower n to one of higher n undergoes a 180°phase change upon reflection. There is no phase change upon reflection when going from a medium of higher n to one of lower n. The transmitted wave in either case undergoes no change in phase.

2. The wavelength of light in a medium of index of refraction n is given by $\frac{\lambda}{n}$, where λ is the wavelength in vacuum.

According to rule (1), the light undergoes a phase change of 180° at surface A, but not a surface B. Therefore, the condition for maximum intensity in the reflected light is that the path difference between rays 1 and 2 in Figure 9–4(*a*) be equal to $\frac{1}{2}\lambda_m$, $\frac{3}{2}\lambda_m$, and so forth, where λ_m is the wavelength of light in that medium. Since the path difference is 2d for normal incidence, we can write the condition for *maximum intensity* as

$$2d = \left(m + \frac{1}{2}\right)\lambda_m \qquad (m = 0, 1, 2, \ldots) \tag{9.10}$$

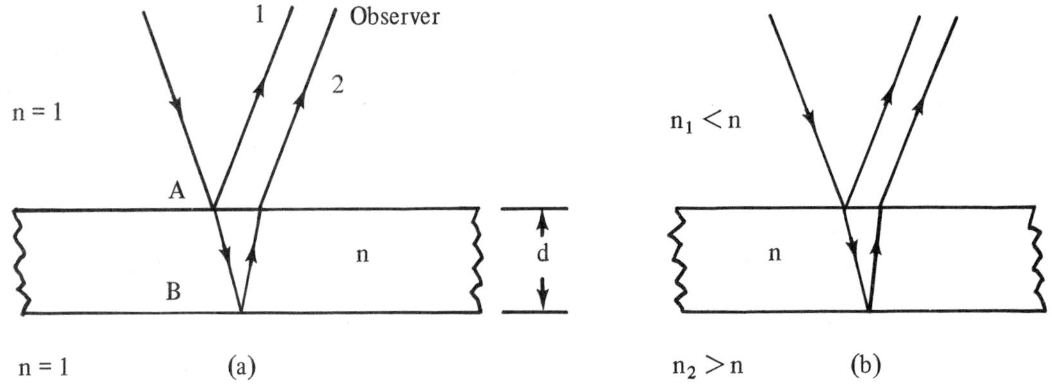

Figure 9–4 (*a*) Interference of light resulting from reflections at two surfaces of a thin film of thickness d, index of refraction n. (*b*) A thin film between two different media.

where

$$\lambda_m = \frac{\lambda}{n} \tag{9.11}$$

The condition for *minimum intensity* in the reflected light is

$$2d = m\lambda_m \qquad (m = 0, 1, 2, \ldots) \tag{9.12}$$

In reality, films are not uniform in thickness; therefore, some parts of the film cause constructive interference and others destructive interference. As a result, the film is multicolored, corresponding to the various wavelengths present in ordinary light.

Note that Equations (9.10) and (9.12) apply only when the film has an index of refraction which is greater or less than the indices of refraction of the media on both sides of the film. If a film is surrounded by two media, one of lower index of refraction and one of higher index, as in Figure 9–4(*b*), the conditions for minima and maxima in the reflected light are different. In this case, there is a phase change of 180° in the reflected light at both interfaces; hence, the net change in phase is zero for the two reflections. Therefore, the condition for an intensity *minima* in the reflected light is given by Equation (9.10), while the condition for intensity *maxima* is given by Equation (9.12). (The proof of this is left as a problem for the student – see Problem 10.)

Example 9.2

Calculate the thicknesses of an oil film (n = 1.46) which will result in constructive interference (in the reflected light) using light of wavelength 5000 Å in vacuum.

Solution

To get the *minimum* thickness, we take m = 0 in Equation (9.10), and note that λ_m is less than λ by the factor n = 1.46.

$$d_0 = \frac{1}{4}\lambda_m = \frac{1}{4} \times \frac{5000 \text{ Å}}{1.46} = 856 \text{ Å}$$

Consequently, film thicknesses of $3d_0$, $5d_0$, $7d_0$, and so forth will also produce constructive interference.

9.3 DIFFRACTION OF LIGHT

The deviation of light from a straight-line path is known as *diffraction.* Diffraction results from interference of waves from many coherent sources. The "smearing out" of shadows is a diffraction phenomenon. If light is incident on a screen with a slit in it, as in Figure 9–5, the "shadow" of the slit at the observing screen B will not be perfectly sharp. When the observing screen is far from the slit, that is, when $L \gg d$, and when parallel rays are used, the diffraction pattern observed is called *Fraunhofer diffraction.* Parallel rays incident on the slit can be produced with lenses, as in Figure 9–5(*a*), or with a distant source. A bright fringe is observed at $\theta = 0$, with alternating bright and dark fringes on either side of the central bright fringe.

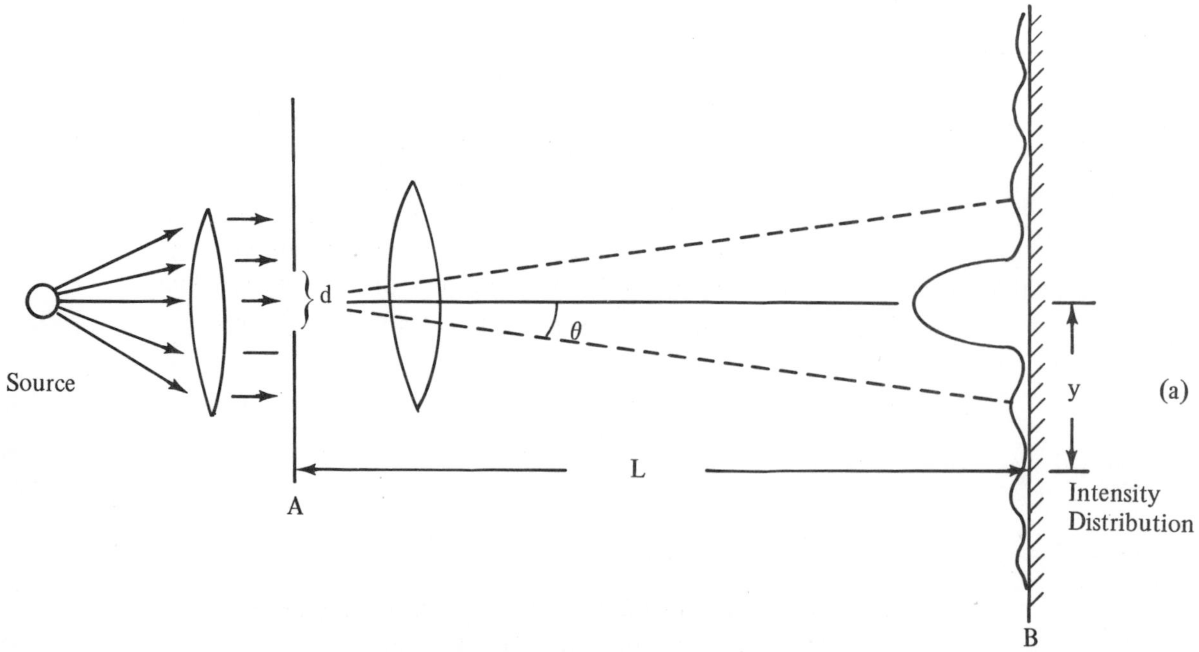

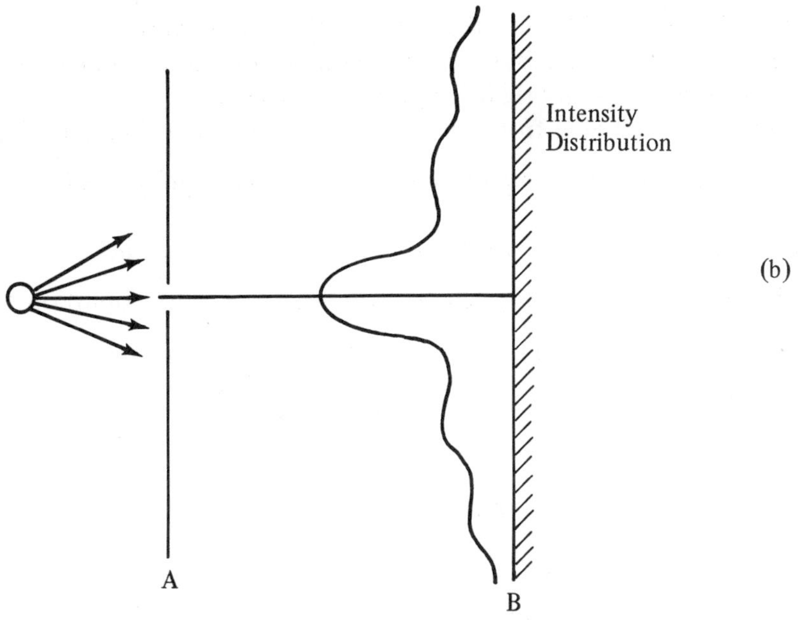

Figure 9-5 Schematic representations of (*a*) Fraunhofer diffraction, where parallel rays are incident on the slit and L ≫ d. (*b*) Fresnel diffraction, where the incident rays are not parallel and the observing screen is at a finite distance from the slit.

When the observing screen is at a finite distance from the slit, and parallel rays are not used, the diffraction pattern observed is called *Fresnel diffraction* (see Fig. 9–5). We will restrict our discussion to the first case of Fraunhofer diffraction.

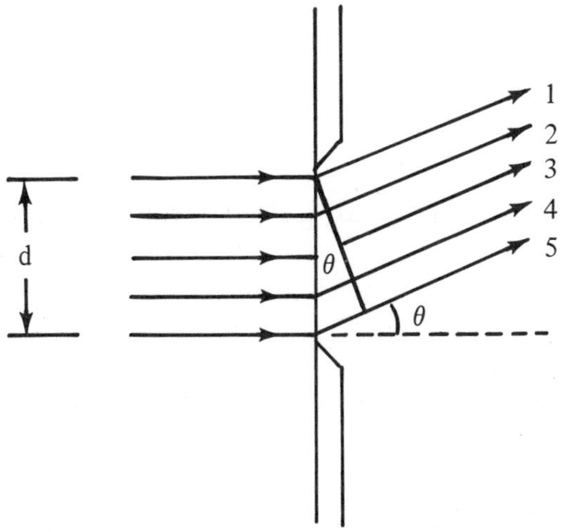

Figure 9-6 Diffraction of light by a narrow slit of width d.

Now let us examine the single slit Fraunhofer diffraction in more detail. A magnified picture of the slit is shown in Figure 9–6. Huygens' principle says that each portion of the slit acts as a source of waves. Hence, light from one portion of the slit can interfere with light from another portion, and the intensity of light on a particular point on the screen will depend on the angle of observation, θ. At the center of the screen, where $\theta = 0$, all rays arriving at the screen will be in phase and, when focused at the screen, a central bright fringe will be observed.

Now consider the light emerging from the slit at some angle θ as in Figure 9–6. *Destructive* interference results between any two rays whose path difference δ is $n\frac{\lambda}{2}$, where n is an odd integer. Therefore, for rays 1 and 3, which originate at points separated by $\frac{d}{2}$, the path difference is given by $\delta_{1,3} = \frac{1}{2} d\sin\theta$. The condition for destructive interference of rays 1 and 3 (or any other rays originating at points separated by $\frac{d}{2}$ is

$$\delta_{1,3} = \frac{1}{2} d\sin\theta = n\frac{\lambda}{2} \qquad (n = \pm 1, \pm 3, \ldots)$$

or

$$d\sin\theta = n\lambda$$

Now consider two rays originating at points separated by $\frac{d}{4}$, such as rays 1 and 2. The condition for destructive interference of these waves, or any pair of waves originating at points separated by $\frac{1}{4} d\sin\theta$, is

$$\delta_{1,2} = \frac{1}{4} d\sin\theta = n\frac{\lambda}{2} \qquad (n = \pm 1, \pm 3, \ldots)$$

or,

$$d\sin\theta = m\lambda \qquad (m = \pm 2, \pm 6, \ldots)$$

If we continue this analysis, we find that, in general, *destructive* interference occurs at angles satisfying the condition

$$d\sin\theta = n\lambda \qquad (n = \pm 1, \pm 2, \pm 3, \ldots) \tag{9.13}$$

Thus, the points on the screen which have *zero* intensity occur at the angles for which

$$\sin\theta = \pm\frac{\lambda}{d}, \pm\frac{2\lambda}{d}, \pm\frac{3\lambda}{d}, \ldots \tag{9.14}$$

A simple formula such as Equation (9.14) is not available for the positions of the points of maximum intensity; however, they do lie approximately halfway between the minima. The general features of the intensity distribution are as shown in Figure 9–7. A broad central bright area is observed, flanked by alternate minima and maxima.

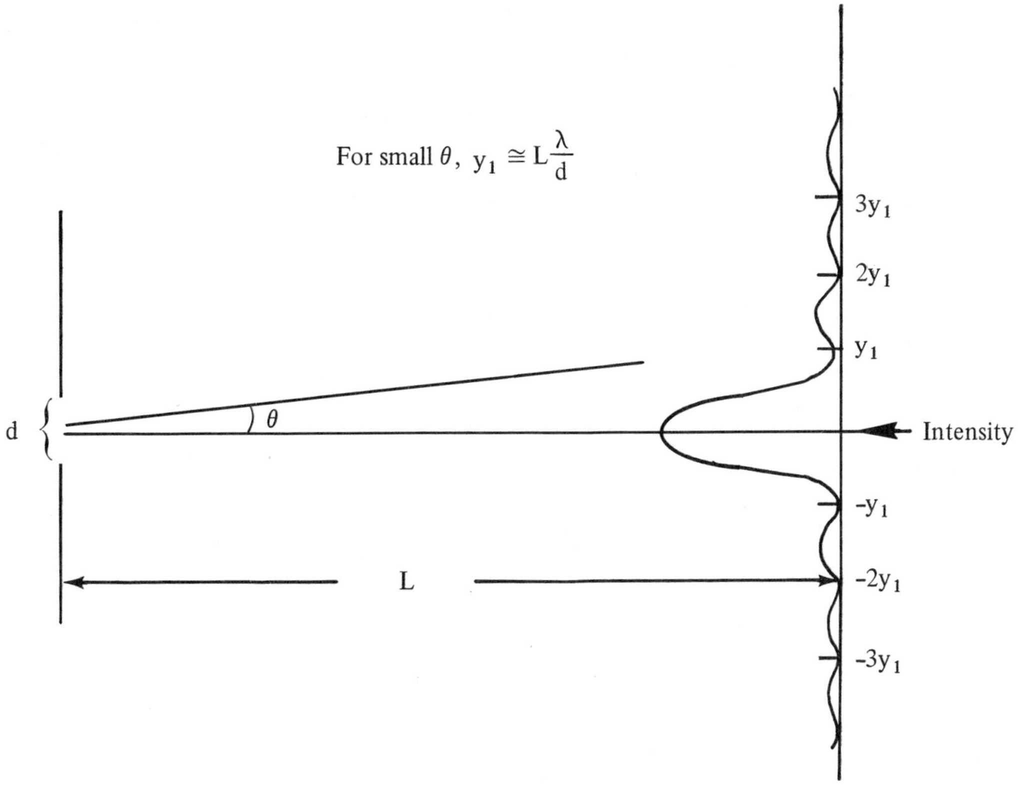

Figure 9–7 Intensity distribution of a single slit diffraction pattern.

Example 9.3

Light of wavelength 6500 Å is incident on a slit of width 0.05 mm. The observing screen is placed 2 m from the slit. Find the width of the central bright fringe.

Solution

The width of the central fringe corresponds to the distance between the two minima on either side of the axis. Using Equation (9.14), we see that these first minima occur at the angles

$$\sin\theta = \pm\frac{\lambda}{d}$$

or,

$$\sin\theta = \pm\frac{6.5 \times 10^{-7}}{5 \times 10^{-5}} = \pm\ 1.3 \times 10^{-2}$$

Since θ is very small, the positions of these minima measured from the central axis are approximately

$$|y_1| \cong L\sin\theta \cong L\frac{\lambda}{d} = 2.6 \times 10^{-2}\ \text{m}$$

The width of the central bright fringe is therefore $2|y_1| = 5.2 \times 10^{-2}$ m = 52 mm. That is, the central maximum is much wider than the slit. If the slit width is *increased,* the central maximum would *decrease* in width and would ultimately reduce to the size of the slit. That is, for large slit widths of ~ 1 mm, the width of the central maximum would correspond to the spread of the geometric shadow of the slit. Try a calculation for d = 1 mm. You should get a width of approximately 2.6 mm.

The *intensity distribution* of the single slit diffraction pattern can be calculated by the phasor method, which is graphical and direct, or by summation of the wave amplitudes arriving at a given point on the observing screen. The result which is derived in most texts is

$$I = I_0 \frac{\sin^2 \gamma}{\gamma^2} \tag{9.15}$$

where

$$\gamma = \frac{\pi d \sin\theta}{\lambda} \tag{9.16}$$

and I_0 is the intensity of the central maximum. Equation (9.15) is in good agreement with the observed intensity distribution as shown in Figure 9–7. Note that as $\theta \rightarrow 0$, the function $\sin\gamma \approx \gamma$, and $I \rightarrow I_0$. Also, I goes through zeroes at $\gamma = \pm\pi$, $\pm 2\pi, \ldots$; corresponding to the values of $\sin\theta$ given by Equation (9.14).

Diffraction patterns are also observed for other types of apertures. For example, the diffraction pattern of a *small circular hole* of diameter d for wavelengths $\lambda \ll d$ is a series of concentric rings, which alternate in intensity as in Figure 9–8. The central region is a bright circular spot when $L \gg d$, and the *first minimum* is a dark circle which occurs at an angle θ given by

$$\sin\theta = 1.22 \frac{\lambda}{d} \tag{9.17}$$

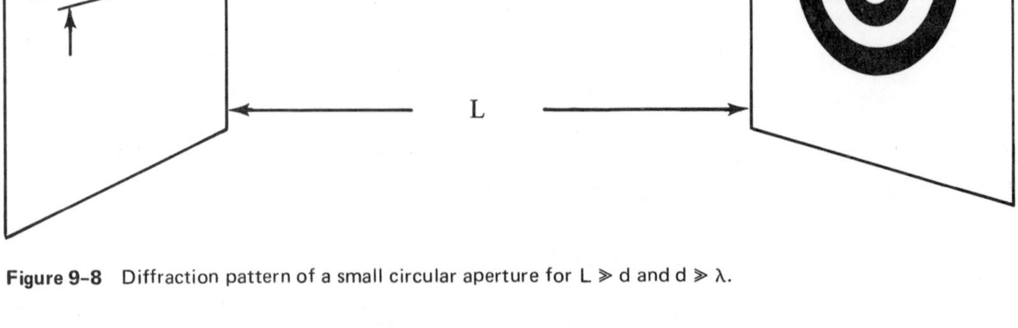

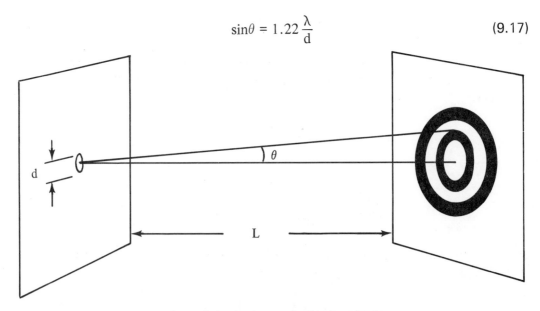

Figure 9–8 Diffraction pattern of a small circular aperture for $L \gg d$ and $d \gg \lambda$.

9.4 DIFFRACTION GRATINGS

A *diffraction grating* consists of a large number of slits of the same width, equally separated by some distance a as in Figure 9–9. For two slits, we found that constructive interference, or maximum intensity fringes, occurs, when the *path difference,* δ, is some integral multiple of the wavelength. That is, the condition for intensity maxima can be written as

$$\delta = a\sin\theta = n\lambda \qquad (n = 0, \pm 1, \pm 2, \ldots) \tag{9.18}$$

Adding more slits doesn't change the condition for the maxima. However, as the number of slits N increases, the number of primary maxima increases. These maxima increase in sharpness and in brightness as N increases. In practice, N is several thousand lines per inch.

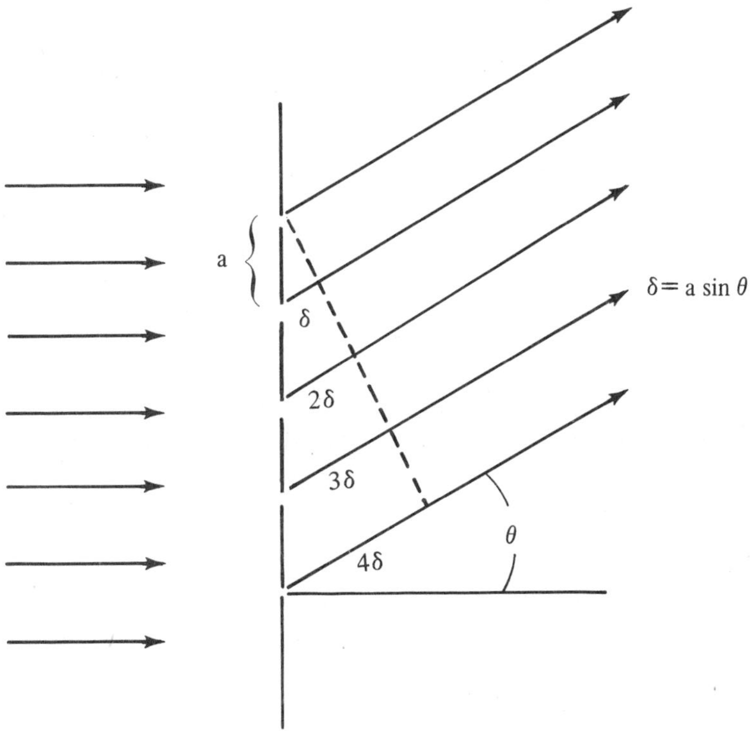

Figure 9–9 Diffraction from a grating.

Therefore, for normal incidence of light as in Figure 9–9, the positions of the primary maxima for the diffraction grating can be calculated from Equation (9.18). The value of n is called the *order number* of the principal maximum. Note that the position of the nth-order principal maximum as determined from Equation (9.18) depends only on the ratio $\frac{\lambda}{a}$, and is independent of N. A sketch of an intensity distribution pattern for the case N = 5 is shown in Figure 9–10. Note that there are very weak *secondary maxima* lying between the primary maxima. The number of such secondary maximary is N – 2, and their intensity relative to the primary maxima decreases as N is increased. Therefore, when N is very large, the secondary maxima will be practically zero in intensity. Note that the number of primary maxima depends on the ratio $\frac{\lambda}{a}$. Since the maximum value of $\sin\theta$ is *one,* we can calculate the largest possible value of n if λ and d are known.

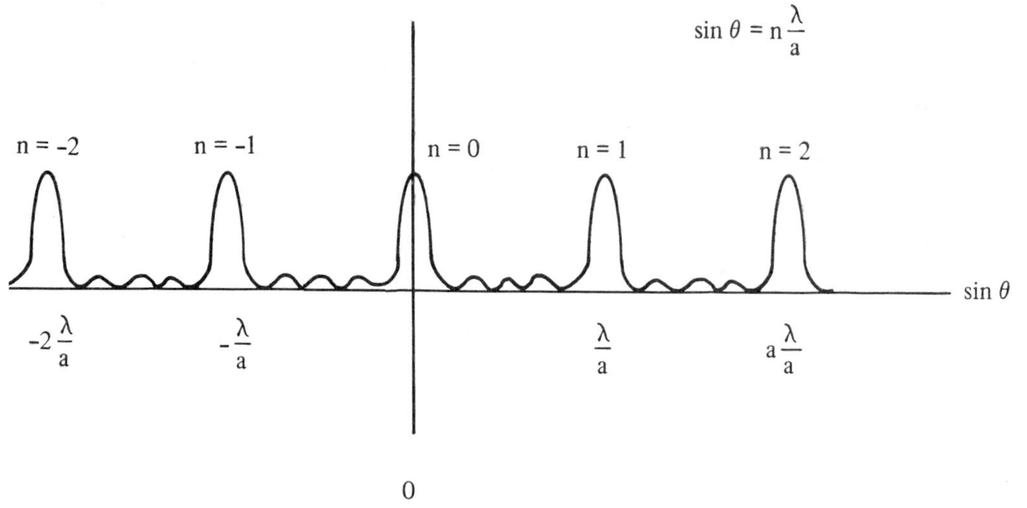

Figure 9–10 Intensity distribution for a diffraction grating containing five slits.

Example 9.4

A diffraction grating consists of 5000 slits over a distance of 2 cm. Light of wavelength 6000 Å is incident normal to the grating. (a) Calculate the angles at which the first- and second-order maxima occur.

Since there are 2500 slits per centimeter, the spacing, a, between each slit is $\frac{1}{2500}$ cm $= 4 \times 10^{-4}$ cm $= 4 \times 10^{-6}$ m. Therefore, the first-order maximum corresponding to n = 1 in Equation (9.18) occurs at

$$\theta_1 = \sin^{-1}\left(\frac{\lambda}{a}\right) = \sin^{-1}\left(\frac{6 \times 10^{-7}}{4 \times 10^{-6}}\right)$$

$$\theta_1 = \sin^{-1}(0.15) \cong 8.9^\circ$$

The second-order maximum corresponding to n = 2 occurs at

$$\theta_2 = \sin^{-1}\left(\frac{2\lambda}{a}\right) = \sin^{-1}(0.30) \cong 17.9^\circ$$

(b) Determine the total number of principal maxima that can be observed for this grating.

Again, using Equation (9.18) and noting that the highest value of n corresponds to the maximum value of $\sin\theta$, which is *one,* gives

$$n_{max} = \frac{a}{\lambda} = \frac{4 \times 10^{-6}\ \text{m}}{6 \times 10^{-7}\ \text{m}} = 6.7$$

But since n must be an integer, the largest value of n is 6. This corresponds to $2n + 1 = 13$ primary maxima, that is, the zeroth-order maximum ($n = 0$) and six on either side of the zeroth-order maximum.

9.5 DISPERSION AND RESOLVING POWER OF A DIFFRACTION GRATING

Consider *two* monochromatic waves incident on a diffraction grating, whose difference in wavelengths $\Delta\lambda$ is *small* compared to λ. A useful measure of the *angular separation* between the nth-order maximum corresponding to the two wavelengths is a quantity called the *dispersion,* D, given by

$$D = \frac{d\theta}{d\lambda} \tag{9.19}$$

Differentiating Equation (9.18) gives

$$a\cos\theta\, d\theta = n\, d\lambda$$

Substituting this into Equation (9.19), we get

$$D = \frac{n}{a\cos\theta} \tag{9.20}$$

where we see from Equation (9.19) that D has units of angle per unit length. Therefore, the dispersion varies with the order of the primary maxima. For example, the student should show that the dispersion of the grating described in Example 9.4 for $n = 2$ is given by $D = 5.3 \times 10^5 \dfrac{\text{rad}}{\text{m}}$.

The *resolving power* R of a diffraction grating is the ratio of the mean wavelength of two spectral lines that are barely resolved divided by their wavelength separation $\Delta\lambda$. A grating with a high resolving power has a large number of slits per unit length. The resolving power can be written as

$$R = \frac{\lambda}{\Delta\lambda} = Nn \tag{9.21}$$

From Equation (9.21), we see that the number of slits required in a grating to resolve two spectral lines decreases as the order n increases. Note that $R = 0$ for $n = 0$, which signifies that *all wavelengths* are *indistinguishable* for the zeroth-order maximum. However, suppose one considers the second-order maximum, $n = 2$, for the grating in Example (9.4) which has 5000 rulings. The resolving power of this grating from Equation (9.21) is given by $5000 \times 2 = 10{,}000$. Therefore, the *minimum* wavelength separation between two spectral lines that can be barely resolved in the second-order principal maximum (say, for a mean wavelength $\lambda = 6000$ Å) is given by $\Delta\lambda = \dfrac{\lambda}{R} = 0.6$ Å. For the third-order principal maximum, $R = 15{,}000$ and $\Delta\lambda = 0.4$ Å, and so on.

9.6 POLARIZED LIGHT

Electromagnetic waves are *transverse* waves in that the electric and magnetic vectors associated with the waves are at right angles to the direction of propagation. In the *plane polarized* wave shown in Figure 9–11, the electric field vectors vibrate in a plane perpendicular to the plane of vibration of the magnetic field vectors. An *unpolarized* transverse wave contains a random orientation of such planes of vibration as in Figure 9–12. However, it is possible to polarize such a wave by inserting a sheet of polarizing material in front of the wave. Polarizing materials exist which will only transmit waves whose E vectors are parallel to a certain direction, and absorb those components of the wave whose E vectors are perpendicular to this direction. (This is usually achieved with plastics containing very long molecules oriented in a specific direction.)

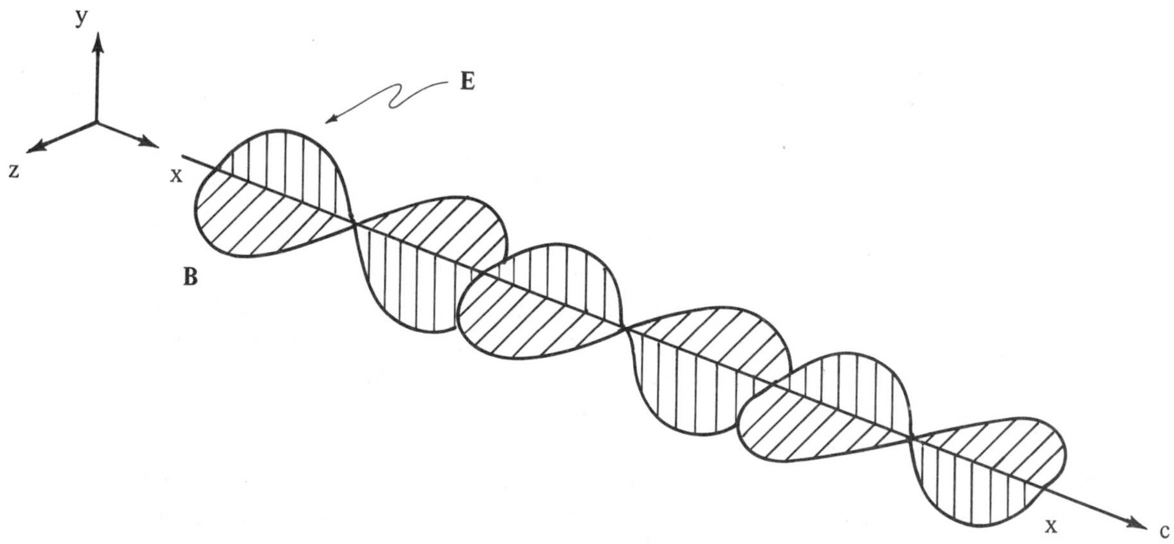

Figure 9–11 Representation of a plane polarized electromagnetic wave. Note that the planes of vibration of **E** and **B** are perpendicular to each other and to the direction of propagation of the wave.

Let us denote the amplitude of the wave that passes through the polarizer as E_0 (the first Polaroid sheet in Fig. 9–12). The amplitude of the wave that passes through the second Polaroid sheet, called the analyzer, is given as $E_0 \cos\theta$, where θ is the angle between the polarizing directions of the two sheets. Since the *intensity* of the transmitted wave is proportional to E^2, the transmitted intensity emerging from the analyzer is given as

$$I = I_0 \cos^2 \theta \quad (9.22)$$

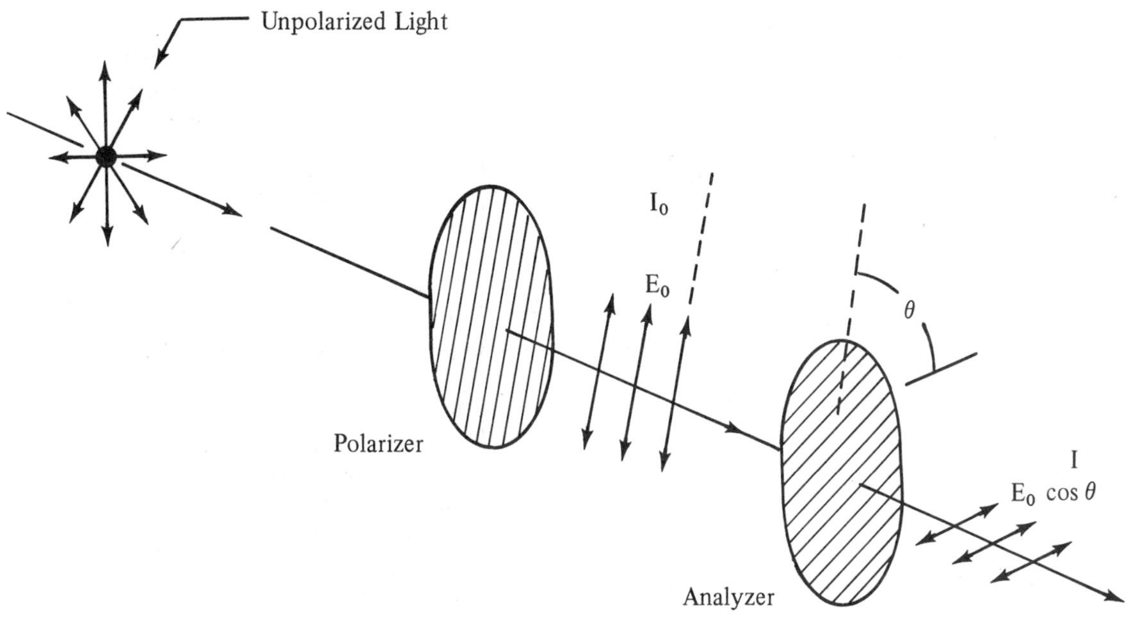

Figure 9-12 Only a fraction of the light is transmitted through the analyzer when θ lies between 0 and 180°.

where I_0 is the intensity of the wave incident on the analyzer. Therefore, we conclude that the maximum intensity is transmitted when $\theta = 0$ or π, that is, when the polarizers are parallel. However, no intensity is transmitted when $\theta = \frac{\pi}{2}$ or $\frac{3\pi}{2}$, since $I = 0$ for these angles. This can be observed using a pair of crossed polarizing sheets. For certain relative orientations of the two sheets, practically no light is transmitted. If, for example, $\theta = 84°$, then $\frac{I}{I_0} \cong (0.1)^2 = 0.01$. That is, only 1 per cent of the light is transmitted.

Light can also be polarized by *reflection* for a *particular* angle of incidence, θ_p. If an unpolarized light wave traveling in air is incident on a refracting medium of index of refraction n, the *reflected* light shown in Figure 9-13 will be completely polarized at the angle θ_p given by

$$\tan\theta_p = n \tag{9.23}$$

At this angle, the reflected and refracted light are at right angles to each other $\left(\theta_p + \theta_2 = \frac{\pi}{2}\right)$. It is left as an exercise to show that this fact, together with Snell's law of refraction, gives Equation (9.23). This expression is known as *Brewster's law,* and the angle θ_p is sometimes referred to as *Brewster's angle,* or the polarizing angle. Note that the reflected light is *plane polarized,* while the refracted light is unpolarized. As an example, the polarizing angle for water ($n = 1.33$) is given by arctan $(1.33) = 53°$.

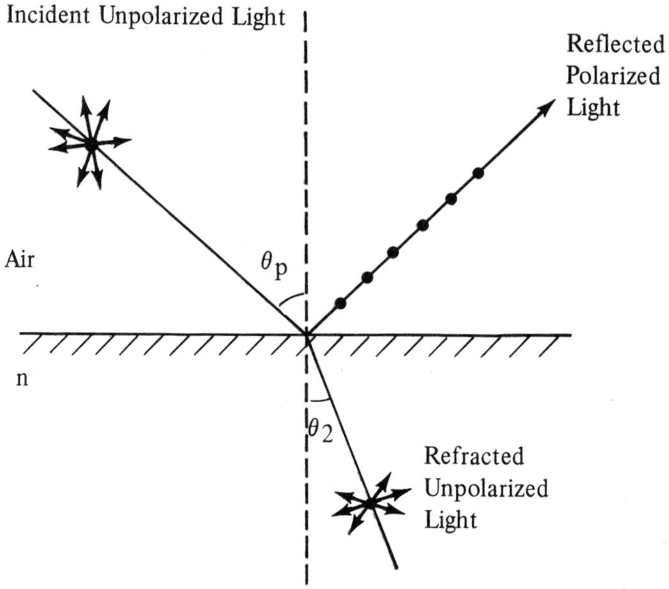

Figure 9-13 Light reflected from a surface is completely polarized at the angle θ_p given by $\tan \theta_p = n$.

9.7 PROGRAMMED EXERCISES

1.A

Parallel monochromatic light is incident on a pair of slits separated by a distance a as shown below. What is the path difference δ between the two rays that arrive at P?

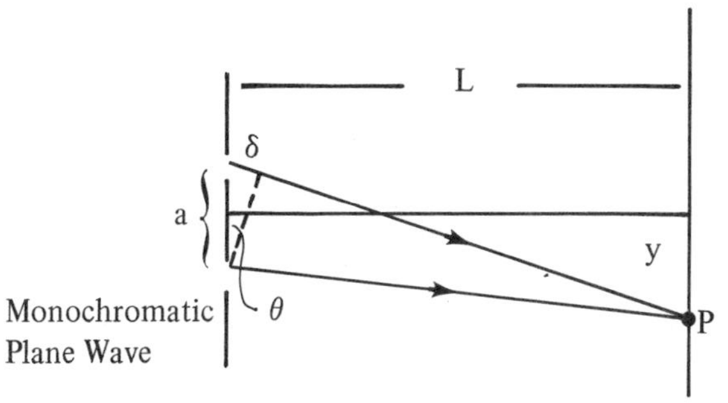

From the small right triangle, we see that

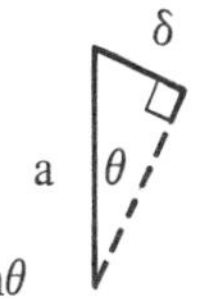

$$\delta = a\sin\theta \qquad (1)$$

1.B

Suppose the wavelength of the incident light is λ. The condition for a maximum in intensity at the point P is δ = ________.

This type of interference is called ________ interference.

For an *intensity maximum,*

$$\delta = a\sin\theta = n\lambda \qquad (2)$$

$$(n = 0, 1, 2, 3, \ldots)$$

This corresponds to *constructive* interference.

1.C

Dark fringes (or intensity minima) occur at P if the path difference is given by δ = ________. This type of interference is called ________ interference.

When δ is *odd* multiples of $\frac{\lambda}{2}$, the wave amplitudes from the two slits cancel each other. Thus,

$$\delta = a\sin\theta = \left(n + \frac{1}{2}\right)\lambda \qquad (3)$$

$$(n = 0, 1, 2, 3, \ldots)$$

is the condition for intensity minimum. This corresponds to *destructive* interference.

1.D

Show that if θ is *small*, the approximation $\sin\theta \cong \frac{y}{L}$ is valid.

For small θ, $\sin\theta \approx \tan\theta$. But $\tan\theta = \frac{y}{L}$ from the right triangle in frame 1.A.

$$\therefore \quad \sin\theta \cong \frac{y}{L} \tag{4}$$

1.E

Now use (2), (3) and (4) to find the positions of the bright and dark fringes on the screen measured from the central axis as shown in frame 1.A.

Using (2) and (4) gives

$$y_n = \frac{\lambda L}{a} n \tag{5}$$

for the positions of the *bright* fringes. From (3) and (4), we get

$$y_n = \frac{\lambda L}{a}\left(n + \frac{1}{2}\right) \tag{6}$$

for the positions of the *dark* fringes.

1.F

Find the separation between two adjacent maxima.

Using (5), we have

$$\Delta y = y_{n+1} - y_n = \frac{\lambda L}{a}(n+1) - \frac{\lambda L}{a} n$$

$$\Delta y = \frac{\lambda L}{a} \tag{7}$$

Note that for small θ, the fringes are evenly spaced.

1.G

The *intensity* of the resultant wave at P is given by $I_p = I_0 \cos^2 \frac{\phi}{2}$, where ϕ is the phase difference between the two waves arriving at P. What is ϕ in terms of the path difference δ? Write an expression for I in terms of δ.

$$\frac{\phi}{2\pi} = \frac{\delta}{\lambda}$$

$$\therefore \quad \phi = \frac{2\pi}{\lambda}\delta \tag{8}$$

$$I_p = I_0 \cos^2 \frac{\phi}{2} = I_0 \cos^2\left(\frac{\pi}{\lambda}\delta\right) \tag{9}$$

1.H

Write an expression for I_p in terms of y. Make use of (1), (4) and (9).

$$I_p = I_0 \cos^2 \left(\frac{\pi}{\lambda} a \sin\theta \right)$$

$$I_p = I_0 \cos^2 \left(\frac{\pi a}{\lambda L} y \right) \quad (10)$$

1.I

Show that (10) is equivalent to (5) in predicting the intensity maxima.

$\cos\gamma$ is a maximum for $\gamma = n\pi$, $n = 0, 1, 2, \ldots$. Thus, from (10) we have

$$\frac{\pi a}{\lambda L} y = n\pi$$

or,

$$y = \frac{\lambda L}{a} n$$

which agrees with (5).

2.A

When monochromatic light passes through a narrow slit of width d, the pattern formed on a screen placed behind the slit is a series of maxima and minima, which is called a

__________ pattern. Can this phenomenon be explained by geometrical optics?

Diffraction.

No. Geometrical optics predicts only a shadow of the slit. The wave properties of light must be invoked to explain the effect.

2.B

If the width of the slit is d, and the observing screen is at a finite distance from the screen, the diffraction is called

__________. If the source and screen are very far from the slit, so that the rays incident on the slit are parallel, the diffraction is called __________.

Fresnel Diffraction

Fraunhofer Diffraction

2.C

Describe in a sketch how the case of Fraunhofer diffraction can be readily obtained in the laboratory using converging lenses.

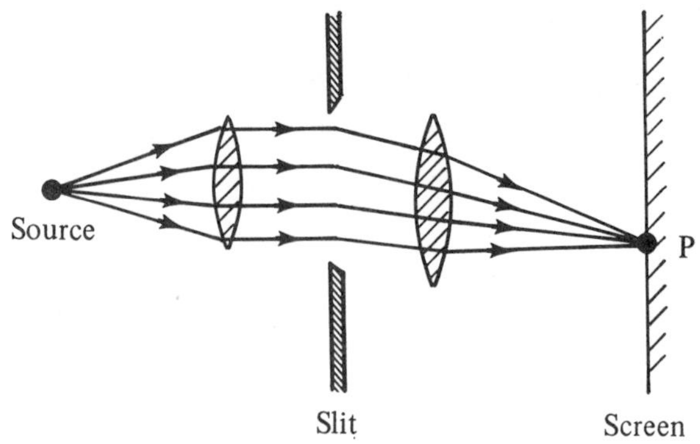

2.D

The rays incident on the slit in frame 2.C are parallel rays. The wave associated with these rays is called a ________ wave.

plane wave

2.E

Each part of the slit acts as a new source of waves. This is called ________ principle.

Huygens'

2.F

Explain qualitatively the reason for bright and dark areas on the screen.

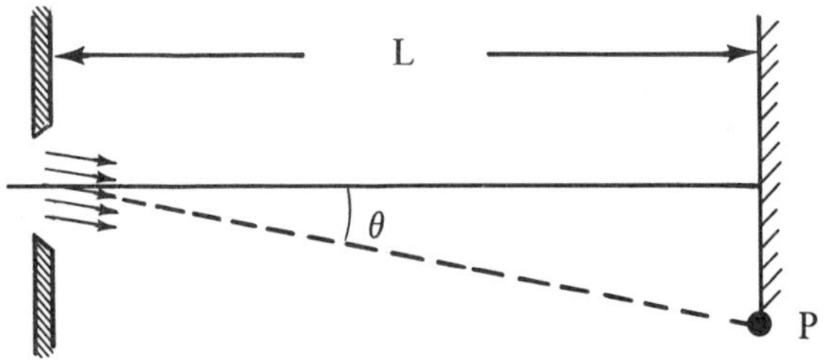

Light from one portion of the slit interferes with light from another portion. The intensity at P is a superposition of all such portions, and it depends on the angle of observation, θ.

2.G

An intensity maximum is observed at the central point, $\theta = 0$. This occurs because all waves arriving at this point are ________ .

in phase (or have the same path length). Such waves interfere constructively.

2.H

If λ is the wavelength of the monochromatic light, then *destructive* interference results at P if $d\sin\theta$ = ________ .

$$d\sin\theta = n\lambda \qquad (1)$$

where $n = \pm 1, \pm 2, \pm 3, \ldots$. These values of θ correspond to points of zero intensity at P.

2.I

Determine the angles for which the *first* points of *zero intensity* occur on the screen. Take $\lambda = 4000$ Å and $d = 0.1$ mm.

For $n = \pm 1$, (1) gives

$$\sin\theta = \pm \frac{\lambda}{d}$$

$$\sin\theta = \pm \frac{4 \times 10^{-7}}{1 \times 10^{-4}} = \pm 4 \times 10^{-3}$$

Since θ is very small,

$$\sin\theta \approx \theta \cong 4 \times 10^{-3} \text{ rad}$$

2.J

For $L \gg d$, calculate the positions of the points of *zero* intensity measured from the central axis.

Note that $\tan\theta = \frac{y}{L}$. But for $L \gg d$, θ is small and $\sin\theta \cong \tan\theta$.

$\therefore$ Using this approximation and (1), we get

$$d\sin\theta \cong d\frac{y}{L} = n\lambda$$

or,

$$y = \frac{n\,L}{d} \qquad (2)$$

2.K

Make a sketch of the diffraction pattern showing the positions of the intensity minima.

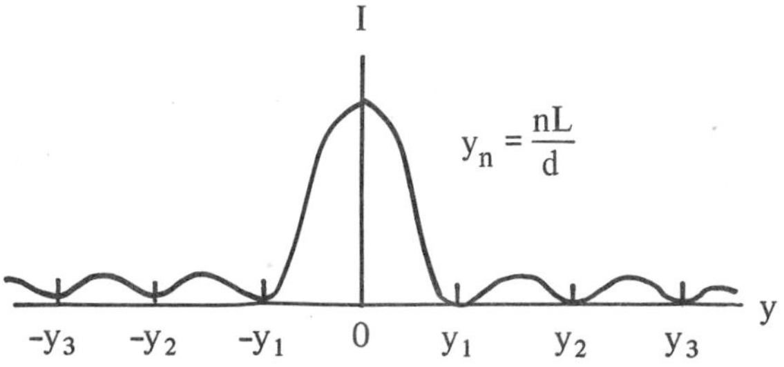

2.L

Show that the "width" of the central maximum is twice the width of the second maxima.

From the sketch in frame 2.K, we see that the width of the central maximum is $2|y_1|$, or

$$2|y_1| = 2\frac{\lambda L}{d}$$

The widths of the second maxima are given by

$$|y_2 - y_1| = 2\frac{\lambda L}{d} - \frac{\lambda L}{d} = \frac{\lambda L}{d}$$

2.M

What expression predicts the intensity distribution of the single slit diffraction pattern sketched in frame 2.K?

$$I = I_0 \frac{\sin^2 \gamma}{\gamma^2} \tag{3}$$

where

$$\gamma = \frac{\pi d \sin\theta}{\lambda}$$

Note that *minima* are predicted by (3) at $\gamma = n\pi$, where $n = \pm 1, \pm 2, \pm 3, \ldots$, which agrees with (1).

3.A

A diffraction grating consists of N slits equally separated by a distance a as shown below. What is the path difference δ between rays 1 and 2?

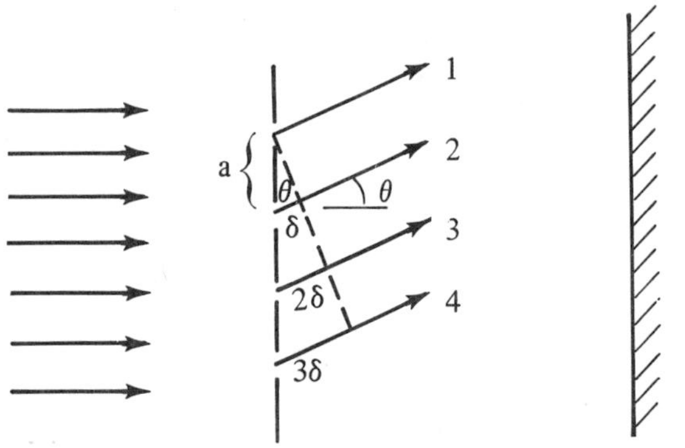

$$\delta = a\sin\theta$$

Note that the path difference between rays 1 and 3 is 2δ, between 1 and 4 it is 3δ, and so on.

3.B

What is the condition for an *intensity maximum*?

The path difference between rays from adjacent slits must equal an integral multiple of the wavelength.

$$\delta = a\sin\theta = n\lambda \quad (n = 0, \pm 1, \pm 2, \ldots) \tag{1}$$

3.C

What is n called?

n is the *order number*.

3.D

Suppose $\lambda = 0.1$ a. How many orders will be observed for this grating? How many primary maxima does this predict?

From (1), we get

$$\sin\theta = n\frac{\lambda}{a} = 0.1\ n$$

Since $(\sin\theta)_{max} = 1$, the maximum number of orders will be 10. The number of primary maxima = 2n +1 = 21.

3.E

Sketch the intensity distribution of a diffraction grating for large N (that is, neglect the secondary maxima). Label the positions of the primary maxima.

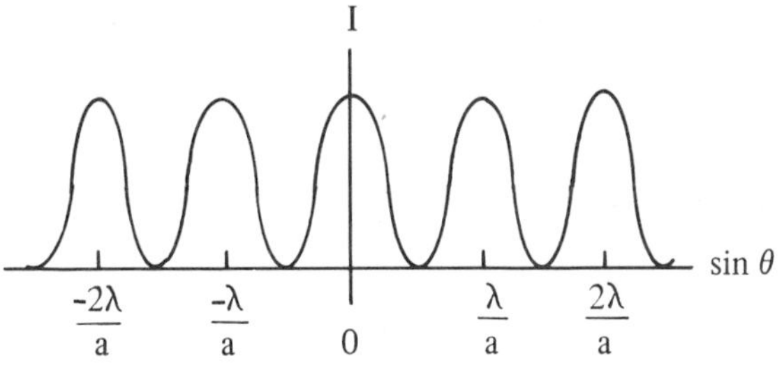

3.F

Does the ratio $\frac{\lambda}{a}$ affect the intensities of the primary maxima? Explain.

No. We see from (1) that only the positions of the maxima are affected by $\frac{\lambda}{a}$. However, the intensities are a function of $\frac{\lambda}{d}$, where d is the slit width. This follows from the treatment of the single slit system (see Exercise 2).

---

†3.G

The *dispersion* of a diffraction grating is given by

$$D = \frac{d\theta}{d\lambda} \qquad (2)$$

Use this definition and (1) to show that $D = \frac{n}{a\cos\theta}$.

Differentiating (1) gives

$$d\,(a\sin\theta) = n d\lambda$$

$$a\cos\theta\, d\theta = n d\lambda$$

$$\therefore \quad D = \frac{d\theta}{d\lambda} = \frac{n}{a\cos\theta} \qquad (3)$$

Note: Since $\frac{n}{a} = \frac{\sin\theta}{\lambda}$, we can also write (3) as $D = \frac{\tan\theta}{\lambda}$.

3.H

The dispersion of a grating is a measure of the angular separation between the nth-order maxima. A particular grating has a dispersion of $3 \times 10^5 \frac{\text{rad}}{\text{m}}$ for the third-order maximum. What is the angular separation in the third-order maximum of two waves separated in wavelength by 20 Å?

From (2), we can write

$$\Delta\theta = D\Delta\lambda$$

$$\Delta\theta = 3 \times 10^5 \frac{\text{rad}}{\text{m}} \times 20 \times 10^{-10} \text{ m}$$

$$\Delta\theta = 6 \times 10^{-4} \text{ rad}$$

or

$$\Delta\theta \cong 0.034°$$

3.I

Define the resolving power of a diffraction grating.

$$R = \frac{\lambda}{\Delta\lambda} \tag{4}$$

where $\Delta\lambda$ is the wavelength separation of two monochromatic waves that are barely distinguishable and λ is their mean wavelength. *Or,*

$$R = Nn \tag{5}$$

where N is the number of rulings in the grating, and n is the *order number.*

3.J

Determine the resolving power in the third-order for a grating which has 8000 rulings.

From (5), for n = 3 we get

$$R = 8000 \times 3 = 24{,}000$$

3.K

If two waves of mean wavelength 4000 Å are incident on this grating, what is the minimum separation between their wavelengths such that they are barely resolved in the third order?

Using (4) and the results to 3.J gives

$$\Delta\lambda = \frac{\lambda}{R} = \frac{4000 \text{ Å}}{24{,}000}$$

$$\Delta\lambda = \frac{1}{6} \text{ Å} \cong 0.17 \text{ Å}$$

3.L

Repeat the calculation in 3.K for the second-order maxima and the first-order maxima.

For n = 2, R = 8000 × 2 = 16,000.

$$\therefore \quad \Delta\lambda = \frac{\lambda}{R} = \frac{4000}{16{,}000} = 0.25 \text{ Å}$$

Likewise, for n = 1, R = 8000 and $\Delta\lambda = 0.50$ Å. That is, the resolving power *decreases* as n decreases.

9.8 SUMMARY

In the double slit interference experiment, the positions of the bright and dark fringes are given by

$$y_n = \frac{\lambda L}{a} n \qquad \text{Bright Fringes} \qquad (9.4)$$

$$y_n = \frac{\lambda L}{a} \left(n + \frac{1}{2}\right) \qquad \text{Dark Fringes} \qquad (9.5)$$

where n = 0, ±1, ±2, . . ., λ is the wavelength of light, a is the separation between the slits, L is the distance between the slits and observing screen and y is measured from the central axis. The intensity at a point P on the screen whose coordinate is y is given by

$$I_P = I_0 \cos^2 \left(\frac{\pi a}{\lambda L} y\right) \qquad (9.9)$$

In the single slit diffraction experiment, the condition for *destructive* interference is

$$d\sin\theta = n\lambda \qquad (9.13)$$

where n = ±1, ±2, ±3, . . ., d is the width of the slit and θ is the angle of observation. The intensity at a point on the observing screen is given by

$$I = I_0 \frac{\sin^2 \gamma}{\gamma^2} \qquad (9.15)$$

where

$$\gamma = \frac{\pi d\sin\theta}{\lambda} \qquad (9.16)$$

The condition for intensity maxima for a *diffraction grating* whose rulings are separated by a distance a is

$$a\sin\theta = n\lambda \qquad (9.18)$$

where n = 0, ±1, ±2, . . . is called the *order number*. The *dispersion* of a diffraction grating is given by

$$D = \frac{n}{a\cos\theta} \qquad (9.20)$$

and the *resolving power* of a diffraction grating is

$$R = \frac{\lambda}{\Delta\lambda} = Nn \qquad (9.21)$$

where N is the total number of rulings and $\Delta\lambda$ is the wavelength separation between two waves that are just distinguishable.

If I_0 is the intensity of light which passes through one polarizer, and a second polarizing material (analyzer) is placed behind the first, the intensity transmitted through the analyzer is given by

$$I = I_0 \cos^2 \theta \tag{9.22}$$

where θ is the angle between the polarizing directions of the two polarizers.

If unpolarized light is incident on a dense medium of refractive index n, the reflected light will be completely polarized when the angle of incidence equals *Brewster's* angle θ_p, given by

$$\tan\theta_p = n \tag{9.23}$$

9.9 PROBLEMS

1. In a double slit interference experiment, the slits are illuminated with light of wavelength 6800 Å. If the second bright fringe is measured to be 3.5 cm from the central line, and the slits are 2 m from the observing screen, calculate (a) the slit separation and (b) the position of the second dark fringe.

2. Monochromatic light of unknown wavelength is used in Young's interference experiment. The slit separation is 0.1 mm, and the observing screen is placed 1.5 m from the slits. If the separation between adjacent bright fringes near the central line is 6.0 mm, what is the wavelength of the light?

3. Two sinusoidal vectors of the same amplitude A and frequency ω have a phase difference ϕ. Calculate the resultant amplitude of the two vectors both graphically and analytically if ϕ equals (a) 0, (b) 60° and (c) 90°.

†4. Two sinusoidal waves of unequal amplitudes E_1 and E_2 vary in time according to the expressions $E_1 \sin\omega t$ and $E_2 \sin(\omega t + \phi)$. Show that if these two waves are superimposed, the resultant *maximum* intensity is given by

$$I_P = \frac{1}{2}[E_1{}^2 + E_2{}^2 + 2E_1E_2]$$

Hint: To get the factor $\frac{1}{2}$, you must take the time average of the instantaneous value of I_P. (See Section 9.1.)

5. An oil film in air whose thickness is 5000 Å is illuminated with white light in the direction perpendicular to the film. What wavelengths will be strongly reflected in the range 3000 Å to 7000 Å? (Take n = 1.46 for oil.)

6. Repeat Problem 5 if the oil film is placed on a thick piece of glass (n = 1.50). (Refer to the end of Section 9.2 for this situation.)

7. A material having an index of refraction of 1.30 is used to coat a piece of glass (n = 1.50). What should be the minimum thickness of this film in order to minimize reflected light at a wavelength of 5000 Å?

8. (a) A uniform film of index of refraction n and thickness d is surrounded on either side by air. For normal incidence of light, an intensity minimum is observed in the reflected light at λ_2, and an intensity maximum is observed at λ_1, where $\lambda_1 > \lambda_2$. If there are no intensity minima observed between λ_1 and λ_2, show that the integer m which appears in Equations (9.10) and (9.12) is given by

$$m = \frac{\lambda_1}{2(\lambda_1 - \lambda_2)}$$

(b) Determine the thickness of the film if n = 1.40, λ_1 = 5000 Å and λ_2 = 3750 Å.

9. A piece of transparent material having an index of refraction n is cut into the shape of a wedge as shown in Figure 9-14. The angle of the wedge is small, and monochromatic light of wavelength λ is normally incident from above. If the height of the wedge is h, and the width is l, show that bright fringes occur at the positions

$$x = \frac{\lambda l \left(m + \frac{1}{2}\right)}{2hn}$$

and dark fringes occur at the positions

$$x = \frac{\lambda l m}{2hn}$$

where m = 0, 1, 2, ...

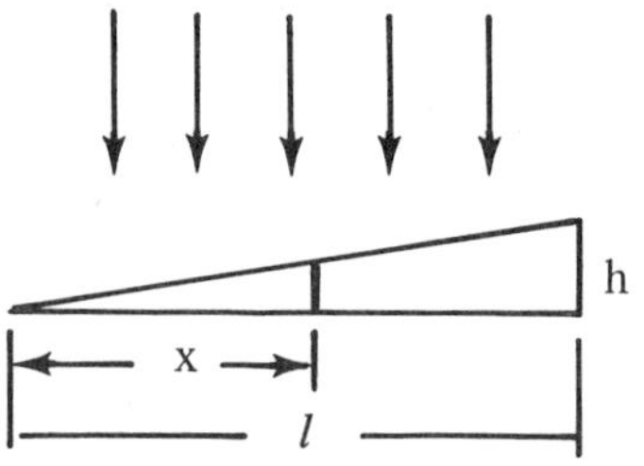

Figure 9-14

10. Show that if a thin film is between two different media as in Figure 9-4(*b*), the condition for a minimum in the reflected light intensity is given by Equation (9.10), while the condition for a maximum is given by Equation (9.12).

11. A screen is placed 50 cm from a single slit which is illuminated with light of wavelength 6900 Å. If the distance between first and third minima in the diffraction pattern is measured to be 3.0 mm, what is the width of the slit?

12. A single slit of width 0.20 mm is illuminated with light of wavelength 5000 Å. The observing screen is placed 80 cm from the slit. (a) Calculate the widths of the central bright fringe and the secondary maxima of the diffraction pattern. (b) What is the distance between the first and fourth minima?

13. A circular aperture of diameter 0.3 mm is illuminated with monochromatic light of unknown wavelength as in Figure 9-8. If the aperture is 2.0 m from the screen, and the radius of the first dark circle is 5 mm, determine the wavelength of the light.

14. A diffraction grating has 8000 rulings per centimeter. When monochromatic light is incident normal to the grating, the first-order maximum is observed at an angle of 15°. (a) What is the wavelength of the light? (b) In what direction will the second-order maximum be observed?

15. Light of wavelength 5000 Å is incident normally on a diffraction grating. If the third-order maximum of the diffraction pattern is observed at an angle of 32°, (a) what is the number of rulings per centimeter for the grating? (b) Determine the total number of primary maxima that can be observed in this situation.

16. Calculate the dispersion of the diffraction grating described in Problem 14 for the first-order maximum and second-order maximum.

17. A diffraction grating of length 4 cm contains 6000 rulings over a distance of 2 cm. (a) What is the resolving power of this grating in the first three orders? (b) If two monochromatic waves incident on this grating have a mean wavelength of 4000 Å, what is their wavelength separation if they are just resolved in the third order?

18. Calculate Brewster's angle for (a) ice, (b) fused quartz, (c) crown glass and (d) diamond. Use the data given in Table 8–1 in your calculations.

APPENDIX 1

TABLE OF TRIGONOMETRIC FUNCTIONS

Angle θ		sin θ	cos θ	tan θ	Angle θ		sin θ	cos θ	tan θ
Degree	Radian				Degree	Radian			
0	0.0000	0.0000	1.0000	0.0000					
1	0.0175	0.0175	0.9998	0.0175	46	0.8029	0.7193	0.6947	1.0355
2	0.0349	0.0349	0.9994	0.0349	47	0.8203	0.7314	0.6820	1.0724
3	0.0524	0.0523	0.9986	0.0524	48	0.8378	0.7431	0.6691	1.1106
4	0.0698	0.0698	0.9976	0.0699	49	0.8552	0.7547	0.6561	1.1504
5	0.0873	0.0872	0.9962	0.0875	50	0.8727	0.7660	0.6428	1.1918
6	0.1047	0.1045	0.9945	0.1051	51	0.8901	0.7771	0.6293	1.2349
7	0.1222	0.1219	0.9925	0.1228	52	0.9076	0.7880	0.6157	1.2799
8	0.1396	0.1392	0.9903	0.1405	53	0.9250	0.7986	0.6018	1.3270
9	0.1571	0.1564	0.9877	0.1584	54	0.9425	0.8090	0.5878	1.3764
10	0.1745	0.1736	0.9848	0.1763	55	0.9599	0.8192	0.5736	1.4281
11	0.1920	0.1908	0.9816	0.1944	56	0.9774	0.8290	0.5592	1.4826
12	0.2094	0.2079	0.9781	0.2126	57	0.9948	0.8387	0.5446	1.5399
13	0.2269	0.2250	0.9744	0.2309	58	1.0123	0.8480	0.5299	1.6003
14	0.2443	0.2419	0.9703	0.2493	59	1.0297	0.8572	0.5150	1.6643
15	0.2618	0.2588	0.9659	0.2679	60	1.0472	0.8660	0.5000	1.7321
16	0.2793	0.2756	0.9613	0.2867	61	1.0647	0.8746	0.4848	1.8040
17	0.2967	0.2924	0.9563	0.3057	62	1.0821	0.8829	0.4695	1.8807
18	0.3142	0.3090	0.9511	0.3249	63	1.0996	0.8910	0.4540	1.9626
19	0.3316	0.3256	0.9455	0.3443	64	1.1170	0.8988	0.4384	2.0503
20	0.3491	0.3420	0.9397	0,3640	65	1.1345	0.9063	0.4226	2.1445
21	0.3665	0.3584	0.9336	0.3839	66	1.1519	0.9135	0.4067	2.2460
22	0.3840	0.3746	0.9272	0.4040	67	1.1694	0.9205	0.3907	2.3559
23	0.4014	0.3907	0.9205	0.4245	68	1.1868	0.9272	0.3746	2.4751
24	0.4189	0.4067	0.9135	0.4452	69	1.2043	0.9336	0.3584	2.6051
25	0.4363	0.4226	0.9063	0.4663	70	1.2217	0.9397	0.3420	2.7475
26	0.4538	0.4384	0.8988	0.4877	71	1.2392	0.9455	0.3256	2.9042
27	0.4712	0.4540	0.8910	0.5095	72	1.2566	0.9511	0.3090	3.0777
28	0.4887	0.4695	0.8829	0.5317	73	1.2741	0.9563	0.2924	3.2709
29	0.5061	0.4848	0.8746	0.5543	74	1.2915	0.9613	0.2756	3.4874
30	0.5236	0.5000	0.8660	0.5774	75	1.3090	0.9659	0.2588	3.7321

Angle θ		sin θ	cos θ	tan θ	Angle θ		sin θ	cos θ	tan θ
Degree	Radian				Degree	Radian			
31	0.5411	0.5150	0.8572	0.6009	76	1.3265	0.9703	0.2419	4.0108
32	0.5585	0.5299	0.8480	0.6249	77	1.3439	0.9744	0.2250	4.3315
33	0.5760	0.5446	0.8387	0.6494	78	1.3614	0.9781	0.2079	4.7046
34	0.5934	0.5592	0.8290	0.6745	79	1.3788	0.9816	0.1908	5.1446
35	0.6109	0.5736	0.8192	0.7002	80	1.3963	0.9848	0.1736	5.6713
36	0.6283	0.5878	0.8090	0.7265	81	1.4137	0.9877	0.1564	6.314
37	0.6458	0.6018	0.7986	0.7536	82	1.4312	0.9903	0.1392	7.115
38	0.6632	0.6157	0.7880	0.7813	83	1.4486	0.9925	0.1219	8.144
39	0.6807	0.6293	0.7771	0.8098	84	1.4661	0.9945	0.1045	9.514
40	0.6981	0.6428	0.7660	0.8391	85	1.4835	0.9962	0.0872	11.430
41	0.7156	0.6561	0.7547	0.8693	86	1.5010	0.9976	0.0698	14.301
42	0.7330	0.6691	0.7431	0.9004	87	1.5184	0.9986	0.0523	19.081
43	0.7505	0.6820	0.7314	0.9325	88	1.5359	0.9994	0.0349	28.636
44	0.7679	0.6947	0.7193	0.9657	89	1.5533	0.9998	0.0175	57.290
45	0.7854	0.7071	0.7071	1.0000	90	1.5708	1.0000	0.0000	∞

APPENDIX 2

TABLE OF NATURAL LOGARITHMS (BASE e)

x	ln x	x	ln x	x	ln x	x	ln x
10^{-9}	-20.723	1.05	0.0488	4.2	1.4351	16	2.773
10^{-6}	-13.816	1.10	0.0953	4.4	1.4816	17	2.833
10^{-5}	-11.513	1.15	0.1398	4.6	1.5261	18	2.890
10^{-4}	-9.210	1.20	0.1823	4.8	1.5686	19	2.944
10^{-3}	-6.908	1.25	0.2231	5.0	1.6094	20	2.996
0.01	-4.6052	1.30	0.2624	5.2	1.6487	22	3.091
0.02	-3.9120	1.35	0.3001	5.4	1.6864	24	3.178
0.03	-3.5066	1.40	0.3365	5.6	1.7228	26	3.258
0.04	-3.2189	1.45	0.3716	5.8	1.7579	28	3.332
0.05	-2.9957	1.50	0.4055	6.0	1.7918	30	3.401
0.06	-2.8134	1.55	0.4383	6.2	1.8245	32	3.466
0.07	-2.6593	1.60	0.4700	6.4	1.8563	34	3.526
0.08	-2.5257	1.65	0.5008	6.6	1.8871	36	3.584
0.09	-2.4079	1.70	0.5306	6.8	1.9169	38	3.638
0.10	-2.30259	1.75	0.5596	7.0	1.9459	40	3.689
0.12	-2.1203	1.80	0.5878	7.2	1.9741	42	3.738
0.14	-1.9661	1.85	0.6152	7.4	2.0015	44	3.784
0.16	-1.8326	1.90	0.6419	7.6	2.0281	46	3.829
0.18	-1.7148	1.95	0.6678	7.8	2.0541	48	3.871
0.20	-1.6094	2.00	0.69315	8.0	2.0794	50	3.912
0.22	-1.5141	2.1	0.7419	8.2	2.1041	55	4.007
0.24	-1.4271	2.2	0.7885	8.4	2.1282	60	4.094
0.26	-1.3471	2.3	0.8329	8.6	2.1518	65	4.174
0.28	-1.2730	2.4	0.8755	8.8	2.1748	70	4.248
0.30	-1.2040	2.5	0.9163	9.0	2.1972	75	4.317
0.35	-1.0498	2.6	0.9555	9.2	2.2192	80	4.382
0.40	-0.9163	2.7	0.9933	9.4	2.2407	85	4.443
0.45	-0.7985	2.8	1.0296	9.6	2.2618	90	4.500
0.50	-0.6931	2.9	1.0647	9.8	2.2824	95	4.554
		3.0	1.0986	10.0	2.30259	100	4.605

x	ln x	x	ln x	x	ln x	x	ln x
0.55	−0.5978	3.1	1.1314	10.5	2.3514	200	5.298
0.60	−0.5108	3.2	1.1632	11.0	2.3979	300	5.704
0.65	−0.4308	3.3	1.1939	11.5	2.4423	400	5.991
0.70	−0.3567	3.4	1.2238	12.0	2.4849	500	6.215
0.75	−0.2877	3.5	1.2528	12.5	2.5257	600	6.397
0.80	−0.2231	3.6	1.2809	13.0	2.5649	10^3	6.908
0.85	−0.1625	3.7	1.3083	13.5	2.6027	10^4	9.210
0.90	−0.1054	3.8	1.3350	14.0	2.6391	10^5	11.513
0.95	−0.0513	3.9	1.3610	14.5	2.6741	10^6	13.816
1.00	0.0000	4.0	1.3863	15.0	2.7081	10^9	20.723

APPENDIX 3

USEFUL MATHEMATICAL RELATIONS

Roots of a Quadratic Equation:

$$ax^2 + bx + c = 0 \qquad x = \frac{-b \pm \sqrt{b^2 - 4\,ac}}{2\,a}$$

Trigonometric Relations:

$$\sin\theta = \frac{y}{r} \qquad \cos\theta = \frac{1}{\sin\theta} = \frac{r}{y}$$

$$\cos\theta = \frac{x}{r} \qquad \sec\theta = \frac{1}{\cos\theta} = \frac{r}{x}$$

$$\tan\theta = \frac{y}{x} \qquad \cot\theta = \frac{1}{\tan\theta} = \frac{x}{y}$$

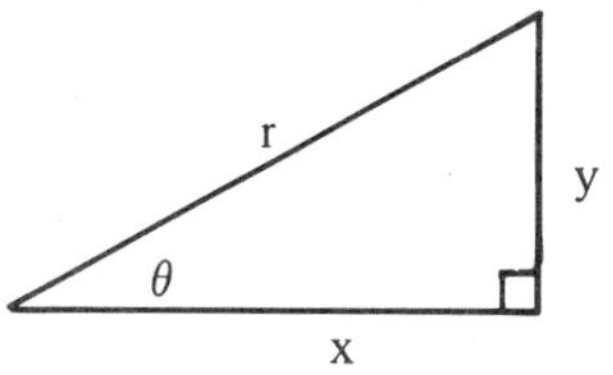

$$\sin^2\theta + \cos^2\theta = 1 \qquad x^2 + y^2 = r^2$$

$$\sin 2\theta = 2\sin\theta\cos\theta \qquad \sin(-\theta) = -\sin\theta$$

$$\cos 2\theta = \cos^2\theta - \sin^2\theta = 1 - 2\sin^2\theta \qquad \cos(-\theta) = \cos\theta$$

$$\sin(\theta \pm \psi) = \sin\theta\cos\psi \pm \cos\theta\sin\psi \qquad \tan(-\theta) = -\tan\theta$$

$$\cos(\theta \pm \psi) = \cos\theta\cos\psi \mp \sin\theta\sin\psi$$

Law of Cosines:

$$c^2 = a^2 + b^2 - 2\,ab\cos\gamma$$

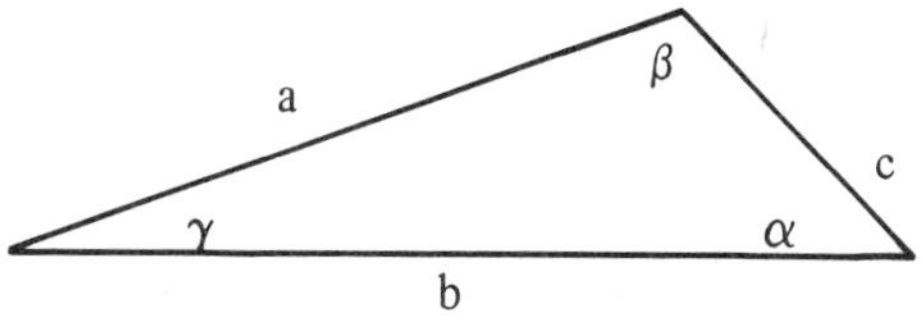

Law of Sines:

$$\frac{\sin\alpha}{a} = \frac{\sin\beta}{b} = \frac{\sin\gamma}{c}$$

Series Expansions:

$$e^x = 1 + x + \frac{x^2}{2!} + \frac{x^3}{3!} + \ldots$$

$$\sin x = x - \frac{x^3}{3!} + \frac{x^5}{5!} - \frac{x^7}{7!} + \ldots$$

$$\cos x = 1 - \frac{x^2}{2!} + \frac{x^4}{4!} - \frac{x^6}{6!} + \ldots$$

$$ln\,(1 \pm x) = \pm x - \frac{1}{2}x^2 \pm \frac{1}{3}x^3 - \frac{1}{4}x^4 + \ldots$$

$$(1 \pm x)^n = 1 \pm nx + \frac{n(n-1)x^2}{2!} \mp \frac{n(n-1)(n-2)x^3}{3!} + \ldots$$

APPENDIX 4

SOME CONVERSION FACTORS

Length:	1 m = 39.37 in = 3.28 ft = 100 cm
	1 ft = 30.48 cm 1 in = 2.54 cm
	1 mi = 5280 ft = 1.609 km
	1A = 10^{-8} cm = 10^{-10} m
	1μ (micron) = 10^{-4} cm = 10^{-6} m
Mass:	1 slug = 14.59 kg = 32.2 lb (mass)
	1 g = 10^{-3} kg = 6.85×10^{-5} slug
Time:	1 year = 365 days = 3.16×10^{7} sec
	1 day = 24 hrs = 1.44×10^{3} min = 8.64×10^{4} sec
Area:	1 cm^2 = 0.155 in^2 1 m^2 = 10.76 ft^2
	1 in^2 = 6.452 cm^2 1 ft^2 = 144 in^2 = 0.0929 m^2
Volume:	1 m^3 = 10^6 cm^3 = 10^3 liters
	= 35.3 ft^3 = 6.1×10^4 in^3
	1 ft^3 = 2.83×10^{-2} m^3 = 28.32 liters
Velocity:	1 mi/hr = 1.47 ft/sec = 0.447 m/sec
	1 m/sec = 100 cm/sec = 3.281 ft/sec
	1 mi/min = 60 mi/hr = 88 ft/sec
Acceleration:	1 m/sec^2 = 3.28 ft/sec^2 = 100 cm/sec^2
	1 ft/sec^2 = 0.3048 m/sec^2 = 30.48 cm/sec^2
Force:	1 N = 10^5 dyne = 0.2247 lb
Pressure:	$1\,\frac{\text{N}}{\text{m}^2} = 10\,\frac{\text{dyne}}{\text{cm}^2} = 1.45 \times 10^{-4}$ lb/in^2
Energy and Power:	1 J = 10^7 ergs = 0.239 cal = 0.738 ft lb
	1 eV = 1.6×10^{-19} J = 1.6×10^{-12} ergs
	1 cal = 4.18 J
	1 hp = 745 W = 550 ft lb/sec

APPENDIX 5

SOME FUNDAMENTAL CONSTANTS*

Universal gravitational constant	G	6.67×10^{-11} Nm^2/kg^2
Speed of light in a vacuum	c	2.9979×10^8 m/sec
Avogadro's number	N_0	$6.02 \times 10^{23} \dfrac{\text{molecules}}{\text{mole}}$
Universal gas constant	R	8.314×10^3 J/mole °K
Boltzmann's constant	$k = R/N_0$	1.38×10^{-23} J/°K
Charge of electron	e	-1.6×10^{-19} C
Rest mass of electron	m_e	9.1×10^{-31} kg
Rest mass of proton	m_p	1.67×10^{-27} kg
Planck's constant	h	6.626×10^{-34} J sec
Permittivity of space	ϵ_0	8.85×10^{-12} C^2/Nm^2
Permeability constant	μ_0	$4\pi \times 10^{-7}$ N/A^2
Mass of earth		5.98×10^{24} kg
Mean radius of earth		6.37×10^6 m

*See B. N. Taylor, et al., Fundamental Constants and Quantum Electrodynamics, *in* Reviews of Modern Physics Monographs, 375 (1969) for more complete list of constants.

APPENDIX 6

ANSWERS TO PROBLEMS

Chapter 1

1. (a) 4 cm (b) π cm (c) $\frac{3}{2\pi}$ Hz (d) $\frac{2\pi}{3}$ sec

2. (a) 2 cm, 0.5 sec
   (b) $v_y = 12\pi \sin(\pi x - 4\pi t)\ \frac{cm}{sec}$;
   $a_y = -48\pi^2 \cos(\pi x - 4\pi t)\frac{cm}{sec^2}$
   (c) $v_y = 6\sqrt{2}\,\pi\ \frac{cm}{sec}$; at t = 0, x = 0.25
   $a_y = -24\sqrt{2}\,\pi^2\ \frac{cm}{sec^2}$, at t = 0, x = 0.25
   (d) $(v_y)_{max} = 12\pi\ \frac{cm}{sec}$; $(a_y)_{max} = 48\pi^2\ \frac{cm}{sec^2}$

3. (a) 6×10^9 Hz (b) 3.3×10^{-8} sec

4. $180\ \frac{m}{sec}$

5. (a) $20\ \frac{cm}{sec}$ (b) $y = 2.0\sin\left(\frac{\pi}{2}x - 10\pi t - \phi\right)$cm

7. $32\ \frac{m}{sec}$

8. (a) 0 (b) $4\ \frac{cm}{sec}$ (c) $v_y = -40\ \frac{cm}{sec}$; $a_y = 0$
   (d) $y' = 2.0\sin(5.0\ x + 20\ t)$, that is, a wave traveling to the left with the same amplitude, frequency and wavelength as y.

9. 775 N

10. (a) $99\ \frac{m}{sec}$ (b) 6.2 Hz, 12.4 Hz, 18.6 Hz (c) Nodes at $x = 0, \frac{8}{3}$ m, $\frac{16}{3}$ m, 8 m;
    Antinodes at $x = \frac{4}{3}$ m, 4 m, $\frac{20}{3}$ m, where x is measured from one end.

12. $760\ \frac{m}{sec}$, about $2.3\ v_{air}$

13. (a) $5100 \frac{m}{sec}$ (b) 32 sec

14. (a) $1400 \frac{m}{sec}$ (b) 4.3 sec

15. $5900 \frac{m}{sec}$

17. 8×10^{-12} m $\leqslant \xi_m \leqslant 1.1 \times 10^{-5}$ m

18. (a) 690 Hz (b) 36 cm, 60 cm, 84 cm

19. (a) 55 cm (b) 600 Hz, 900 Hz

20. 315 Hz, 945 Hz, 1575 Hz

21. 2.5 m

23. 28 Hz, 814 Hz

24. (a) 835 Hz (b) 770 Hz

25. 537 Hz to 703 Hz

26. $13 \frac{m}{sec}$ towards the man

27. (b) 515 Hz

28. $605 \frac{m}{sec}$, Mach number 1.83

Chapter 2

1. (a) 5.4 N (b) $1.8 \times 10^6 \frac{N}{C}$

2. (a) $5.3 \times 10^{11} \frac{N}{C}$ (b) 8.5×10^{-8} N

3. (a) $\mathbf{F} = (-3.6\mathbf{i} + 6.8\mathbf{j})$ N (b) $\mathbf{E} = (-1.2\mathbf{i} + 2.3\mathbf{j}) \times 10^6 \frac{N}{C}$

4. $E_x = k\frac{16qb}{(a^2 + b^2)^{3/2}}$; $E_y = 0$

5. (a) $\mathbf{E} = (0.6\,\mathbf{i} + 9.6\,\mathbf{j}) \times 10^6 \frac{N}{C}$ (b) $\mathbf{F} = (1.2\,\mathbf{i} + 19.2\,\mathbf{j})$ N

6. (a) $5.6 \times 10^{-11} \frac{N}{C}$ in the downwards direction.

(b) $1.0 \times 10^{-7} \frac{N}{C}$ in the upwards direction.

7. (a) 1.1×10^{-8} C (b) 5.5×10^{-3} N

8. (a) $4.4 \times 10^{5} \frac{N}{C}$ (b) $3.9 \times 10^{-6} \frac{C}{m^2}$

9. (0, 0.08) m

11. $\mathbf{E} = -k \frac{\sqrt{2}\lambda}{R} \mathbf{j}$ (By symmetry, $E_x = 0$.)

12. $\mathbf{E} = \frac{\sigma}{2\epsilon_0} \left[1 - \frac{y}{(R^2 + y^2)^{1/2}}\right] \mathbf{j}$ (By symmetry, $E_x = 0$.)

13. $\mathbf{E} = k\lambda_0 \left[ln\left(\frac{l + d}{d}\right) - \frac{l}{l + d}\right] \mathbf{i}$ ($E_y = 0$ at *all* points on the x axis.)

14. (a) $1500\,\pi \frac{Nm^2}{C}$ (b) 0 (c) $750\sqrt{2}\,\pi \frac{Nm^2}{C}$

15. (a) $9 \times 10^{4} \frac{N}{C}$ (b) $3.6\pi \times 10^{2} \frac{Nm^2}{C}$

18. (a) $\sigma_a = -3\sigma, \sigma_b = \sigma$ (b) $E(I) = \frac{3\sigma}{\epsilon_0}$, $E(II) = 0$, $E(III) = \frac{\sigma}{\epsilon_0}$

19. (a) $E = \frac{\rho r}{3\epsilon_0}$ $(r < a)$, $E = k\frac{Q}{r^2}$ $(b > r > a)$

(b) $\sigma_b = -\frac{Q}{4\pi b^2}$, $\sigma_c = \frac{Q}{4\pi c^2}$

20. (a) -4×10^{-9} C (b) $Q_b = 4 \times 10^{-9}$C, $Q_c = 5.5 \times 10^{-9}$C

(c) $\sigma_b = 8 \times 10^{-9} \frac{C}{m^2}$, $\sigma_c = 7 \times 10^{-9} \frac{C}{m^2}$

21. (a) $4.5 \times 10^{5} \frac{N}{C}$ radially inwards towards the cylinder.

(b) 1.4×10^{-2} N, towards the wire.

22. (a) $\lambda_b = -\lambda$, $\lambda_c = -2\lambda$

(b) $\mathbf{E} = 0$ $(r < a)$, $\mathbf{E} = 2k\frac{\lambda}{r}\hat{r}$ $(a < r < b)$,

$\mathbf{E} = 0$ $(b < r < c)$, $\mathbf{E} = -4k\frac{\lambda}{r}\hat{r}$ $(r > c)$.

23. (a) $2 \times 10^{-6} \frac{C}{m}$ (b) $\lambda_b = -2 \times 10^{-6} \frac{C}{m}$, $\lambda_c = -6 \times 10^{-6} \frac{C}{m}$

(c) $-5.4 \times 10^{5} \frac{N}{C}$, directed radially inwards.

24. $E = \frac{\rho_0 R^5}{5\epsilon_0 r^2}$ $(r > R)$, $E = \frac{\rho_0 r^3}{5\epsilon_0}$ $(r < R)$

25. (a) 0 (b) $\frac{\rho}{3\epsilon_0}\left(\frac{r^3 - a^3}{r^2}\right)$ (c) $\frac{\rho}{3\epsilon_0}\left(\frac{b^3 - a^3}{r^2}\right)$

26. (a) $-\frac{\rho q r}{3\epsilon_0}$ (towards the center) (c) 5.8×10^{-15} sec

27. $\mathbf{E} = -\frac{kp}{x^3}\mathbf{i}$ at (x,0); $\mathbf{E} = \frac{2kp}{y^3}\mathbf{j}$ at (0,y)

28. (a) $2.9 \times 10^{13} \frac{m}{sec^2}$ to the left (b) $1.5 \times 10^6 \frac{m}{sec}$ (c) 5.3×10^{-8} sec

30. (a) $3.1 \times 10^7 \frac{m}{sec}$ (b) 5.3×10^{-9} sec (c) 13 cm

Chapter 3

1. 9×10^4 V

2. (a) 4.5×10^5 V (b) (−0.17,0) m

3. (a) $6\sqrt{2}\frac{kq}{a}$ (b) $6\sqrt{2}\frac{qq'}{a}$

4. $6\sqrt{3}\frac{kq}{b}$

5. $11\frac{kq^2}{b}$

6. (a) 3.6×10^5 V (b) 0

7. 7.2 eV

8. 5.7×10^3 V

9. (a) $V = k\frac{Q}{R}$ for $r \leqslant R$, $V = k\frac{Q}{r}$ for $r > R$ (b) $C = \frac{R}{k} = 4\pi\epsilon_0 R$

10. (a) 1.13×10^3 V (b) $15\frac{m}{sec}$

11. 8.3×10^{-5} C

12. (a) $2kq\left[\frac{1}{\sqrt{r^2 + a^2}} - \frac{1}{r}\right]$ (b) $-\frac{kqa^2}{r^3}$

13. $\frac{\pi k \lambda}{2}$

14. $\frac{\lambda R}{2\epsilon_0} \frac{d}{(d^2 + R^2)^{3/2}}$

15. (a) $V = \frac{\sigma}{2\epsilon_0} [(y^2 + R^2)^{1/2} - y]$ (b) $\mathbf{E} = \frac{\sigma}{2\epsilon_0} \left[1 - \frac{y}{(y^2 + R^2)^{1/2}}\right] \mathbf{j}$

16. (a) $V = k\lambda_0 \left[l - d \ln\left(\frac{d + l}{d}\right)\right]$ (b) $\mathbf{E} = k\lambda_0 \left[\ln\left(\frac{d + l}{d}\right) - \frac{l}{l + d}\right] \mathbf{i}$

17. (a) $\lambda = \frac{V}{2k \ln\left(\frac{b}{a}\right)}$ (b) $C = \frac{l}{2k \ln\left(\frac{b}{a}\right)}$

18. (a) $\frac{kQ}{r}$ for $r > R$ (b) $\frac{\rho_0}{12\epsilon_0} (4R^3 - r^3)$ for $r < R$

19. (a) -2×10^3 V (b) 10^{-2} J

20. (a) $E_x = -8 - 24xy$, $E_y = -12x^2 + 40y$, $E_z = 0$
(b) $E_x = -152 \frac{V}{m}$ $E_y = 72 \frac{V}{m}$ (c) 8.8×10^{-5} J

21. $\frac{2kq_1 q_2}{m{v_0}^2}$

22. (a) $1.4 \times 10^5 \frac{m}{sec}$ (b) $6 \times 10^6 \frac{m}{sec}$

23. (a) 1.64μF (b) 4.92×10^{-4} C (c) 164 V, 82 V, and 54 V

24. (a) 18μF (b) 9×10^{-4} C, 1.8×10^{-3} C, and 2.7×10^{-3} C
(c) Obviously, the series combination should be the choice; since C_{eq} is small, the charge stored is the least, therefore the "zap" would be minimum for a given potential difference.

25. (a) 34.8 pF (b) 122 pF

26. (a) 1.1 pF (b) 3.5×10^{-9} J

27. (a) 3.3μF (b) 10^{-4} C on the 3μF and 6μF capacitors; 6.7×10^{-5} C on the 2μF and 4μF capacitors (c) 33 V across the 3μF and 2μF capacitors; 17 V across the 6μF and 4μF capacitors

28. (a) 1.32×10^{-4} C (b) 13.2 V (c) 1.45×10^{-3} J (d) 5.8×10^{-4} J. The wires heat up when charge is transferred from one capacitor to the other.

29. (a) 0 (b) 29μC on the 3μF; 115μC on the 12μF

32. (b) 11 N

33. (a) $V_b = k\left(\frac{Q - q}{b}\right)$; $V_a = k\left(\frac{Q - 2q}{b} + \frac{q}{a}\right)$
(b) Inner sphere has zero charge, outer sphere has charge q - Q.
(c) *Zero.* Both spheres are at the same potential when all the charge is on the outer sphere.

35. (a) $\frac{\epsilon_0 l}{d}[x(\kappa - 1) + l]$ (b) $\frac{\epsilon_0}{2d}[x(\kappa - 1) + l]V^2$ (c) $\frac{\epsilon_0 l V^2}{2d}(\kappa - 1)$ to the *right*
(d) 1.55×10^{-3} N, equivalent to ≈ 0.16 g

Chapter 4

1. 2.2×10^{15} electrons

2. (a) 0.25 Ω (b) 7.5 V (c) $1.5 \times 10^7 \frac{A}{m^2}$

3. (a) 2.6 Ω (b) $0.78 \frac{N}{C}$ (c) 39 V

4. (a) 33 Ω (b) 7.7 A (c) 360 W

5. (a) 8.3 Ω (b) $I_1 = 3$ A, $I_2 = 2$ A, $I_3 = 1$ A (c) 75 W

6. (a) I = 3 + 10t (b) I = 33 μA (c) 54 μC

7. (a) 14 Ω (b) 56 W (c) 2 A

8. (a) 0.55 A in 60 W, 0.91 A in 100 W
(b) 200 Ω for 60 W, 120 Ω for 100 W

9. (a) 15 Ω (b) 5.5 A

10. (b) 6.1 Ω

11. $I_1 = 0$, $I_2 = 0.5$ A, $I_3 = 0.5$ A

12. $I_1 = 2.4$ A, $I_2 = 2.6$ A, $\mathcal{E} = 15.8$ V

13. (a) $I_1 = 5$ A, $I_2 = I_3 = 2.5$ A
(b) $V_4 = 42.5$ V, $V_8 = 20$ V, $V_{10} = 0$
(c) $q_4 = 170\ \mu C$, $q_8 = 160\ \mu C$, $q_{10} = 0$
(d) $U_4 = 3.6 \times 10^{-3}$ J, $U_8 = 1.6 \times 10^{-3}$ J, $U_{10} = 0$

14. (a) 5×10^4 Ω (b) 5 W

15. 10 Ω

19. (a) 10^{-4} A (b) 4×10^{-5} C (c) 0.4 sec (d) 1.4×10^{-5} A (e) 3.5×10^{-5} C
(f) 1.2×10^{-3} J

20. (a) 1.5×10^{-5} C (b) 30 V (c) 30 V (d) No. $\mathcal{E} = 0$, so Kirchhoff's second rule gives $\frac{q}{C} - IR = 0$. This is consistent with our results.

Chapter 5

1. 1.6×10^{-13} **k** N

2. (a) 0.28 m (b) $3.8 \times 10^5 \dfrac{\text{rad}}{\text{sec}}$

3. $2.4 \times 10^5 \mathbf{j} \dfrac{\text{N}}{\text{C}}$

5. (a) Zero, since ds × **B** = 0 for each element. (b) 7.1×10^{-2} N *out* of the paper. (c) Zero, since the force on the section from b to c is also 7.1×10^{-2} N, but *into* the paper.

6. (a) $\dfrac{\pi}{2}$IRB directed upwards. (b) Zero

7. $B = \dfrac{\mu_0 I}{4\sqrt{2}\,\pi a}$, directed into the paper.

8. (a) 0.628 T (b) 3.15×10^{-3} Wb

9. (a) 9×10^{-6} T *out* of the paper.
   (b) Yes. **B** = 0 along a line parallel to both wires, 7.6 cm *below* the top wire. This line lies in the plane of the two wires.

10. (a) Zero, since ds is || to $\hat{r}$, so ds × $\hat{r}$ = 0 for these elements.
    (b) $\dfrac{\mu_0 I}{12b}$ *into* the paper.
    (c) $\dfrac{\mu_0 I}{12}\left[\dfrac{1}{a} - \dfrac{1}{b}\right]\mathbf{k}$, where **k** is a unit vector *out* of the paper.

11. 3.8×10^{-2} T or 380 G

12. (a) $\left(\dfrac{ke^2}{mr_0{}^3}\right)^{1/2}$ (b) $\dfrac{e^2}{\pi r_0}\left(\dfrac{k}{mr_0}\right)^{1/2}$ (c) $\dfrac{\mu_0 e^2}{2\pi r_0{}^3}\left(\dfrac{k}{mr_0}\right)^{1/2}$

13. $\phi = \dfrac{\mu_0 Ib}{2\pi}\, ln\left(\dfrac{a+d}{d}\right)$

14. (a) $2.3\pi \times 10^{-6}$ T *into* the paper. (b) 7.2 cm

15. 0.59 T or 5900 G

Chapter 6

1. (a) 40π mV (b) 80π mA

2. 2.5 V

5. 1.8×10^{-4} V

6. (a) 0.81 μH (b) 4.1 μV

7. $\frac{\mu_0 I_0 b}{2\pi\tau} ln\left(\frac{a+d}{d}\right) e^{-t/\tau}$

8. (b) 3 μJ

9. 4.2 μC

10. (a) 0.5 msec (b) 0.25 A (c) 0.35 msec

11. 0.46 msec

12. $I_0 \mathcal{E} e^{-t/\tau} (1 - e^{-t/\tau})$, where $\tau = \frac{L}{R}$

13. (a) $\frac{\mu_0 n R^2}{2r} \frac{dI}{dt}$ (b) $\frac{\mu_0 nr}{2} \frac{dI}{dt}$

Chapter 7

1. (a) $2500 \frac{rad}{sec}$ (b) 0.48 A

2. 6.3 mH

4. (a) 1.3 msec (b) 2×10^{-4} C (c) 1 A (d) 5×10^{-3} J (e) $\frac{\pi}{2} \times 10^{-4}$ sec

5. (a) $1.7 \times 10^3\ \Omega$, 33 Ω (b) 600 Ω, 3000 Ω

6. (a) $\sqrt{2} \times 10^3\ \Omega$ (b) $\frac{\pi}{4}$, I lags V since $X_L > X_C$ (c) 3.2 W

7. (a) $V_R = 40\sqrt{2}$ V, $V_L = 48\sqrt{2}$ V, $V_C = 8\sqrt{2}$ V
   (b) $v_R = 40\sqrt{2} \sin\left(2000\,t - \frac{\pi}{4}\right)$, $v_L = 48\sqrt{2} \cos\left(2000\,t - \frac{\pi}{4}\right)$,
   $v_C = -8\sqrt{2} \cos\left(2000\,t - \frac{\pi}{4}\right)$

8. (a) 5A, 0 (b) 150 Ω, $-87°$

10. (a) $5 \times 10^4 \frac{rad}{sec}$ (b) 18 W

13. 3.9×10^{26} W

14. (a) 600 kW (b) $2 \times 10^{-8} \frac{Wb}{m^2}$

15. $6.4 \times 10^{-7} \frac{N}{m^2}$

16. (a) 72 J (b) 3.6×10^{-7} kg $\frac{m}{sec}$

17. 2500 m^2

18. 2.4×10^3 years, opposite to the motion of the vehicle. (Obviously, he would be better off using a can of spray deodorant.)

Chapter 8

1. (a) $1.24 \times 10^8 \frac{m}{sec}$ (b) 2440 Å

2. 1.24

3. 22.4°

4. (a) 50° (b) 43° (c) 32°

5. 1.34

6. (a) 45 cm in front of the mirror, $M = \frac{1}{2}$
   (b) 60 cm behind the mirror, M = -3

7. (a) 22.5 cm behind the mirror, M = -0.25
   (b) 12 cm behind the mirror, M = -0.6

8. 48 cm

9. (a) $\frac{20p}{p - 20}$ (b) -6.67 cm, -20 cm, ∞, 40 cm, 30 cm

10. 5.7 cm in front of the mirror. The image is real and inverted.

12. (a) M = -0.4, 12 cm behind the mirror.
    (b) M = -0.25, 15 cm behind the mirror.
    (c) Erect.

13. n = 2

14. 8.6 cm from the surface, or 6.4 cm to the left of center.

15. 9.7 cm from the surface, or 5.3 cm to the left of center.

16. (a) 18 cm to the right of V (b) 54 cm to the right of V (c) 6 cm to the left of V

17. 53 cm to the right of V

18. (a) $M = -\frac{2}{3}$ (b) $M = \frac{4}{3}$

19. (a) $d' = \frac{2}{3}d \cong 23$ cm (b) 1.5 cm

21. (a) 33 cm (b) 60 cm (c) -20 cm (d) -0.66, -2, 2, inverted in cases (a) and (b). Erect in case (c).

22. (a) -14.3 cm (b) -12 cm (c) -6.7 cm (d) 0.29, 0.40, 0.67. Erect in all cases.

23. (a) $\frac{4}{3}f$ (b) $\frac{3}{4}f$ (c) -3, 4

24. (a) $\frac{3}{2}f$ (b) $-f$ (virtual object)

25. (a) 10 cm from the second lens, or 60 cm from the first lens of focal length 20 cm.
(b) 2 cm (real)

26. (a) 4 cm (b) 20 cm (real image)

27. (a) 6.4 cm (b) -0.25 (c) converging

28. (a) 16 cm (c) 8 cm (c) 4 cm (d) $\frac{8}{3}$ cm (e) 1 cm (f) 0

29. The final image is virtual, erect and 5.9 cm to the left of the diverging lens. Its height is 0.84 cm.

30. $d = 14$ cm

Chapter 9

1. (a) 7.8×10^{-2} mm (b) 2.63 cm

2. 4000 Å

3. (a) 2A (b) $\sqrt{3}$ A (c) $\sqrt{2}$ A

5. 5840 Å, 4170 Å, 3240 Å

6. 4870 Å, 3650 Å

7. 960 Å

8. (b) 2680 Å

11. 0.23 mm

12. (a) 4 mm, 2 mm (b) 6 mm

13. 6150 Å

14. (a) 3235 Å (b) 31°

15. (a) 3450 rulings per cm (b) 5

16. $8.28 \times 10^5 \frac{\text{rad}}{\text{m}}$ in first-order. $1.87 \times 10^6 \frac{\text{rad}}{\text{m}}$ in second-order.

17. (a) 12.000 24,000 36,000 (b) 0.11 Å

18. (a) 52.6° (b) 55.5° (c) 56.4° (d) 67.5°

INDEX

Page numbers in *italic* type refer to illustrations.
Page numbers followed by (t) refer to tables.
Page numbers preceded by I refer to Volume I and II to Volume II.